T0324213

MECHANICS OF CREEP BRITTLE MATERIALS
2

Proceedings of the International Colloquium 'Mechanics of Creep Brittle Materials 2' held at the University of Leicester, UK, 2–4 September 1991.

MECHANICS OF CREEP BRITTLE MATERIALS

2

Edited by

A. C. F. COCKS

Department of Engineering, Cambridge University, UK

and

A. R. S. PONTER

Department of Engineering, Leicester University, UK

ELSEVIER APPLIED SCIENCE
LONDON and NEW YORK

ELSEVIER SCIENCE PUBLISHERS LTD
Crown House, Linton Road, Barking, Essex IG11 8JU, England

Sole Distributor in the USA and Canada
ELSEVIER SCIENCE PUBLISHING CO., INC.
655 Avenue of the Americas, New York, NY 10010, USA

WITH 19 TABLES AND 186 ILLUSTRATIONS

© 1991 ELSEVIER SCIENCE PUBLISHERS LTD
© 1991 NUCLEAR ELECTRIC PLC—pp. 90–99

British Library Cataloguing in Publication Data
European Mechanics Colloquium 'Mechanics of Creep
Brittle Materials 2' (1991 : University of Leicester)
Mechanics of creep brittle materials.
I. Title II. Cocks, A. C. F.
III. Ponter, A. R. S.
620.11233

ISBN 1-85166-701-6

Library of Congress CIP data applied for

Preface

Mechanics of Creep Brittle Materials—1 was published in 1989 as the proceedings of a Colloquium held in Leicester in the summer of 1988. The Colloquium examined the creep response of a wide range of materials, including metals, engineering ceramics and ice, with the aim of determining similarities in the response of these materials and the way in which their behaviour is modelled. The proceedings were structured so as to reflect the interdisciplinary nature of the Colloquium, with papers grouped together largely on the basis of the phenomena being examined, rather than by class of material.

Mechanics of Creep Brittle Materials—2 was held in Leicester in September 1991 to discuss advances made in our understanding of the response of creep brittle materials since the first Colloquium. The scope of the Colloquium was extended to include mineral salts, concrete and composite systems. These proceedings are once more structured so that the reader can readily compare the response of different material systems and evaluate the suitability of the range of models presented to the materials he is interested in. In fact a number of papers directly compare the behaviour of a range of different materials with the aim of identifying general strategies for the testing and modelling of creeping materials.

The proceedings are divided into two main sections. The first considers the propagation of cracks in creeping materials, with papers concerned with crack stability, transient and steady state crack growth and crack growth under cyclic loading conditions. The second section is largely concerned with continuum processes, with papers describing micromechanical and phenomenological models of creep deformation and failure and the application of these models in the design of high temperature components. It is often difficult to determine the continuum response of creep brittle materials, and this section concludes with a series of papers which investigate the use of indentation techniques to obtain appropriate properties.

We would like to take this opportunity to thank all the people involved in *Mechanics of Creep Brittle Materials—2*. We are grateful to Leicester University for allowing us to hold the Colloquium in Beaumont Hall and the University Conference Office and Beaumont Hall staff for providing a welcoming and relaxed environment. We are once more indebted to Jo Denning for all the time and effort she has put into making the arrangements for the Colloquium and into the preparation of these proceedings.

A. C. F. Cocks
University of Cambridge, UK
A. R. S. Ponter
University of Leicester, UK

Contents

1. Crack Propagation in Creeping Bodies

A STUDY OF CREEP CRACK GROWTH IN ENGINEERING MATERIALS

ALAN C.F. COCKS[*] & JULIAN D.J. DE VOY[+]

[*] Department of Engineering, Cambridge University,
Trumpington Street, Cambridge CB2 1PZ, UK.

[+] Department of Engineering, Leicester University,
University Road, Leicester LE1 7RH, UK.

ABSTRACT

From an examination of the different types of stress and strain-rate fields that can develop ahead of a growing crack in an elastic / creeping material a simple map is developed for presenting creep crack growth data. The map can also be used to determine the most suitable, if any, crack tip parameter to correlate a particular set of data and to assess the range of applicability of theoretical models describing the process of crack growth.

INTRODUCTION

Structural components operating at high temperatures can often fail as the result of the time dependent propagation of flaws which have either been introduced into the component during manufacture or nucleate early in the life of the component. Assessment procedures are currently being developed for metallic components used in the power generating industry which addresses this particular problem [1]. This type of failure is not, however, limited to metallic components. Ceramic components invariably contain flaws which are introduced into the component during cooling from the sintering temperature as a result of variations in the elastic and thermal properties along different crystallographic planes or due to the presence of inhomogeneities in the compact. There are broad similarities between the response of these two classes of materials and the theoretical models that have been developed to describe their behaviour [2,3].

If a crack grows sufficiently quickly then the stress and strain-rate fields ahead of the growing crack can be expressed in terms of the stress intensity factor K and it is found that K is a suitable parameter for correlating the crack growth behaviour. Alternatively if the crack grows slowly then the near crack tip fields are characterised by C^*, which, under these conditions, represents a more suitable parameter for correlating creep crack growth data. In this paper we examine the conditions under which K and C^* determine the component response and present a simple map where the range of applicability of each parameter can be identified. Material data can be plotted directly onto this map without making any prior judgement about how this data will be correlated. The predictions of a wide range of theoretical models can also be plotted onto the map, allowing the assumptions used in the development of these models to be readily assessed.

CRACK TIP FIELDS

In this section we describe the nature of the stress and strain-rate fields which develop ahead of a growing crack in a material which deforms according to the relationship

$$\dot{\varepsilon} = \frac{\dot{\sigma}}{E} + \dot{\varepsilon}_0 \left(\frac{\sigma}{\sigma_0}\right)^n \tag{1}$$

in uniaxial tension, where $\dot{\varepsilon}$ is the strain-rate at a stress σ and stress rate $\dot{\sigma}$ for a material of Young's modulus E which creeps at a rate $\dot{\varepsilon}_0$ under a constant stress σ_0, with creep exponent n.

It is instructive initially to consider a stationary crack. On initial fast loading the component responds elastically and the stress field, σ_{ij}, ahead of the crack is [4]

$$\sigma_{ij} = \frac{K}{\sqrt{2\pi r}} \tilde{\sigma}_{ij}(\theta) \tag{2}$$

with respect to a polar co-ordinate system centred on the crack tip, where $\tilde{\sigma}_{ij}(\theta)$ is a dimensionless function of the angular co-ordinate θ. A small zone within which creep strains dominate subsequently forms about the crack tip which spreads out with time until the stresses have completely relaxed. The stress field in the vicinity of the crack is then given by [5,6]

$$\sigma_{ij} = \sigma_0 \left[C^* / \dot{\varepsilon}_0 \sigma_0 I_n r \right]^{1/(n+1)} \tilde{\sigma}_{ij}(\theta, n) \qquad (3)$$

where $\tilde{\sigma}_{ij}$ is now a function of n and θ and I_n is well approximated by [7]

$$I_n = \frac{n+1}{n} \pi$$

Ainsworth [8] has demonstrated that C^* is adequately approximated by

$$C^* = \dot{\varepsilon}_R \sigma_R \lambda \qquad (4)$$

where $\dot{\varepsilon}_R$ is the uniaxial strain-rate at a reference stress

$$\sigma_R = \frac{P}{P_L} \sigma_y$$

where P_L is the limit load for a perfectly plastic material of yield strength σ_y and P is the applied load; and λ is a characteristic length for the component,

$$\lambda = K^2 / \sigma_R^2$$

At intermediate times the stress field in the creep zone is given by an eqn with the form of eqn (2) with C^* replaced by a quantity C[t] [9], the detailed form of which need not concern us here.

If the crack is now allowed to grow at a rate $\dot{a}$ then a further zone forms immediately ahead of the crack tip where, for n > 3, there is a strong coupling between the elastic and creep response. The stress field within this zone is given by [10]

$$\sigma_{ij} = \left(\frac{\alpha_n \sigma_R^n \dot{a}}{\dot{\varepsilon}_R E r} \right)^{\frac{1}{n-1}} \tilde{\sigma}_{ij}^{HR}(\theta, n) \qquad (5)$$

where $\tilde{\sigma}_{ij}^{HR}$ is again a dimensionless function of θ and n and α_n is well approximated by [11]

$$\alpha_n = \frac{1}{n-1}$$

Hawk and Bassani [12] have determined the full transient field ahead of a growing crack for small scale creep. They demonstrate that the field consists of the HR field of eqn (5) surrounded the transient C[t] creep field of eqn (3), which, in turn, is surrounded by the elastic K field of eqn (2). The extent of each of these fields is obtained by simply determining the value of r where the von Mises effective stress for the two adjoining fields are the same. Under steady state conditions the situation is even simpler; the field then consists of two zones: an inner HR field surrounded by the K field. The extent of the HR zone, r_c can be obtained by simply equating eqns (2) and (5) when $\tilde{\sigma}_e = \tilde{\sigma}_e^{HR} = 1$. We obtain

$$\frac{r_c}{\lambda} = \frac{1}{2\pi} \left(\frac{n-1}{2\pi} \frac{E}{\sigma_R} \frac{\dot{\varepsilon}_R \lambda}{\dot{a}} \right)^{\frac{2}{n-3}} \tag{6}$$

This is effectively the size of zone within which creep effects become important in an otherwise elastic material. This zone gradually increases in size as the crack growth rate decreases. If the crack velocity is very slow creep effects will dominate in the remote field and the region where elastic effects become important is confined to a small region surrounding the crack tip. The extent of this zone can be obtained by equating eqns (2) and (3):

$$\frac{r_e}{\lambda} = \frac{n}{(n+1)\pi} \left(\frac{(n+1)\pi}{n(n-1)} \frac{\sigma_R}{E} \frac{\dot{a}}{\dot{\varepsilon}_R \lambda} \right)^{\frac{n+1}{2}} \tag{7}$$

Now the size of the HR zone decreases with increasing crack velocity. The transition from K to C* dominating in the remote field can be determined roughly by equating eqns (6) and (7), ie when

$$\frac{\sigma_R}{E} \frac{\dot{a}}{\dot{\varepsilon}_R \lambda} = f(n) \tag{8}$$

where $f(n) \simeq (n+1)/(n-1)$ for $n > 3$.

It is important to note the form of the above expressions. We describe the load applied to the body in terms of the reference stress σ_R and all stresses and modulii are normalised with respect to this quantity. Similarly all rates are suitably normalised using the reference strain-rate $\dot{\varepsilon}_R$ and all quantities with the dimension of length are normalised by the characteristic length λ. The full significance of these normalisations will become apparent when we examine a number of micromechanical models of the crack growth process.

The material model used to obtain these results is a rather simplified description of how a material responds in practice. As a crack grows the material immediately ahead of the crack tip can suffer time independent plastic straining. Also there must be a mechanism of crack growth. In general this mechanism involves the nucleation and growth of voids, which link with the dominant crack, causing it to grow. The presence of this damage ahead of the crack can significantly affect the elastic and creep properties of the material in these regions and therefore the local stress and strain-rate fields immediately surrounding the crack tip. If, however, the process zone associated with the damaging process is small compared with λ but larger than r_c then the boundary conditions for this zone can be expressed in terms of K which can, therefore, be used to characterise the crack growth process. Similarly if the process zone is larger than r_e the crack growth rate can be correlated using C^*.

Without detailed examination of the state of the material ahead of a growing crack or detailed micromechanical modelling of the evolution of the damage zone it is not possible to determine whether the above conditions are met; but a consequence of the above requirements is that for K controlled growth r_c must be much less than λ and for C^* controlled growth r_e must be much less than λ. In the following we arbitrarily assume that these requirements are met if $r_c/\lambda < 0.02$ and $r_e/\lambda < 0.02$ respectively. From eqns (6) and (7) we find that if

$$\frac{\dot{a}}{\dot{\varepsilon}_R \lambda} > \frac{g_1(n)}{\sigma_R/E} \tag{9}$$

K controls crack growth, and if

$$\frac{\dot{a}}{\dot{\varepsilon}_R \lambda} < \frac{g_2(n)}{\sigma_R/E} \tag{10}$$

C^* controls crack growth, where

$$g_1(n) = \frac{(n-1)}{50} \left(\frac{25}{\pi} \right)^{\frac{n-1}{2}} \quad \text{and} \quad g_2(n) = \frac{(n-1)}{50^{2/(n+1)}} \left(\frac{n+1}{n} \ \pi \right)^{\frac{n-1}{n+1}}$$

Eqn (8) and inequalities (9) and (10) are plotted in Fig 1 for n=9 using axes of $\dot{a}/\dot{\varepsilon}_R\lambda$ and σ_R/E. The range of dominance of K and C^* are identified on this figure. The shaded band represents the range of values of $\dot{a}$ over which there is a transition from K to C^*, and neither of these represent suitable parameters for correlating creep crack growth.

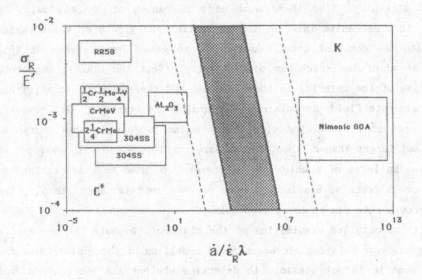

Figure 1. A creep crack growth map showing the range of dominance of C^* and K and test data for a number of engineering materials.

The above analysis applies for situations where n > 3, but we can readily demonstrate that the general features of this analysis are also applicable for smaller values of n. Consider the situation where n=1; the stress field ahead of a moving crack tip is then simply the field of eqn (2) at all times, where the co-ordinate system moves with the crack tip. The rate of change of stress of a material element which is instantaneously a distance r ahead of the crack is

$$\dot{\sigma}_{ij} = \frac{K}{2\sqrt{2\pi r}} \, \tilde{\sigma}_{ij}(0) \, \frac{\dot{a}}{r} \tag{11}$$

Substituting this into eqn (1) together with eqn (2) yields the total strain-rate in the material element. It is immediately evident that the elastic strain-rate will dominate immediately ahead of the crack tip. The extent of this elastic zone can be evaluated by determining the position where the elastic and creep strain-rates are equal:

$$\frac{r_e}{\lambda} = \frac{1}{2} \frac{\dot{a}}{\dot{\varepsilon}_R \lambda} \tag{12}$$

If this zone is large then the response is mainly elastic and if it is small the elastic deformation is negligible and we can treat the material as a linear creeping material, using C^* to characterize the crack growth rate. As before we assume that if $r_e / \lambda < 0.02$ C^* controls the growth and if it is greater than 1 K is the appropriate characterizing parameter. These inequalities can be expressed in the form of eqns (9) and (10) with $g_1(1) = 0.5$ and $g_2(1) = 0.04$. These limits are perhaps academic when one is interested only in choosing a suitable parameter to correlate creep crack growth data, but they become important when assessing the suitability of models of the crack growth process.

APPLICATION OF MAPS

In this section we examine ways in which the maps can be used to assess crack growth data, to determine the suitability of micromechanical material models and to identify the significance of creep crack growth in the design and assessment of structural components.

Presentation of Material Data

Material data can readily be plotted onto the maps presented in the last section. To do this a knowledge of the uniaxial creep properties are required in addition to information about the rate of crack growth. Nikbin et al [13] have assembled all the necessary data for a range of steels and the aluminium alloy RR58; Riedel [14] provides data for Nimonic 80A; and Blumenthal et al [15] have studied the response of alumina. Limit loads and stress intensity factors for specimen geometries used in these studies are documented by De Voy and Cocks [11]. The data from each of these studies is plotted in Fig 1, for the situation where σ_R and $\dot{\varepsilon}_R$ represent the plane strain reference quantities. Similar plots assuming plane stress conditions are given by De Voy and Cocks [11]. As noted earlier the shaded band on this figure represents the transition region from K to C^* controlled growth for n=9. The dashed line to the left of this band represents the line below which C^* controls crack growth for n=1 and the dashed line to the right of the band represents the lower limit for K controlled growth for n=13; these values of n correspond to those for alumina [15] and Nimonic 80A [14] respectively.

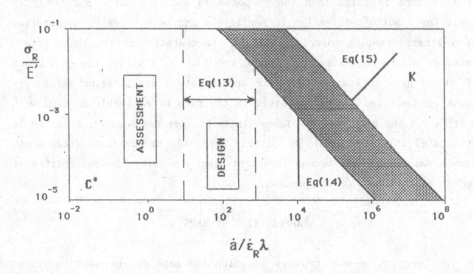

Figure 2. A creep crack growth map showing regions associated with the design and assessment of high temperature plant and the predictions of a number of models.

Examination of Fig 1 quickly reveals that C^* is the controlling parameter for each of the above materials over the range of loads and temperatures used in the testing programmes apart from the Nimonic 80A, where K is a more suitable parameter; a result which is consistent with experiment in each case.

Design and Assessment

Design rules for structures which operate in the creep range limit the mean creep strain, ε_d, accumulated during the design life, t_d, to 1%. If the structure can tolerate a critical flaw size, a_c, which is much greater than the initial flaw size, and a crack is permitted to grow to this size in a time t_d, then the mean crack length over the design life is $a_c/2$. If we now assume that the critical crack length is much smaller than other structural dimensions, then the mean value of λ is of the order of a_c giving a mean value of $\dot{a}/\dot{\varepsilon}_R\lambda$ of 10^2. This result is represented as the design box in Fig 2 for situations where the design reference stress is less than $10^{-3}E'$. If the experimental data lies to the left of this box then failure due to the steady growth of a crack will not occur during the design life.

The maps can also be used as a guide to the assessment of defective structures. If a crack forms early in the design life of a component that is a significant fraction of the critical crack length, a_c, then the extent of crack growth that can be tolerated during the remainder of the design life is substantially reduced, thus reducing the level of crack growth rate that can be tolerated. The acceptable level of normalised growth rate is shown diagrammatically in Fig 2. As before, the component will achieve its design life provided the material data is to the left of the assessment box. If the data is to the right or too close to the box for comfort then the component either needs to be repaired, replaced or the design load needs to be reduced, thus reducing $\dot{\varepsilon}_R$ and moving the assessment box to the right and away from the material data.

Models of Creep Crack Growth

A large number of material models have been developed to explain and correlate creep crack growth in engineering materials. De Voy and Cocks [11] have examined a range of models and demonstrated how the maps of Figs 1 and 2 can be used to assess the relative merits and predictive capabilities of each model. Perhaps the simplest model in the literature is that for C^* controlled growth due to Cocks and Ashby [16] who considered the accumulation of strain in an element of material as it moves towards the crack tip and assumed that a critical strain ε_f is accumulated as it links with the crack. Webster has further extended this model to take into account the different ductilities under plane stress and plane strain conditions [2] and to evaluate the influence of damage accumulated in the remote field [17]. Thouless [3] has extended the model to determine the rate of creep crack growth in engineering ceramics. The basic model [16] can be written in the following dimensionless form:

$$\dot{\bar{a}} = \frac{n+1}{\varepsilon_f} \left(\frac{\bar{R}}{I_n} \right)^{\frac{1}{n+1}} \tag{13}$$

where $\dot{\bar{a}} = \dot{a}/\dot{\varepsilon}_R \lambda$ and $\bar{R} = R/\lambda$ is the normalized size of the process zone. For given values of n, ε_f and $\bar{R}$ this equation plots as a vertical line on the maps. The position of these lines is not very sensitive to the value of $\bar{R}$, particularly for large values of n and it is generally sufficient to set $\bar{R}/I_n$ equal to 1. The resulting range of lines for n=9 and ε_f in the range 0.01 to 1 are shown in Fig 2.

In the above model the rate of growth of damage in the material is assumed to be related to the rate of creep deformation, so that there is only one characteristic rate for the crack growth process. Thouless [3] and De Voy and Cocks [11] describe a set of models where the functional dependence of the damage growth rate and deformation rate with stress within the process zone differ from that in the surrounding material, ie the reference displacement rate $\dot{u}_R$ for the process zone is not related to $\dot{\varepsilon}_R$. An example of this is when the deformation within the process zone results from the plating out of material onto the grain boundaries as material diffuses away from the growing voids, while the surrounding material either deforms elastically or creeps according to eq (1). Here we consider a class of models where the process zone is confined to a plane directly ahead of the crack tip and the crack advances when a void directly ahead of the crack tip reaches a critical size (a certain fraction of the void spacing), or, equivalently, when the crack tip opening displacement reaches a critical value, δ_c. De Voy and Cocks [11] present a number of methods for analysing this situation. A particularly useful technique involves evaluating the C^*-integral around the process zone for situations where creep effects dominate or evaluating the J-integral for conditions where the component only deforms elastically in the remote field. In these models the form of the displacement field in the process zone is required. For assumed fields of the form

$$\dot{u}_y = f\left(\frac{x}{R}\right) \qquad \text{or} \qquad u_y = f\left(\frac{x}{R}\right)$$

for C^* or K controlled growth respectively, where x is the distance directly ahead of the crack tip and u_y and $\dot{u}_y$ are the displacement and displacement rates normal to the line of the crack around the process zone, we find

$$\dot{\bar{a}} = c_1 \frac{\bar{R}}{\bar{\delta}_c} \dot{\bar{u}}_R^{1/2} \tag{14}$$

for C^* controlled growth and

$$\dot{\bar{a}} = c_2 \frac{\bar{R}}{\bar{\delta}_c^2} \dot{\bar{u}}_R \frac{\sigma_R}{E} \tag{15}$$

for K controlled growth, where $c_1 = 0.54$ and $c_2 = 0.079$ for linear fields, $\bar{\delta}_c = \delta_c/\lambda$ and $\bar{\dot{u}}_R = \dot{u}_R/\dot{\varepsilon}_R\lambda$. These equations are plotted in Fig 2 for $\bar{R} = 0.05$ $\bar{\delta}_c = 5.10^{-5}$ and $\bar{\dot{u}}_R = 100$, where the range of applicability of each model can readily be identified. At low stresses the creep model of eq (14) is appropriate, while at high stresses the rate of creep is too slow and eq (15) provides a more accurate description of the component response. High values of $\bar{\dot{u}}_R$ and low values of $\bar{\delta}_c$ promote brittle behaviour, pushing both lines to the right on Fig 2.

CONCLUSIONS

In this paper we have presented a simple map which readily allows the performance of defective components to be assessed by directly comparing material data with design and assessment criteria. The predictions of models for the process of creep crack growth can also be plotted directly onto these maps, providing a means of assessing the range of applicability of the models.

Acknowledgement

JDJDV acknowledges financial support from the SERC through the award of a studentship.

REFERENCES

1. Ainsworth, R.A., Chell G.G., Coleman, M.C., Goodall, I.W., Gooch, D.J., Haigh, J.R., Kimmins, S.T. and Neate, G.J., CEGB assessment procedure for defects in plant operating in the creep range, Fat. Fract. Eng. Mat. Struc., 1987, **10**, 115-27.

2. Webster, G.A., Modelling of creep crack growth. In Mechanics of Creep Brittle Materials I, eds. A.C.F. Cocks and A.R.S. Ponter, Elsevier Applied Science Publishers, London, 1989, pp 36-49.

3. Thouless M.D., Modelling creep crack growth processes in ceramic materials. In Mechanics of Creep Brittle Materials I, eds. A.C.F. Cocks and A.R.S. Ponter, Elsevier Applied Science Publishers, London, 1989, pp 50-62.

4. Williams, M.L., J. Appl. Mech., On the stress distribution at the base of a stationary crack, 1957, **24**, 109.

5. Hutchinson, J.W. Singular behaviour at the end of a tensile crack in a hardening material, J. Mech. Phys. Solids, 1968, **16**, 13-31.

6. Rice, J.R. and Rosengren, G.F., Plane strain deformation near a crack tip in a power law hardening material, J. Mech. Phys. Solids, 1968, **16**, 1-12.

7. Cocks, A.C.F. and Ashby, M.F., Cleavage cracks in elastic-creeping solids, 1983, Cambridge University Engineering Department Report CUED/C/MATS/TR.93.

8. Ainsworth, R.A., Some observations on creep crack growth, Int J. Fracture, 1982, **20**, 147-59.

9. Riedel, H. and Rice, J.R., Tensile cracks in creeping solids, in Fracture Mechanics: Twelfth Conference, ASTM STP 700, 1980, pp 112-30.

10. Hui, C.Y. and Riedel, H., The asymptotic stress and strain field near the tip of a growing crack under creep conditions, Int. Jnl. Fracture, 1981, **17**, 409-25.

11. De Voy, J.D.J. and Cocks, A.C.F., Creep crack growth, to appear.

12 Hawk, D.E. and Bassani, J.L., Transient crack growth under creep conditions, Jnl. Mech. Phys. Solids, 1986, **34**, 191-212.

13. Nikbin, K.M., Smith, D.J. and Webster, G.A., Prediction of creep crack growth from uniaxial data, Proc. Roy. Soc., 1984, **A396**, 183-197.

14. Riedel, H., Fracture at High Temperature. Springer-Verlag, Berlin, 1987.

15. Blumenthal, W. and Evans, A.G., High temperature failure of polycrystalline alumina: II, Creep crack growth and blunting, J. Am. Ceram. Soc., 1984, **67**, 751-59.

16. Cocks, A.C.F. and Ashby, M.F., The growth of a dominant crack in a creeping material, Scripta Met., 1982, 16, 109-14.

17. Nishida, K. and Webster, G.A., Interaction between build up of local and remote damage on creep crack growth. In Creep and Fracture of Engineering Materials and Structures, eds. B. Wilshire and R.W. Evans, Inst. Metals, London, 1990, pp 703-14.

TRANSITION EFFECTS IN CREEP-BRITTLE MATERIALS

K M NIKBIN
Department of Mechanical Engineering
Imperial College
London SW7 2BX

ABSTRACT

The definition for a creep-brittle material is considered in the light of material and geometric constraints that are imposed at the crack tip. It is assumed that creep-brittle fracture is achieved in plane strain and the damage at the crack tip is localized regardless of the creep ductility of the material. Data from four representative engineering alloys are considered. The state of stress local to the singularity is described in terms of the elastic and creep stresses present at initial loading. The rate of transition from the elastic field to the steady state creep field is dictated by the rate of stress redistribution that occurs at the crack tip but this transition time does not in itself explain the initial cracking rate. It has been found that the transient or stage one of crack growth in geometries, behaving in a creep-brittle manner, can be described in terms of a model based on C* incorporating the creep uniaxial ductility. The model assumes a process zone in which the transition time to accumulate sufficient crack tip creep strains dictate the initial cracking rate.

INTRODUCTION

Many engineering components in gas turbines, power plants and nuclear reactors need to be made to precision and are not expected, during their operating life at high temperatures, to deform excessively by creep. On the other hand design against creep cracking may dictate a choice of a crack-resistant creep ductile material. The mechanical and geometric constraints used in the design of components, regardless of the creep ductility of the material, invariably produce creep-brittle crack growth and failures in which creep damage is localized since the permitted design tolerances for the deformation of key components during service need to be restrictive.

The creep fracture and failures in components are usually related to initiation and growth of cracks in areas of high stress concentration, heat affected zone, and in degraded materials due to overaging and creep damage. Localized embrittlement in a creep ductile material may also fail in a creep-brittle manner. Therefore it is possible to induce a creep-brittle crack in a material with a high uniaxial creep ductility if the crack tip damage is contained locally by means of geometry or some form of material degradation.

The aim of this paper is to show the applicability of linear and non-linear fracture mechanics crack growth model based on K and C* to the initial stages of creep crack growth in creep-brittle modes of cracking. Before creep redistribution occurs K describes the crack tip linear stresses but with the advance of damage with time C* will be applicable. Sample high temperature tests results from four engineering materials, consisting of a 1%Cr 1%Mo 1/4%V Ferritic steel [1], a 10%Cr Martensitic steel FV448 [2], a 31%Ni 20%Cr Nickel-Chrome alloy 800H [1] and a 14%Cr 17%Co 5%Mo isostatically pressed powder nickel alloy AP1 [3] are presented. The crack growth data are compared to the predicted calculations for the initial cracking rates using the present transient model. The characterization of the crack growth rates in the initial stage of damage accumulation at first loading is described in terms of initial drop in the crack growth rate followed by the gradual increase in the rate to coincide with the steady state cracking rate.

CRACK GROWTH MODEL BASED ON A CREEP PROCESS ZONE

a) Steady State Behaviour

In correlating creep crack growth data the following linear elastic based relationship has been shown to have had limited success [4-5]

$$\dot{a} = C K^m \qquad (1)$$

where $\dot{a}$ is the crack growth rate, C and m are material constants. However by assuming that the creep strain rate $\dot{\varepsilon}$ is governed by Norton's creep law given as

$$\dot{\varepsilon} = A \sigma^n \qquad (2)$$

where A is the proportionality factor, n is the creep index and σ is the applied stress, a correlating parameter C* has been developed [6-7] to describe the elevated temperature crack growth rate in terms of the creep stress singularities present at the crack tip. From both theoretical and experimental [7] standpoints a relevant and an appropriate correlating relationship has been put forward between C* and creep crack growth in the form of

$$\dot{a} = D_0 C^{*\phi} \qquad (3)$$

where $\phi = (n/n+1)$, D_0 is a material constant, and C* is the non-linear creep parameter[6-7] describing the state of stress ahead of a crack tip in a creeping body. It has been extensively shown [7-8] from experimental tests that C* has been found to correlate data in a range of materials where the creep ductility is high but where creep-brittle behaviour is sustained by crack tip constraints. In general both K and C* describe the creep-brittle steady state crack growth adequately but in the initial stages of loading and damage accumulation both parameters fail to describe adequately the process of crack initiation and growth. Figure 1 shows a representative example of the 'tail' that exists in the data of 1CrMoV steel specimens tested at 550°C when correlating crack growth rate versus C*. The initial transition tail constitutes a substantial part of test times and the modelling of this feature should improve crack growth rate predictions using the C* parameter.

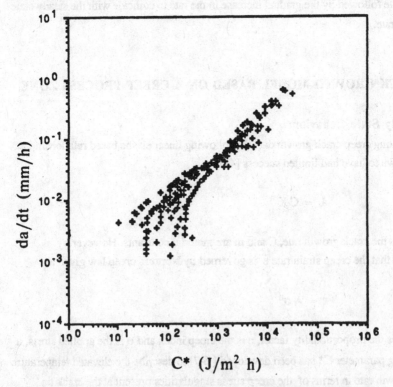

Figure 1. Example of the transient effect in the correlation of crack growth versus C* for a 1CrMoV steel compact tension specimens tested at 550 °C.

In order to predict crack growth at high temperatures a process zone (Figure 2a) can be postulated at the crack tip [9] where cracking proceeds when an element of material experiences damage and rupture stressed according to the local magnified state of stress. This non-linear stress singularity determines the rate at which the element of material accumulates damage and failure occurs when the creep ductility ε_f^* appropriate to the state of stress (or to the extent of constraint) at the crack tip is exhausted. For the case of plane stress $\varepsilon_f^* = \varepsilon_f$, where ε_f is the uniaxial creep ductility, The NSW model [9] of steady state creep crack growth gives $\mathring{a}_s$ in the form

$$\mathring{a}_s = \{(n+1)/\varepsilon_f^*\} \left[C^*/I_n \right]^{n/(n+1)} (A\, r_c)^{1/(n+1)} \qquad (4)$$

where I_n is a non-dimensional function of n, r_c is the creep process zone size over which each element sees the appropriate stress history. This expression assumes the development of steady state damage distribution in the region of the process zone. It assumes that zero damage exists at $r=r_c$ and that progressively more damage is accumulated as the crack tip is approached. Therefore the model assumes that little extra strain is required to break a ligament dr at the crack tip since it will be almost broken before the crack reaches it.

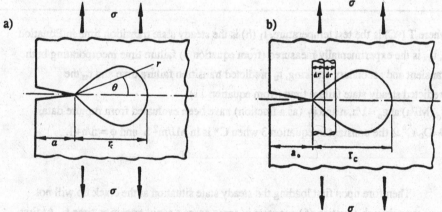

Figure 2a. Model of a creep process zone ahead of a crack tip.
   2b. Ligament damage development in a creep process zone.

b) Transient analysis

However experimental evidence as shown in figure 1 suggests that the initial loading produces a 'tail' in the crack growth. This phenomenon has also been described mathematically in [10-11]. This effect would be mainly attributed to two factors. Primary creep would induce an initial slowing down of crack growth and the transition times t_1 which is described as the time taken to go from an elastic state of stress at the crack tip to a steady state creep stress is given as [6]

$$t_1 = G/\{(n+1) C^*\} \tag{5}$$

where G is the elastic strain energy release rate. Table 1 gives typical range of values for four materials at the particular testing temperature. By eliminating the data points up to the point t_1 the actual periods of the tail that have been measured are still apparent in the test data. It is thus clear that the values of t_1 are very much less than the periods of the measured tails.

TABLE1

Geometry = Compact Tension, Material properties, and failure times.

Material	T	n	σ_0	D	ϕ	ε_f	t_1	t_{ss}/t_{tr}	t_{trx}	t_{trx}/t_{tr}
1 CrMOV	550	6.5	964	3	.87	.15	10-50	0.63	280	.95
FV448	550	9	361	5	.89	.1	1-10	0.59	9	.6
AP1	700	8	1100	5	.88	.15	<1	0.6	59	.85
Alloy 800H	800	6.5	170	6	.87	.3	<<1	0.59	940	1.04

where T (°C) is the test temperature, t_1 (h) is the steady state transition time in Equation 5, t_{trx} is the experimentally measured (from equation 3) failure time incorporating both transient and secondary cracking, t_{tr} predicted transition failure time and t_{ss} the predicted steady state failure time (from equation 11).
σ_0 (MPa) at $\varepsilon_0 = 1/h$, n and ε_f (as a fraction) have been evaluated from rupture data, $D = D_0/\varepsilon_f^*$ is the constant in equation 3 when C^* is in MJ/m^2 h and $\phi = n/n+1$.

Therefore upon first loading the steady state situation at the crack tip will not exist even though equation (5) suggests in some cases a rapid transition time t_1. At first loading a stable distribution of damage will need to built up ahead of the crack before steady state crack growth begins. The small ligament dr (Figure 2b) will not have suffered any creep strain and failure will not occur until a time dt has elapsed, given by

$$\varepsilon_f^* = \dot{\varepsilon} \, dt \tag{6}$$

where ε is the creep displacement rate. This leads to an initial creep crack growth rate $\mathring{a}_0$ giving

$$\mathring{a}_0 = dr/dt \tag{7}$$

$$\mathring{a}_0 = (1/\varepsilon_f^*) \left[C^*/I_n \right]^{n/(n+1)} (A\, dr)^{1/(n+1)} \tag{8}$$

Equation (8) is very similar to that derived from the steady state damage conditions (equation (4)) . It results in the relation

$$\mathring{a}_0 = (1/n+1) \left(dr/r_c\right)^{1/(n+1)} \mathring{a}_s \tag{9}$$

the ligament dr can be chosen to be a suitable fraction of r_c . However since dr/r_c is raised to a small power in equation (9)

$$\mathring{a}_0 \approx (1/n+1)\ \mathring{a}_s \tag{10}$$

For most engineering materials therefore the initial crack growth rate is expected to be approximately an order of magnitude less than that predicted from the steady state analysis. The cracking rate will progressively reach the steady state cracking rate as damage is accumulated. The cracking rate will increase incrementally corresponding to

$$\mathring{a}_i = \{1/ (\varepsilon_f^*/ \varepsilon_u^*)\} \left[C^*/I_n \right]^{n/(n+1)} (A\, dr)^{1/(n+1)} \tag{11}$$

where i is an integer >1, ε_u^* is the ductility already used up in the ligament prior to arrival of the crack.

Numerical integration is required to evaluate equation (11). A computer program has been developed, using incremental crack extension, to evaluate the transition period resulting from the development and the accumulation of damage in accordance with equations 4 and 11.

RESULTS AND ANALYSIS

Experimental results for the representative materials in Table1 are shown in figures 3-6. C* has been calculated from the general relationship

$$C^* = F \{ (P\dot{\Delta} / B_n W) \} \tag{12}$$

where F is a non-dimensional factor which can be obtained from limit analysis techniques [5], $\dot{\Delta}$ is the loadline creep displacement rate, B_n is the net thickness of the specimen with side-groove and W is the width. The results were obtained on standard compact tension specimens and the data were correlated versus C* which was evaluated by using equation 12.

The predicted steady state cracking rates were obtained from equations 3 and 4 . The predictions for the transient behaviour were made using the data from table 1 and numerically evaluating equation 11. The sample results for the four alloys are shown in figures 3-6 and the predicted failure times are tabulated in table 1. The value of I_n was taken from the tabulation by Hutchinson [12]. The values of r_c and dr in equations 4 and 11, which are given as orders of grain sizes, respectively are raised to a small power and it has been found in the integration that the results in figures 3-6 are insensitive to their choice.

DISCUSSION

The comparison of the experimental cracking rates in figures 3-6 and the predicted results show that the trends at the initial stages of the cracking rates, where a tail exists in the experimental data, are predicted by the model satisfactorily. The values of D_0 and ϕ used in equation 3 determine the accuracy with which the steady state crack growth rates are predicted. There are no visible differences in trends comparing the four alloys. All show varying degrees of an initial tail and the model within a factor of two or less describe the transition to steady state. The reason may be due to the differing constraint conditions, either due to temperature geometry or loading, that are

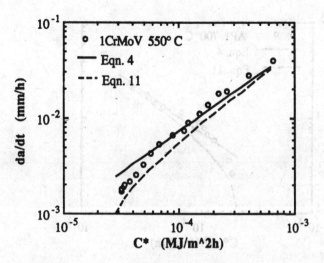

Figure 3. Comparison of the test and predicted results for 1CrMoV compact tension tested at 550° C at constant load.

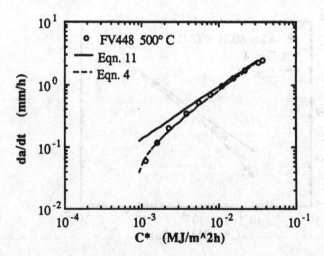

Figure 4. Comparison of the test and predicted results for FV448 compact tension tested at 550° C at constant load.

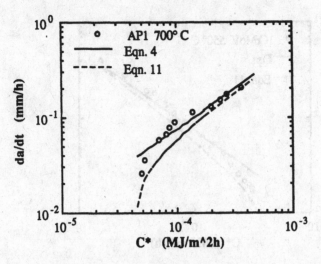

Figure 5. Comparison of the test and predicted results for AP1 compact tension tested at 700° C at constant load.

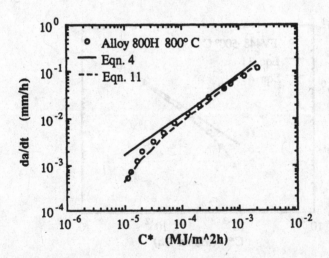

Figure 6. Comparison of the test and predicted results for Alloy 800H compact tension tested at 800° C at constant load.

differing constraint conditions, either due to temperature geometry or loading, that are imposed on the four materials in order that they would behave in a creep-brittle manner under plane strain conditions.

Considering the total experimental failure times t_{trx} and the predicted total failure times t_{tr} (which includes the transition and steady state time to failure) and the predicted steady state failure time t_{ss} from Table 1, it is clear that the longer the test time the nearer the value of t_{trx} / t_{tr} to unity and therefore the better the prediction of the total time to failure. The ratio of the predicted steady state failure time to the total transient time to failure t_{ss} /t_{tr} suggests that approximately 40% of the test time is taken up in the transient stage of crack growth. This proportion is a significant amount of the life of the specimen and could describe the time needed for initiation and growth of the crack to a steady state stage.

CONCLUSION

Creep-brittle material crack growth behaviour is defined as a situation in which a crack is sufficiently constrained (ie. plane strain) within the structure to allow it to grow under as a single dominant crack and with the creep damage and deformation contained locally. After the initial elastic loading of the geometry creep redistribution occurs at the crack tip to allow the non-linear state of stress to dominate. It has been found that this transition period to steady state C* is not sufficient to explain the 'tail' that exists in the creep crack growth data. Expressions developed to predict initial ligament damage accumulation at the crack tip have been applied to four representative engineering alloys. These alloys can be described as having a creep-brittle crack growth behaviour due to the constraints that have been imposed on them by way of geometry and crack-tip containment.

The predicted results show that initial cracking rate $\mathring{a}_0$ could be upto to approximately 1/5 of the steady state crack growth rate $\mathring{a}_s$. This value is consistent with the experimental data for the four alloys considered. The period over which the transition takes place in the predictions has also been found to be about 40% of total life and it compares well with the first stage crack initiation and growth times found experimentally. Since the ideal secondary steady state does not exist in the laboratory data and primary and tertiary effects are prevalent to varying degrees the present predictions, using the ligament damage development model, show the right trends but would be dependent on the accuracy of the C* estimation procedure used.

Improvements in the method of evaluating of C* will further improve the model's ability to predict the early stages of cracking.

REFERENCES

1.   Djavanroodi, F., Webster, G. A., Comparison between numerical   and experimental estimates of the creep fracture mechanics parameter C*, Proc. 22nd National Symp. on Fracture Mechanics. ASTM Committe E-24 , Houston June 1990.

2.   Nikbin, K.M., Nishida, K., Webster, G.A., Creep/Fatigue crack growth in a 10% Cr martensitic steel, to be presented at ICM6 , Japan, August, 1991.

3.   Winstone, M.R., Nikbin, K.M., Webster, G.A., Models of failure under creep/fatigue loading of a nickel-based superalloy, J. of Mat. Sci., 1985, **20**, pp.2471-2476.

4.   Ellison, E.G., Creep behaviour of components containing cracks - a critical review.J. Strain Anal., 1978, **13**, 35-51.

5.   Webster, G.A., Crack growth at high temperature. In Engineering approaches to high temperature design, eds. B. Wilshire, D.R.J. Owen, Pineridge Press,Swansea, 1983, pp. 1-56.

6.   Riedel, H., Rice, J.R., Tensile cracks in creeping solids, Fracture Mechanics. ASTM STP 700, 1980, pp.112-130.

7.   Nikbin, K.M., Smith, D.J., Webster, G.A., An engineering approach to the prediction of creep crack growth, J. Eng. Mat. Tech., 1986, **108**, pp.186-191.

8.   Ainsworth, R.A., Goodall, I.W., Defect assessment at elevated temperatures, J. of Press. Vessel Tech., 1983, **105**, pp.263-268.

9.   Nikbin, K.M., Smith, D.J., Webster, G.A., Influence of creep ductility and state of stress on creep crack growth, in Advances in Advances in life prediction methods at elevated temperatures, eds. D.A. Woodford, J.R. Whitehead, 1983, ASME, New York, pp. 249-258.

10.   Kubo, s., Ohji, k., Ogura, k., An analysis of creep crack propagation on the basis of the plastic singular stress field, Eng. Fract. Mech., **11**, 1979, pp. 315-329.

11.   Riedel, H., The extension of a macroscopic crack at elevated temperature by the growth and coalescence of microvoids, Creep in Structures, eds. A.R.S. Ponter, D.R.Hayhurst, Springer-Verlag, Berlin, 1981, pp.514-519.

12.   Hutchinson, J.W.,Singular behaviour at the end of a tensile crack in a work-hardening material, J. Mech. Phys. of Solids, 1968, **16**, pp.13-31.

CRACK GROWTH STABILITY IN SALINE ICE

Samuel J. DeFranco and John P. Dempsey
Department of Civil and Environmental Engineering
Clarkson University, Potsdam, N.Y. 13699-5710

ABSTRACT

Crack initiation and propagation in saline ice were investigated in this study in an attempt to understand the processes of crack growth under low loading rates and variable temperatures. As has been previously observed in sea ice and freshwater ice, crack growth occurred in initiation/arrest increments. The energetic stability criterion of crack growth are examined and crack growth is characterized in terms of the fracture resistance–stress intensity factor K_R. Salient features of this study are the development of a new fracture geometry capable of sustained crack growth over longer crack lengths and the presentation of a fracture resistance curve for saline ice.

INTRODUCTION

Ice creeps readily under sustained external loading and while many studies exist [1,2] in the literature concerning loading rate effects on the initiation fracture toughness (K_{Ic}) of ice (for ice, K_{Ic} increases with decreasing loading rate), crack propagation and crack growth stability has received only limited attention [2-5]. In these studies, crack propagation in sea ice [3,4] and freshwater ice [3-5] was observed to be incremental initiation/arrest behavior in the two different test geometries which were utilized. In [3,4] the double torsion [6] constant K geometry was used in an attempt to study subcritical crack growth in ice while a modified double cantilever beam geometry based on the crack ligament model of Kanninen [7] was used in [5]. Both studies found the apparent rate sensitivity of crack propagation in freshwater ice; i.e., the fracture resistance R decreases with increasing crack velocity $\dot{a}$. In sea ice [3,4] similar behavior was observed.

For a crack to propagate stably in a medium, the crack driving energy G must equal the resistance to fracture R. In a purely brittle material, R consists solely of the energy required to create new surface area or $R = 2\gamma$. In ice however, polycrystalline effects as well as creep effects at the crack tip cause the experimentally determined critical energy release rate R to be much higher than the fracture surface energy[1].

While the equality of the driving and resisting energies is a necessary condition for stable crack growth, the rate of change of each energy with respect to crack growth Δa must also equate. Therefore,

$$G = R \tag{1}$$

and

$$\frac{\partial G}{\partial \Delta a} = \frac{\partial R}{\partial \Delta a}. \tag{2}$$

In terms of the test geometry compliance $C(a)$, the energy release rate G can be expressed as:

$$G = \frac{P^2}{2h} \frac{dC}{da} \tag{3}$$

where h is the specimen thickness and a is the crack length.

ENERGETIC STABILITY IN FRACTURE TESTING

Gurney and Hunt [9] noted the functional dependence of G on a greatly influences crack growth stability – by manipulation of (1), (2) and (3) and considering testing machine compliance C_M, they determined that

$$\frac{1}{R}\frac{dR}{da} = \frac{C''}{C'} - 2\frac{C'}{C}\left(\frac{C}{C + C_M}\right) \tag{4}$$

where C, C' and C'' are the compliance and the first and second derivatives of the compliance with respect to the crack length a for the case of displacement control ($\delta u/u > 0$). They classified ideal test geometries as having the desired 'geometric stability factor' (hereafter referred to as the gsf) if $\partial G/\partial a$ was negative over an appropriate crack length interval. In other words, the test geometry must exhibit the gsf to match the fracture resistance behavior in $(dR/da)/R$. As indicated in (4), however, a compliant testing machine increases the gsf and may induce unstable fracture in geometries which are normally stable in a more rigid testing machine.

The application of this concept on ice was first utilized in [4] using the double torsion (DT) specimen; based upon compliance information in [6] and assuming a rigid testing machine, the normalized gsf for this geometry is given as:

$$\frac{L}{R}\frac{dR}{da} = -2\frac{L}{a} \tag{5}$$

where L is a characteristic geometric dimension of the test geometry. As can be seen from (5), stable crack growth should occur for materials which exhibit increasing, flat or slightly decreasing $(dR/da)/R$ behavior.

Preliminary experiments to this study were performed on freshwater columnar ice using the crack line wedge loaded (CLWL) geometry. The CLWL geometry is commonly used in R curve determination of metals and is well known to effect slow crack growth in positive $(dR/da)/R$ materials. An important feature of the CLWL geometry is the

---

[1]For ice, Ketcham and Hobbs [8] give the specific free surface energy for fracture along an easy cleavage plane of a single crystal of freshwater ice at $0°$ as $\gamma_{sc} = 0.109$ J/m^2. The energy for fracture along a grain boundary is given by $\gamma_{gb} = 0.065$ J/m^2.

mechanical advantage achieved through wedge loading; the machine load required for crack propagation is reduced by a factor of up to 5 [10], hence reducing any fracture instabilities created by machine compliance effects [9]. Crack growth in the preliminary freshwater ice experiments, however, consisted mainly of long crack jumps that arrested very near the specimen edge. As pointed out in [11], reflected stress waves from the boundaries of finite sized specimens tends to reduce the validity of quasi–static evaluation of long arrested crack jumps thus limiting the usefulness of the CLWL as a fracture geometry for ice testing.

The motivation for a new test geometry was provided in [12] where the authors showed that reverse tapering (i.e. decreasing specimen width W with increasing crack length a) greatly improved crack growth stability. They also showed that crack stability in reverse tapered geometries was greatly improved from typical forward tapered, constant K fracture geometries. Therefore, a new geometry, the reverse tapered CLWL (RT–CLWL) was developed and utilized for fracture resistance experiments on saline ice at Clarkson.

RT–CLWL FRACTURE GEOMETRY

Figure 1 shows a schematic of the RT–CLWL geometry. Loading is applied via a wedge which is in contact with two sets of split pins made in accordance with [11]. Each split pin consists of a load sector which directly contacts the specimen and a wedge load block which is tapered in order to fully contact the loading wedge (see Figure 2). To ensure proper alignment of the load sectors and hence the load path, the wedge load blocks are machined such that line contact exists. This prevents load eccentricity which would tend to cause shear as well as tensile loading at the crack tip. Figure 2 shows a schematic of the loading fixture. As the wedge is displaced downward by the test frame, loading is applied by the load sectors and wedge load blocks which rest on sliding retainer blocks which in turn rest on a base block. The wedge, wedge load blocks and load sectors were constructed of stainless steel and the base block and sliding retained block were constructed of T-6061 aluminum. To reduce friction between all components, powdered graphite was used as a lubricant. No fluid lubricants were found which were capable of maintaining low viscosity at the test temperatures used for this study.

An important parameter to be determined was the friction which exists between the loading components. To accomplish this, a steel plate (355 mm × 152 mm × 6 mm) was constructed as shown in Figure 3 and loading was applied to the center hole by direct tension and by wedge loading. Stress concentrators (14 mm diameter holes) were placed along the tension line in order to reduce the load needed to cause measurable strain at load levels that would be experienced during an ice fracture experiment. By comparison of load versus strain plots, the ratio of the opening load P to the machine load P_M was found through several load–unload cycles to be:

$$\phi = P/P_M = \begin{cases} 3.52 & \text{graphite} \\ 2.89 & \text{no graphite} \end{cases} \tag{6}$$

All loading cycles were performed in a cold room at $-25°$ C and in each case the load versus strain plots were linear.

In order to determine an expression for the energy release rate, compliance measurements were performed on a specimen made of PMMA (thickness $h = 25.4$ mm, all other

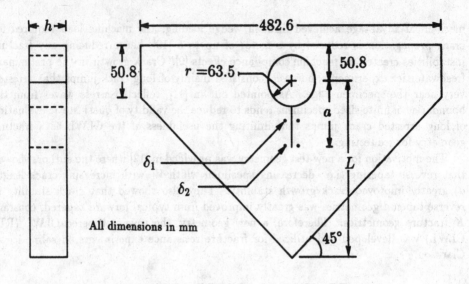

Figure 1. The RT–CLWL fracture geometry.

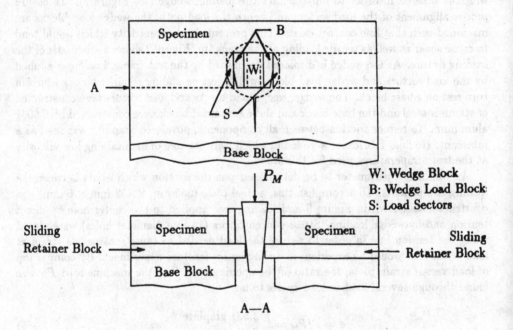

Figure 2. Wedge loading apparatus and setup.

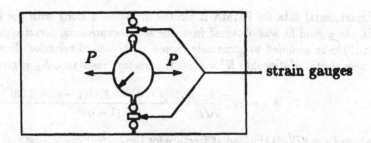

Figure 3. Schematic of device used for friction analysis.

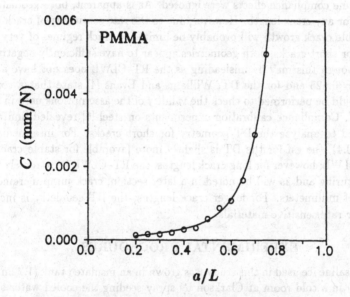

Figure 4. RT–CLWL compliance curve.

dimensions are given in Figure 1) at $0.05a/L$ crack increments for $0.20 \leq a/L \leq 0.75$. Three loading trials were performed for each crack length and the average value of the three was used. Elastic response of the PMMA was ensured by zero to peak loading times of less than 5 seconds and all loading components were liberally coated with powdered graphite. The specimens were not loaded to fracture in any trial and all experiments were performed at $-25°$ C. The load versus load point displacement plots were linear up to the maximum applied load in each case.

The compliance was found by a least squares curve fit to be:

$$C = \frac{10^2}{E'h}\left[3.29\ln\left[(1-\alpha)^{-1}\right] - \frac{4.47}{1-\alpha} + \frac{0.89}{(1-\alpha)^2} + 3.53\right] \tag{7}$$

where E' is the effective elastic modulus (assumed to be 3.93 GPa for PMMA), h is the specimen thickness and $\alpha = a/L$. The form for (7) is similar to equation (14) in [12]; the experimental data for PMMA is plotted in Figure 4 along with the least squares curve fit. As a good fit was obtained from the above expression, derivatives of the compliance in (7) were assumed to accurately represent the actual behavior. From (3), (7) and the linear elastic relationship $K^2 = GE'$, the fracture resistance K_R is given by:

$$K_R = 5\sqrt{2}\frac{P_f}{h\sqrt{L}}\frac{[3.29(1 - \alpha)^2 - 4.47(1 - \alpha) + 1.78]^{1/2}}{(1 - \alpha)^{3/2}} \qquad (8)$$

where $P_f = \phi P_M^f$ is the load at fracture for the current crack length α. P_M^f is the machine load at fracture and ϕ is the mechanical advantage given in (6).

Figure 5 shows the geometric stability factors for the DT and RT–CLWL geometries assuming a rigid testing machine. For the RT–CLWL, preliminary experiments revealed fracture loads for ice specimens that were less than 1% of the testing machine capacity and thus machine compliance effects were ignored. As is apparent, both geometries have negative gsf's for any crack length. However, due to the rate sensitivity of crack propagation in ice, stable crack growth will probably be limited to crack regimes of very negative $(dR/da)/R$. For short cracks, both geometries appear to have sufficiently negative stability factors although this may be misleading as the RT–CLWL does not have a physical crack until $\alpha = 0.125$ and for the DT, Williams and Evans [4] state that a compliance calibration should be performed to check the validity of the assumptions used in the analysis of the DT. Compliance calibration experiments on steel [6] revealed limitations on the theory used to analyze the DT geometry for short cracks. For intermediate crack lengths ($\alpha \approx 0.4$), the gsf for the DT is slightly more favorable for stable crack growth than the RT–CLWL; however for long crack lengths, the RT–CLWL is extremely favorable for stable fracturing and as will be noted in a later section, crack jump increments were on the order of millimeters. For longer crack lengths, the DT geometry is increasingly unfavorable for rate sensitive materials.

EXPERIMENTAL PROCEDURES

The columnar saline ice used in this study was grown in an insulated tank (1.22m × 1.14m × 0.78m deep) in a cold room at Clarkson by spray seeding the cooled water surface to initiate columnar growth. The ice was grown to a depth of approximately 0.15 m and cut into blocks measuring approximately 0.30 m × 0.50 m × 0.15 m thick using a chain saw. The average grain size was approximately 15 mm and the ice had an average density at $-25°$ C of approximately 912 kg/m^3. Salinity was measured using a LabComp SCT microprocessor controlled salinity meter. The bulk salinity was found to be 7.59 ± 1.44 $°/_{oo}$. The ice blocks were immediately moved to an adjacent cold room where they were stored at $-25°$ C for a minimum of 24 hours. After the blocks were uniformly cold, 3 plates were cut from each block and were made parallel by a planer. The edges were formed by a jointer thus ensuring that each plate was a uniform prism. To form the final shape for each RT–CLWL, the dimensions were carefully measured and then cut on a bandsaw. Each finished specimen was approximately 50 mm thick. Drilling the 63.5mm hole in each plate for the loading sectors presented special difficulties in that successively larger holes had to be drilled in order to prevent the specimen from fracturing during the drilling operation. The drilling sequence was: (i) drilling a pilot hole with a 25.4 mm wood spade, (ii) enlarging the hole with a 38.1 mm wood spade and (iii) drilling to the final

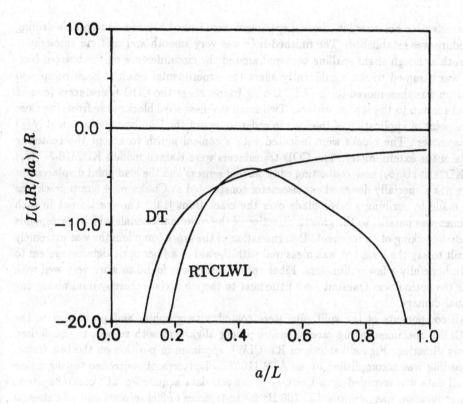

Figure 5.   Energetic stability comparison.

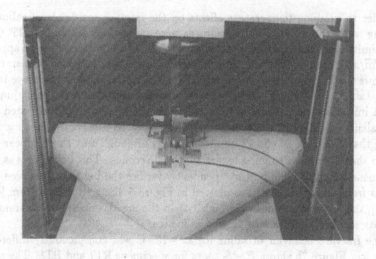

Figure 6.   RT–CLWL specimen ready for testing.

diameter with a Forstner bit. Several specimens were ruined before an acceptable drilling procedure was established. The finished hole was very smooth and uniform throughout its depth although slight spalling occurred around the circumference on the bottom face. This was assumed to not significantly affect the experimental results. Each completed specimen was then moved to the ATS testing frame where the COD transducers (δ_1 and δ_2) were frozen to the top ice surface. Two small stainless steel blocks were frozen as close to the point of application of the load in order to mount the load line displacement (δ_{LL}) extensometer. The blocks were indented with a conical punch to accept the centering points of the extensometer. The COD transducers were Kaman models KD2310-3U (δ_1) and KD2810-1U (δ_2) non–contacting eddy current sensors and the load point displacement gauge was a specially designed extensometer constructed at Clarkson. A sharp crack was then made by scribing a razor blade over the crack front [13]. The crack front in each specimen was parallel to the growth direction of the columnar crystals which corresponds to radial cracking of an ice sheet. Determination of the crack jump lengths was extremely difficult to say the least but was measured with the aid of a fiber optic lighting system to within hopefully a few millimeters. Fiber optic lighting was found to work very well with ice as the optic fibers transmit very little heat to the ice surface thereby minimizing any thermal damage.

All components of the split pins were coated with graphite and positioned in the RT–CLWL specimen taking care to ensure proper alignment with respect to the desired loading direction. Figure 6 shows an RT–CLWL specimen in position on the test frame. All loading was accomplished by an ATS 1105SE displacement controlled testing frame and all data was recorded by a Keithley Series 500 data acquisition and control system. Data acquisition was performed at 100 Hz for tests times of 200 seconds and all data was stored in a Zenith Z248 microcomputer.

RESULTS AND DISCUSSION

In order to examine temperature effects on the fracture resistance of saline ice at slow loading rates, three RT–CLWL specimens were successfully tested at $-25°$ C and due to time limitations, only two were tested at $-15°$ C. Loading rates $\dot{K}$ were approximately 2 kPa$\sqrt{\mathrm{m}}$/s for each experiment. The differences in fracture behavior were startling. Figure 7a shows the $P - \delta_2$ plot for specimen RT5[2]. The initial slope of the curve is linear up to the instability load where crack initiation occurred causing a rapid crack jump of 23 mm. K_R at initiation was calculated from (8) to be 105 kPa$\sqrt{\mathrm{m}}$; the associated arrest value, also calculated from (8), was found to be 95 kPa$\sqrt{\mathrm{m}}$. A total of five crack jumps occurred of lengths 23, 23, 20, 20 and 12 mm respectively although only three appear on Figure 7a due to the experiment running longer than 200 seconds. This behavior was typical of all specimens tested at $-25°$ and independently verifies the behavior observed in [3,4]. The results from all experiments are plotted in Figure 8 along with data from [14][3]. Similar to [14], the initiation values were higher than the arrest values—which simply illustrates the rate sensitivity of fracture in cold saline ice.

The fracture behavior of saline ice at $-15°$ C was conspicuously different from the colder ice. Figure 7b shows $P - \delta_2$ plots for specimens RT7 and RT8. The plots are

---

[2]Note: $\delta_2|_{\Delta a \to 0} \equiv$ CTOD

[3]In Figure 8, only specimens in which three or more crack initiation/arrest events occurred were plotted.

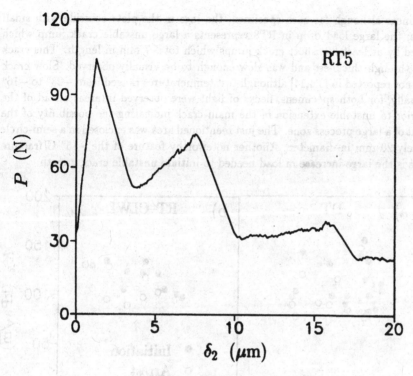

Figure 7a. P vs δ_2 at $-25°$ C.

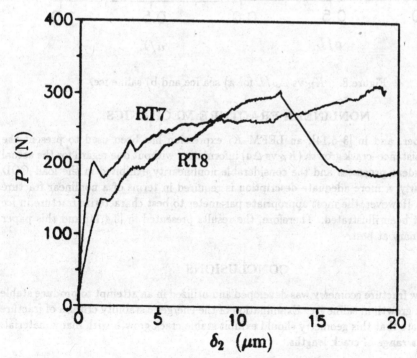

Figure 7b. P vs δ_2 at $-15°$ C.

clearly nonlinear although for each specimen, the jags in the plots coincide with small crack jumps; the large load drop in RT8 represents a large unstable crack jump which was preceded by at least five short crack jumps which total 7 mm in length. This crack growth was through thickness and was slow enough to be visually observed. Slow crack growth was not reported in [3,4,14] although test temperatures ranged from −23° to −16° C. Additionally, for both specimens, flecks of light were observed in areas ahead of the crack tip prior to unstable extension of the main crack indicating the possibility of the development of a large process zone. The just mentioned area was enclosed in a semi-circle approximately 20 mm in diameter. Another noteworthy feature of the −15° C fracture experiments is the large increase in load needed to initiate unstable crack growth.

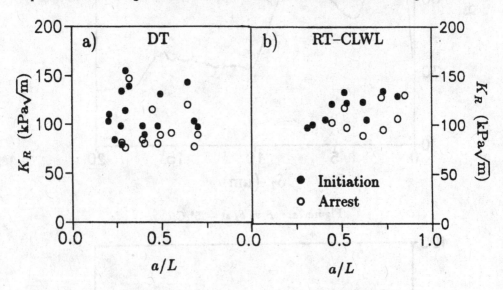

Figure 8. K_R vs $\Delta a/L$ for a) sea ice and b) saline ice.

NONLINEAR FRACTURE MECHANICS

In this paper, and in [3–5,14], an LEFM-K_I expression has been used to present the fracture resistance–crack growth (K_R vs Δa) information without due regard for the actual time dependent processes and the considerable nonlinearity exhibited in the load–COD plots. Clearly, a more adequate description is required in terms of a nonlinear fracture parameter. However, the most appropriate parameter to best characterize fracture in ice has not yet been illustrated. Therefore, the results presented in [3,4,14] and this paper are preliminary at best.

CONCLUSIONS

- A new fracture geometry was developed and utilized in an attempt to produce stable crack growth in saline ice. Examination of the energetic stability criteria of fracture indicate that this geometry should exhibit stable crack growth with many materials over a range of crack lengths.

- Fracture experiments on saline ice at −25° C showed brittle crack initiation/arrest behavior which verified previous experimental results. Conversely, for ice at −15° C, more ductile behavior was observed with load displacement plots exhibiting a much greater degree of nonlinear behavior throughout the fracture experiment. Slow crack growth was observed up to a critical load which coincided with a long unstable crack jump.

- While no unique crack arrest toughness was found, initiation/arrest values were observed to increase with crack length which may well indicate a size/geometry effect in the RT–CLWL, since the DT geometry did not give the same behavior.

- Due to nonlinearities in the load–displacement plots, particularly for warmer ice, it is suggested that LEFM is not an appropriate discipline to treat the fracture of warm ice. A nonlinear fracture mechanics approach may be the only suitable method for treating high temperature fracture of sea ice.

ACKNOWLEDGEMENT

The authors thank Mr. R. Cary, Mr. P. Parker and Mr. Y. Wei for their assistance during the experimental portions of this study. Mr. Cary and Mr. Parker were supported under a grant by the National Science Foundation's Research Experience for Undergraduates Program under Grant No. MSM–86–18798.

This work was supported in part by the U.S. Office of Naval Research under Grant No. N00014–90–J–1360 and in by the U.S. National Science Foundation under Grant Nos. MSM–86–18798 and MSS–90–079291.

REFERENCES

1. Dempsey, J.P., The fracture toughness of ice. In IUTAM/IAHR Symp. Ice/Struct., Springer–Verlag, 1989, (in press).

2. Dempsey, J.P., Wei, Y. and DeFranco, S.J., Fracture resistance to cracking in ice: initiation and growth, In Cold Regions Engineering, ed. D.S. Sodhi, ASCE, 1991, 579–594.

3. Parsons, B.L., Snellen, J.B. and Muggeridge, D.B., The initiation and arrest stress intensity factors of first year columnar sea ice. In 9th IAHR Ice Symp., 1988, 1, 502-512.

4. Parsons, B.L., Snellen, J.B. and Muggeridge, D.B., The double torsion test applied to fine grained freshwater columnar ice, and sea ice. In European Mechanics Colloquium 239, 1989, 188–200.

5. DeFranco, S.J. and Dempsey, J.P., Crack growth stability in S2 ice. 10th IAHR Ice Symp., 1990, 1, 168–181.

6. Williams, D. P. and Evans, A.G., A simple method for studying slow crack growth. J. Test. Eval., 1973, 1, 264-270.

7. Kanninen, M.F., An augmented double cantilever beam model for studying crack propagation and arrest. Int. J. Frac., 1973, 9, 83–92.

8. Ketcham, W.M. and Hobbs, P.V., An experimental determination of the surface energies of ice. Phil. Mag., 1969, 19, 1161–1173.

9. Gurney, C. and Hunt, J., Quasi–static crack propagation, Proc. Roy. Soc., 1967, A361, 254–263.

10. ASTM E561–86, 1987. R-curve determination. ASTM Stds, 03.01, 793–812.

11. Kanninen, M.F. and Popelar, C.H., Advanced Fracture Mechanics, Oxford University Press, New York, 1985, p. 212.

12. Mai, Y–W., Atkins, A.G., and Caddell, R.M., On the stability of cracking in wedge tapered DCB specimens. Int. J. Frac., 1975, 11, 939–953.

13. DeFranco, S.J., Wei, Y., and Dempsey, J.P., Notch acuity effects on the fracture toughness of saline ice. Annals of Glac., 1991, 15, in press.

14. Parsons, B.L., Subcritical crack growth, initiation and arrest in columnar freshwater and sea ice. Doctoral Thesis, Memorial University of Newfoundland, St. John's, Newfoundland, 1989.

FINITE ELEMENT PREDICTION OF CREEP DAMAGE AND CREEP CRACK GROWTH

Yunling Duan*, J.J.Webster and T.H.Hyde.
Department of Mechanical Engineering,
The University, Nottingham, NG72RD, U.K.

*Visiting Scholar from Zhengzhou Institute of Technology, P.R. of China; now at Department of Civil Engineering, University of Bradford, U.K.

ABSTRACT

Eight node plane stress finite elements have been used to simulate the steady load creep behaviour of a tensile loaded strip with a central hole and a crack specimen subjected to mode II loading using continuum damage material behaviour models. The results for the tensile loaded strip are compared with previous constant strain finite element and experimental results and the crack results are compared with test data. The finite element results illustrate some of the unsatisfactory characteristics of this local approach which can be eliminated by using a localisation limiter.

INTRODUCTION

The design and life assessment of plant operating at high temperature requires the prediction of failure of both cracked and uncracked components under creep conditions. Potentially the most powerful and general method for predicting the creep behaviour of components with complex geometry is finite element analysis with a creep continuum damage material behaviour model. This enables the behaviour to be simulated by time marching without regenerating new finite element meshes to model the development and growth of damage zones and cracks during the simulation. Hayhurst et al were the first to apply this approach, initially to tension plates containing holes [1] and subsequently to a number of other uncracked and cracked components made of a variety of materials [2,3]. They established that the approach provided good predictions for component behaviour by comparing their predictions with experimental test results.

Since the mid 1980s the authors and co-workers have used the finite element continuum damage approach to predict the creep behaviour of cracked, 2-D and 3-D components. They found [4-6] that :-

(i)   the solutions were dependent on the finite element mesh;

(ii)   numerical instabilities sometimes developed during the solution;

(iii)   in constrained regions eg. in plane strain and 3-D crack problems, regions of high hydrostatic tension developed which prevented highly damaged material from shedding its load. This problem was overcome by introducing damage into the elastic constants.

At about the same time it was reported eg. [7,8] that other workers were finding similar problems with the approach, both for plasticity problems as well as creep. It was considered possible that these problems arose with higher order elements but not with the 3-node constant strain triangular elements used by Hayhurst et al. [1-3]

In this paper 8-node element solutions are compared with the constant strain solutions, reported by Hayhurst et al. [1], for a tension plate with a central circular hole. In addition some continuum creep damage finite element predictions for plane stress mode II creep crack growth in a 316 stainless steel specimen at $600^{0}C$ are compared with experimental test results.

COMPARISON OF 3-NODE AND 8-NODE ELEMENT SOLUTIONS

Problem definition

The constant strain F.E. and also some experimental results for uniaxially loaded copper tension plate specimens with a central hole are reported in reference [1]. The specimens were approximately 150 mm overall length and 30 mm wide with a central hole of 3.3 mm diameter. The middle (approximately 78 mm) length of the specimen was thinned to provide a uniform thickness plane stress section. Loads were applied to give a stress of 31.03 N/mm^2 on the minimum section; this gives a remote stress of 27.62 N/mm^2. Hayhurst et al's [1] form of the Kachanov - Robotnov coupled creep strain / damage law is :-

$$\dot{\varepsilon}^c = G \left[\sigma_{eq} / (1-\omega)\right]^n t^m$$

(1)

$$\dot{\omega} = M \left\{ \left[\propto \hat{\sigma}_I + (1-\propto) \sigma_{eq} \right]^\chi / \left[(1+\phi) (1-\omega)^\phi \right] \right\} t^m$$

The numerical values of the constants for copper at 250°C are given in Table 1. The elastic constants [2] are Young's modulus = 66.24 kN/mm^2 and Poisson's ratio = 0.3.

TABLE 1

Material behaviour law constants (Units:- Fractional strain, N, mm, h)

Material	G	M	n	χ	m	φ	∝
Copper	3.21 x10^{-12}	1.89 x10^{-7}	5.00	3.19	-0.430	6.000	1.0
316 SS	7.293x10^{-17}	2.029x10^{-16}	5.39	5.76	-0.537	5.668	0.75

Finite element models and solution

One symmetric quadrant of the central 30 mm length of the specimen was modelled. The present mesh of 8-node elements and part of the mesh of 3-node elements may be seen in Figures 1(a) and (b) respectively. Details of the solution for the constant strain elements are given in reference [1]. For the present analysis symmetry and free edge conditions were

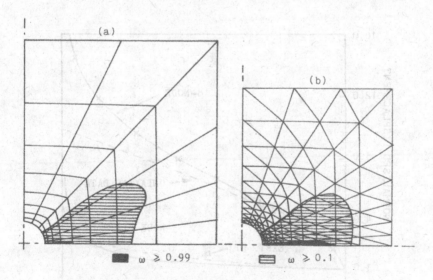

Figure 1. Finite element meshes and damage distributions for:-

(a) Eight node elements and (b) Three node elements.

imposed on the appropriate boundaries and the loaded edge was constrained to have uniform displacement in the load direction, with no lateral constraint. Four Gauss integration points were used for each 8-node element. Details of the solution procedure are given in reference [4]; for this plane stress problem the stresses reduce as the damage increases and material failure is simulated without the neccessity of reducing the local stiffness. The limiting value of damage for the solution was set at 0.99.

Results and discussion

The rupture time obtained from the present solution was 483.5 h. The value obtained from the constant strain element analysis and the average of three experimental values, interpolated from graphical data [1], are 400h and 780h respectively.

Hayhurst et al. [1] compared their F.E. and experimental results for the normalised displacement (defined as the ratio of the total to the elastic displacement) during the life of the component. These results and the present solutions are plotted in Figure 2. The present results are obtained from the F.E. results for the region of the mesh plus the calculated elastic and creep extension for the additional length of the thinned section of the test specimen, assuming that this length is subjected to the remote stress. It may be seen that their are

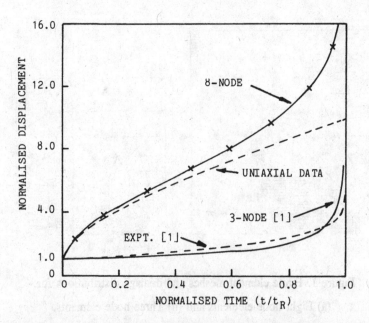

Figure 2. Variation of normalised displacement with normalised time.

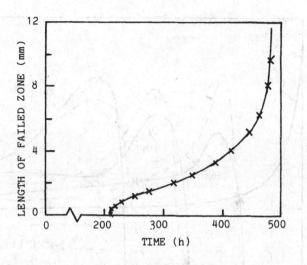

Figure 3. Increase in length of failed material zone with time.

considerable differences between the present and previous results. Also included in the figure is the curve for a uniform tensile stress, equal to the remote stress, determined from the material data. The present results gradually diverge from the uniform tensile stress curve as is expected.

The present results for the failed (0.99 damage) and 0.1 damage contours at 351h are shown in Figure 1a. This time was chosen because the extent of the 0.1 contour is similar to that on the distribution given for the constant strain elements [1] and reproduced in Figure 1b. However the present solution for the 8-node elements has a shorter failed region than the 3-node element solution. The growth of the failed material ($\omega = 0.99$) region through the minimum section for the present analysis is shown in Figure 3.

Distributions of direct stress across the minimum section initially and at three times during the simulation are shown in Figure 4. These times were chosen so that the extent of the failed regions were similar to those for the stress distributions given in reference [1]; these are also included in the figure. For the present simulation the peak stress increases with increase in length of the failed region whereas the peak stress decreases for the previous simulation.

It appears that the constant strain triangular element continuum creep damage solutions for the test specimen differ considerably from the 8-node element solutions. The rupture times are in reasonable agreement but there is no obvious explanation for the differences in the

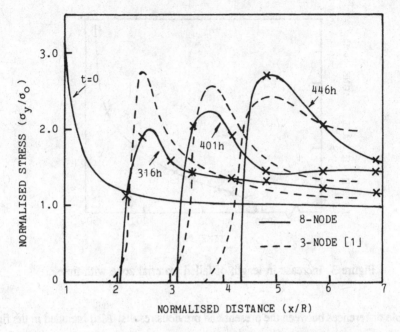

Figure 4. Distributions of normalised direct stress across minimum section.

displacements. Also the differences between the stress distributions and the damage contour plots indicate that the balance between stress redistribution and damage growth differs between the two solutions. A possible explanation is that different material behaviour laws were used. Solutions are given in reference [1] for a range of values of the stress exponents in the material behaviour laws. However the effect of the other constants in these laws and the elastic constants, which also affect the stress distributions, is not discussed. The stress exponents for both the constant strain and the 8-node element stress distributions are those given in Table 1 for copper but the other constants in the laws used to obtain the constant strain element results are not stated explicitly in reference [1].

PREDICTION OF MODE II CREEP CRACK GROWTH

Experimental Results

A Compact Mixed Mode (CMM) [9] specimen, Figure 5, with an a/W ratio of 0.5 and a load angle θ of 106^0 was used to give almost pure mode II loading. The 316 stainless steel specimen was tested at 600^0C with a steady load of P = 4147.7N for 2184 h. The preparation of the specimen, equipment and test procedure are described in reference [10]. Figure 6 is a photograph of the specimen after the test. The right hand crack has grown approximately

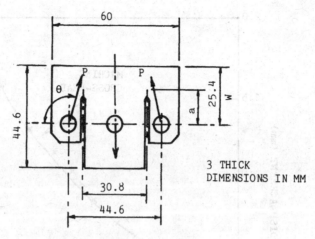

Figure 5. Compact mixed mode crack specimen.

2 mm at 90⁰ to the initial crack direction during the 2184h test. It may also be seen that the left hand crack initially grew at approximately 90⁰ to the initial crack direction but after about 2 mm growth the direction changed to approximately 60⁰. This latter growth occurred in the final stages of the test and it involved considerable local plastic deformation and tearing.

The movement of the cross-head of the testing machine during the test is plotted in Figure 7.

Figure 6. Specimen after creep test.

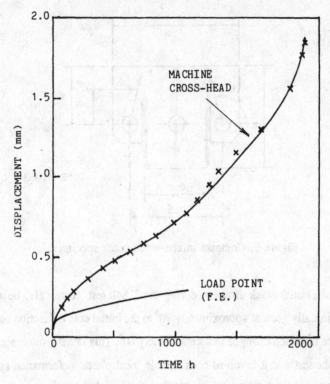

Figure 7. Variation of cross-head and load point displacements with time.

Finite element and material behavior models

The mesh of 8-node isoparametric elements for a symmetric half of the specimen is shown in Figure 8; the crack tip is at the origin. The loading was applied at points A and B. Symmetry conditions were imposed along the top edge with the load point A being completely restrained.

Hayhurst et al's [1] form, equations (1), of the Kachanov-Robotnov coupled creep/damage equations were used to model the material behaviour; the values of the constants, determined from uniaxial tests [11], are included in Table 1. The fit of the law with the uniaxial data is shown in reference [4]. The value of the constant $\propto$, which defines the creep rupture surface, is that used by Hayhurst [2] for 316 SS at 550^0C. The elastic constants are a Young's modulus of 148 GPa and Poisson's ratio of 0.3

Finite Element Results and Discussion

In the finite element simulation there was considerable local deformation at the load points.

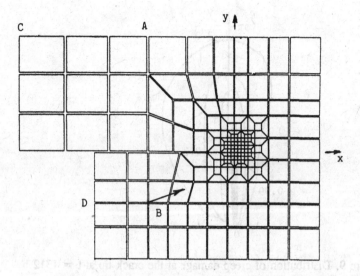

Figure 8. Finite element mesh for the crack specimen.

Examination of the displacements along C-A and D-B, in Figure 8, indicated that a good estimate of the overall deformation of the specimen, excluding the local load point deformation, was obtained from the difference in the diplacements of points C and D. This displacement is included in Figure 8. It may be seen that the cross-head movement is much larger than the F.E. solution due, in part to the deformation of the load shackles, but mainly to the deformation of the load pin holes in the specimen which may be seen in Figure 6.

The distribution of damage at 1312h in the vicinity of the crack is shown in Figure 9; the failed zone is confined to a narrow region, ie. in the form of a crack. The direction of the crack is at 90^0 to the initial crack direction which agrees with the experimental observations. The predicted crack growth (extension of the failed zone) during the simulation upto 1312h is plotted in Figure 10. In the test there was approximately 2 mm crack growth in 2184h indicating that the predicted crack growth rate is somewhat greater than that in the test.

GENERAL DISCUSSION

It is possible that the differences between the 3-node and the 8-node element solutions for the tensile loaded copper plate with the central hole may be due to differences in the material behaviour laws used for the solutions. However, these solutions and also those for the mode

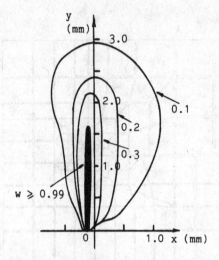

Figure 9. Distribution of creep damage at the crack tip at t = 1312 h.

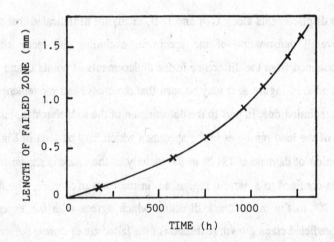

Figure 10. Crack growth (extension of failed zone) during creep simulation.

II crack illustrate some of the inadequacies of the local continuum damage approach. In each solution the failed material zone is restricted to a narrow band of material which is one element thick for the 3-node elements or contains a single Gauss integration point through its thickness for the 8-node elements. This localisation leads to mesh dependency of the solutions. Not all local continuum damage finite element solutions exhibit this localisation, eg. Hayhurst et al's [2] solutions for 316 stainless steel circumfirentially notched tension bars have failure zones which are many elements thick.

Saanouni et al. [12] present solutions for a tensile specimen loaded at constant displacement rate and for a compact tension specimen subjected to constant load. These solutions also show the deficiency of the local continuum damage approach. They propose a non-local damage variable which eliminates the localisation problem. This non-local damage is a weighted volumetric average of the local damage. The weighting or localisation function is defined in terms of a length parameter which defines the proximity effect of local damage on the non-local damage. The length parameter is a material property which was determined for the C.T. specimens as the value for which the solutions for crack initiation times agreed with experimental results. Saanouni et al's [12] results show that the non-local damage approach eliminates the localisation of the failure zone and results in solutions which converge with mesh refinment. Also it eliminates the rapid changes in the stresses which occur as each Gauss point fails; this allows much larger time steps to be used for the simulation and gives a significant reduction in computer time.

Localisation may be restricted by putting a lower limit on the size of the elements in the idealisation. This may be satisfactory when the stresses are non-uniform and the failure path is localised but it would not eliminate localisation in large uniform stress regions. Also the element size limitation may result in inaccurate modelling of fine geometric detail.

Hall and Hayhurst [13] have also suggested a non-local damage variable. In this case the finite element model of the structure is divided into a mesh of cells whose size was taken to be seven grain diameters. The non-local damage for an element is the volumetric weighted average of the local damage for the cell containing the element. This procedure enabled the differences in crack growth due to size effect, which are not predicted by the local approach, to be predicted accurately.

The use of a non-local damage variable eliminates localisation of the failure zone and produces finite element solutions which converge satisfactorily. Further investigation is required to determine the most suitable localisation functions and to develop methods for determining the appropriate length parameter for a material.

REFERENCES

1.   Hayhurst, D.R., Dimmer, P.R. and Chernuka, M.W., Estimates of the creep rupture lifetime of structures using the finite element method. J. Mech. Phys. Solids, 1975, **23**, pp. 335-355.

2. Hayhurst, D.R., Dimmer, P.R. and Morrison, C.J., Development of continuum damage in the creep rupture of notched bars. Phil. Trans. R. Soc. Lond., 1984, A311, pp. 103-129.

3. Hayhurst, D.R., Brown, P.R. and Morrison, C.J., The role of continuum damage mechanics in creep crack growth. Phil. Trans. R. Soc. Lond., 1984, A311, pp. 131-158.

4. Smith, S.D., Webster, J.J. and Hyde, T.H., Three dimensional damage calculations for creep crack growth in 316 stainless steel. In Applied Solid Mechanics-3, ed.I.M. Allison and C. Ruiz, Elsevier Applied Science Publishers, London, 1989, pp. 363-377.

5. Chambers, A.C., Webster, J.J. and Hyde, T.H., Continuum damage predictions of pure and mixed mode I and II creep crack growth. In preparation.

6. Chambers, A.C., Mixed Mode Creep/Fatigue Crack Growth, Ph.D. Thesis, University of Nottingham, 1989.

7. Murakami, S., Kawai, M. and Rong, H., Finite element analysis of creep crack growth by a local approach. Int. J. Mech. Sci., 1988, 30, pp. 491-502.

8. Lemaitre, J., Local approach of fracture. Eng. Fract. Mech., 1986, 25, pp. 523-537.

9. Hyde, T.H. and Chambers, A.C., A compact mixed-mode (CMM) fracture specimen. Jl. Strain Anal., 1988, 23, pp. 61-66.

10. Hyde, T.H. and Chambers, A.C., An experimental investigation of mixed-mode creep crack growth in Jethete M152 at 550°C. Accepted by Matls. at High Temps..

11. Hyde, T.H., Creep crack growth in 316 stainless steel at 600°C. High Temp. Tech., 1988, 6, pp. 51-61.

12. Saanouni, K, Chabochche, J.L. and Lesne, P.M., On the creep crack-growth prediction by a non local damage formulation. Eur. J. Mech., A/Solids, 1989, 8, pp. 437-59.

13. Hall, F.R. and Hayhurst, D.R., Creep continuum damage mechanics studies of size effects and weldments. Fourth IUTAM Symposium, Creep in Structures, Cracow, Poland, 1990.

INFLUENCE OF CONTINUUM DAMAGE ON STRESS DISTRIBUTION NEAR A TIP OF A GROWING CRACK UNDER CREEP CONDITIONS

V.I. ASTAFJEV, T.V. GRIGOROVA and V.A. PASTUKHOV
Department of Mechanics and Mathematics,
State University, Samara, 443011, USSR

ABSTRACT

The asymptotic stress and damage fields near a tip of a growing crack in creep-damaged matarial are definded. The creep-damage interaction is taken into account by Rabotnov-Leckie-Hayhurst constitutive equations. The stress and damage fields analysis is carried out for anti-plane shear, plane strain and plane stress conditions.

A new type of stress and damage fields in process zone near a tip of a growing crack is obtained. It has been shown that in process zone the net stress σ/ψ is bounded, and the stress σ and the continuity ψ fall to zero at the crack tip. Adjoined to the crack surface the fully damaged (failed) zone appears.

The asymptotic fields are completely specified by current crack growth rate, path-independent integral of steady state creep theory, creep and damage material constants. Real crack growth rate can be defined numerically by making egual near and far crack tip fields.

INTRODUCTION

Creep fracture mechanics is one of the most important parts of modern fracture mechanics. It is a result of intensive development of power-building machinery, aerospace industry and other technical branches.

Over the past two decades many experimental and theoretical works were concerned with subcritical creep crack growth. In experimental investigations the crack growth rate has been measured and correlation of measured rate with mechanical parameter (stress intensity factor K_I, C^*- integral or the nominal stress on the orack ligament) hás been sought. A short survey of that investigations on steels, alloys and ceramics has been given by Riedel [1].

Thereafter, theoretical models aimed to describe creep crack growth in metals from a mechanical point of view have been developed. Hayhurst with co-authors [2] and Jansson [3] have discussed the trends in the development of this models.

The main idea of the all theoretical models lies in assumption that crack growth occurs when damage (or strain), defined at a short distance in front of the crack tip, reaches a critical value. The singular HRR-field and Kachanov's damage evolution law were used by Kubo with co-authors [4] and Astafjev [5]. Riedel [6] included more complex damage evolution law based on diffusion and power law creep growth of voids. Cocks and Ashby [7] have developed a similar approach with critical strain failure criterion.

The stress redistribution in elastic-power law creeping material, calculated for stationary crack by Riedel and Rice [8], was used in creep crack growth analysis by Astafjev [9,10]. The role of new type of singularity, discovered for moving crack by Hui and Riedel [11], was investigated by Hui [12]. A more general model based on stress dependent critical damage approach was proposed by Astafjev [13].

As was pointed out by Hayhurst [2] none of the existing models describes the effects of tertiary creep on regions near the crack tip. There are only few works [1,2,14] where the coupled creep-damage theory, proposed by Rabotnov [15] and developed by Leckie and Hayhurst [16,17], was used for finite element calculation of cracked members. The aim of this investigation is to carry out an asymptotical analysis of stress and damage fields near the growing crack tip for Rabotnov - Leckie - Hayhurst coupled creep-damage theory.

STATEMENT OF THE PROBLEM

Crack-tip geometry.   Let us consider the two-dimensional crack problems of anti-plane shear (Mode III crack) and plane-stress or plane-strain tension (Mode I crack).The origin of the Cartesian axes is always at the moving crack-tip.The crack is assumed to be propagating with velocity $\upsilon(t)$ in the x_1 direction. The polar coordinates r, φ are connected with the Cartesian ones x_1, x_2 by $x_1 = r\cos\varphi$, $x_2 = r\sin\varphi$. The material time derivation in moving coordinate system is determined by

$$d/dt = \partial/\partial t - \upsilon\, \partial/\partial x_1 = \partial/\partial t - \upsilon\,(\cos\varphi\, \partial/\partial r - r^{-1}\sin\varphi\, \partial/\partial\varphi)$$

Equilibrium equations. The equilibrium equations under quasi-static conditions (low crack velocity υ with respect to elastic shear-wave speed) and without body forces are

$$\partial\sigma_{ij}/\partial x_j = 0 \qquad (1)$$

where σ_{ij} is the symmetric stress tensor, indexes i,j have range $1 \div 3$ and summation on repeated indices is implied.

Constitutive laws

Multi-axial creep law. The constitutive equations for multi-axial creep used here are those proposed first by Rabotnov [15] and treated extensively by Leckie and Hayhurst [2,16,17]

$$\frac{d\varepsilon_{ij}}{dt} = \frac{1}{2}\left[\frac{\partial \upsilon_i}{\partial x_j} + \frac{\partial \upsilon_j}{\partial x_i}\right] = \frac{3}{2} B \left[\frac{\sigma}{\psi}\right]^{n-1} \frac{s_{ij}}{\psi} \qquad (2)$$

where υ_i is the velocity vector, $s_{ij} = \sigma_{ij} - \sigma_{kk}\,\delta_{ij}/3$ is the stress deviator, $\sigma = (3s_{ij}s_{ij}/2)^{1/2}$ is the effective stress, ψ is Kachanov's continuity parameter ($\omega = 1 - \psi$ is the Rabotnov's damage parameter), B, n are material constants. When $\omega = 0$ ($\psi = 1$) the material is in it's virgin undamaged state and constitutive equations (2) are related to power-law creep equations (Norton's law)

$$\frac{d\varepsilon_{ij}}{dt} = \frac{3}{2} B \sigma^{n-1} s_{ij}$$

When $\omega = 1$ ($\psi = 0$) the strain rates become infinite and the material is failure.

Damage evolution law. The evolution equation for damage state variable introduced for uni-axial tension by Kachanov [18] has been generalized on multi-axial stress states by Leckie and Hayhurst [16,17] and is given by

$$\frac{d\psi}{dt} = - A\left[\frac{\sigma_e}{\psi}\right]^m , \quad \psi(t=0) = 1 \qquad (3)$$

where σ_e is the equivalent stress which in uni-axial loading coincides with tensile stress, A, m are material constants.

Hayhurst [16] has defined an equivalent stress σ_e as a linear combination of maximum principal stress σ_1, the effective stress σ and the hydrostatic stress σ_{kk}

$$\sigma_e = \alpha\sigma_1 + \beta\sigma + (1-\alpha-\beta)\sigma_{kk}$$

To determine material constants A,m and B,n it is necessary to carry out uni-axial tests and to calculate them from minimum strain rate vs stress and time to failure vs stress relations. Parameters α and β can be determined from two-axial tests using the form of isochronous stress curves.

Initial and boundary conditions. The initial conditions at the time t=0 are the next. Continuity parameter ψ is equal to unity and the stress distribution has Hutchinson-Rice-Rosengren (HRR) form for the non-linear viscous media [19,20]

$$\sigma_{ij} = \left[\frac{C^*}{B I_n r}\right]^{1/(n+1)} \bar{\sigma}_{ij}(n,\varphi) \qquad (4)$$

where C^* is the path-independent integral of steady state creep theory , I_n and $\bar{\sigma}_{ij}$ (n, φ) are given by Hutchinson [19].

Boundary conditions are formulated on the traction-free crack surface with unit normal vector n_j

$$\sigma_{ij} n_j = 0 \text{ at } \varphi = \pi \tag{5}$$

and at infinity. For the "small-scale yielding" solution it is sufficient to regard the crack as being of semi-infinite with the boundary conditions (4) at infinity [21].

Governing equations

Dimensionless variables. From equations (1) - (3), initial and boundary conditions (4), (5) one can conclude that the solution at the point r, φ and at the time t is the function of the following set of variables and material parameters : r, φ, t, A, υ, C^*/BI_n . Hence, from standard consideration of dimensional consistency, the stress σ_{ij} and the continuity ψ fields have the next form [21]:

$$\sigma_{ij} = \left[\frac{A \, C^*}{\upsilon \, B \, I_n} \right]^{1/(n+1-m)} \Sigma_{ij}(R, \varphi, T) \tag{6a}$$

$$\psi = \Psi(R, \varphi, T) \tag{6b}$$

where Σ_{ij} and Ψ are the dimensionless stress tensor and the continuity parameter. $R = r/r_o$ and $T = t\upsilon/r_o$ are the dimensionless radial coordinate and the dimensionless time respectively, $p = m/(n+1)$,

$$r_o = \left[\frac{C^*}{B \, I_n} \right]^{p/(p-1)} \left[\frac{A}{\upsilon} \right]^{1/(p-1)}$$

Dimensionless governing equations. Equilibrium equations, compatibility conditions in stresses and damage evolution law for the dimensionless functions Σ_{ij} and Ψ have in polar coordinates the next form

$$\partial \Sigma_{3r}/\partial R + \Sigma_{3r}/R + R^{-1} \partial \Sigma_{3\varphi}/\partial \varphi = 0 \tag{7a}$$

$$\partial \Sigma_{rr}/\partial R + (\Sigma_{rr} - \Sigma_{\varphi\varphi})/R + R^{-1} \partial \Sigma_{r\varphi}/\partial \varphi = 0 \tag{7b}$$

$$\partial \Sigma_{r\varphi}/\partial R + 2 \, \Sigma_{r\varphi}/R + R^{-1} \partial \Sigma_{\varphi\varphi}/\partial \varphi = 0 \tag{7c}$$

$$\frac{\partial}{\partial \varphi} \left[\frac{\Sigma^{n-1} \Sigma_{3r}}{\Psi^n} \right] - \frac{\partial}{\partial R} \left[\frac{R \, \Sigma^{n-1} \Sigma_{3\varphi}}{\Psi^n} \right] = 0 \tag{8a}$$

$$2 \frac{\partial}{\partial R} \left[R \frac{\partial}{\partial \varphi} \left[\frac{\Sigma^{n-1} \Sigma_{r\varphi}}{\Psi^n} \right] \right] - \frac{\partial^2}{\partial \varphi^2} \left[\frac{\Sigma^{n-1} S_{rr}}{\Psi^n} \right] +$$

$$+ R \frac{\partial}{\partial R} \left[\frac{\Sigma^{n-1} S_{rr}}{\Psi^n} \right] - R \frac{\partial^2}{\partial R^2} \left[\frac{R \Sigma^{n-1} S_{\varphi\varphi}}{\Psi^n} \right] = 0 \tag{8b}$$

$$\partial \Psi / \partial T - \cos\varphi \, \partial \Psi / \partial R + R^{-1} \sin\varphi \, \partial \Psi / \partial \varphi = - (\Sigma_e/\Psi)^m \tag{9}$$

where $S_{rr} = - S_{\varphi\varphi} = (\Sigma_{rr} - \Sigma_{\varphi\varphi})/2$, $\Sigma = (S_{\varphi\varphi}^2 + \Sigma_{r\varphi}^2)^{1/2}$
for the plane strain;

$$S_{rr} = 2 (\Sigma_{rr} - \Sigma_{\varphi\varphi}/2)/3, \quad S_{\varphi\varphi} = 2 (\Sigma_{\varphi\varphi} - \Sigma_{rr}/2)/3,$$

$\Sigma = (S_{rr}^2 + S_{\varphi\varphi}^2 + S_{rr}S_{\varphi\varphi} + \Sigma_{r\varphi}^2)^{1/2}$ for the plane stress

and $\Sigma = (\Sigma_{3\varphi}^2 + \Sigma_{3r}^2)^{1/2}$ for the anti-plane shear conditions
(in equations (2) and (4) $3^{(n+1)/2}B$ is replaced by B and
$3^{m/2}A$ by A), $\Sigma_e = \alpha \Sigma_1 + \beta \Sigma + (1-\alpha-\beta)\Sigma_{kk}$.

In dimensionless variable the initial conditions and the boundary conditions at the infinity are

$$\Sigma_{ij} = R^{-1/(n+1)} \bar{\sigma}_{ij}(n,\varphi), \tag{10a}$$

$$\Psi = 1 \tag{10b}$$

ASYMPTOTIC BEHAVIOUR NEAR THE CRACK TIP

For the asymptotic analysis of stress and damage fields let us assume that the solution in the vicinity of the crack tip has the separable form, i.e.

$$\Psi = R^\mu (g^{(1)}(\varphi,T) + R g^{(2)}(\varphi,T) + \dots) \tag{11a}$$

$$\Sigma_{ij}/\Psi = R^\lambda (f_{ij}^{(1)}(\varphi,T) + R f_{ij}^{(2)}(\varphi,T) + \dots) \tag{11b}$$

Substituting (11) in equations (7) – (9) we have the sequence of sets of ordinary differential equations for unknown $(f_{ij}^{(1)}, g^{(1)})$, $(f_{ij}^{(2)}, g^{(2)})$, etc. From equation (9) it can be seen that $\mu = 1+m\lambda$. Let us consider the asymptotic solutions of this equations for anti-plane shear, plane strain and plane stress conditions.

of the $\Lambda(t)$ value by means of modificated path-independent integral will be done.

The numerical solution of this problem is based on straightforward shooting method. In this method a trial value for λ has been assumed. The initial-value problem (12), (13b), (13c), (13d) with chosen λ has been solved by the fourth-order Runge-Kutta method, starting at $\varphi=0$, until $\varphi=\pi$ is reached. The trial value for λ has been varied to meet the boundary condition (13a).

Results of numerical solution. The results of solution by shooting method were curious. No values of λ give the solution of this eugenvalue problem with acceptable accuracy. Moreover, after some value φ_o the function y_1 becomes negative what contradicts physical sence of damage parameter.

Asymptotic analysis near $\varphi = \pi$ shows that y_1 behaves as

$$y_1 = C \; \xi^{\mu} - y_e^m(\pi)/\mu$$

where $\xi = \pi - \varphi$ and C is undefined constant. Thus, there exists some value $\varphi_o < \pi$ where y_1 is equal to zero and becomes negative in range $[\varphi_o, \pi]$.

There are two ways to remove this contradiction and to find the solution of the eugenvalue problem. The first one is to assume additionally that $y_3(\pi) = 0$ together with $y_2(\pi)$. But additional homogeneous boundary condition $y_3(\pi) = 0$ requires an additional undefined constant missing in boundary value problem (12), (13).

The second one is to determine the first value of φ where $y_2 = 0$, and to nullify the solution in range $[\varphi_o, \pi]$. For nullified solution the governing equations (12) and the boundary condition (13a) in range $[\varphi_o, \pi]$ are valid. The continuity requirement for $y_1(\varphi)$ leads to $y_1(\varphi_o) = 0$ (it should be noted that requirement continuity for $y_3(\varphi)$ isn't necessary).

Thus, for governing equations (12) in range $[0, \varphi_o]$ we have the modified eugenvalue problem: to find the eugenvalue λ and eugenfunctions y_1, y_2, y_3 for equations (12) and boundary conditions

$$y_1(0)=1/\mu, \; y_2(0)=1, \; y_3(0)=0, \; y_2(\varphi_o)=0 \qquad (14a)$$

where unknown value φ_o is determined from additional boundary condition

$$y_1(\varphi_o)=0 \qquad (14b)$$

The numerical solution of the modified problem by shooting method with Runge-Kutta integration technique has shown that there exists only one eugenvalue $\lambda=0$ with $\varphi_o = \pi/2$. It's easy to see that for $\lambda=0$ and $\varphi_o=\pi/2$ the problem has the following analytical solution

$$y_1 = \psi_o \cos\varphi, \; y_2 = \cos\varphi, \; y_3 = \sin\varphi, \; y = 1, \; y_e = \alpha+\beta \qquad (15)$$

where $\psi_o=(\alpha+\beta)^m$.

Anti-plane shear analysis

Eugenvalue problem. The anti-plane shear deformation is characterized by two non-zero stress components Σ_{3r} and $\Sigma_{3\varphi}$. Denoting

$$y = g_1^{(1)}, \quad y_2 = f_{3\varphi}^{(1)}, \quad y_3 = f_{3r}^{(1)}, \quad y = (y_2^2 + y_3^2)^{1/2}, \quad y_e = (\alpha + \beta)y$$

equations (7a), (8a) and (9) can be written for $r \rightarrow 0$ as

$$y_1' = (\mu y_1 \cos\varphi - y_e^m)/\sin\varphi \qquad (12a)$$

$$y_2' = -(\lambda + \mu + 1)y_3 - y_2 y_1'/y_1 \qquad (12b)$$

$$y_3' = ((n\lambda + 1)y^2 y_2 - (n-1)y_3 y_2 y_2')/(n y_3^2 + y_2^2) \qquad (12c)$$

where a prime denote the derivative with respect to angle φ.

Equations (12) are the set of third-order non-linear ordinary differential equations. Boundary conditions for this equations follow from the fraction-free condition on the crack surface (5)

$$y_2(\pi) = 0 \qquad (13a)$$

from the symmetry condition with respect to $\varphi = 0$

$$y_3(0) = 0 \qquad (13b)$$

and from the regularity requirement at $\varphi = 0$

$$y_1(0) = y_e^m(0)/\mu \qquad (13c)$$

The system of homogeneous differential equations (12) together with the homogeneous boundary conditions (13) constitute an eugenvalue problem for unknowns eugenvalue λ and eugenfunctions y_1, y_2, y_3. From homogeneity of equations (12) and boundary conditions (13) it follows that $A(t)y_2$, $A(t)y_3$, $A^m(t)y_1$ are the solution too. For the purpose of simplicity a normalizing condition is chosen as

$$y_2(0) = 1 \qquad (13d)$$

The amplitude $A(t)$ cannot be specified by analyzing the asymptotic problem alone. In steady state creep theory without damage ($\psi \equiv 1$) the value of $A(t)$ coincides with amplitude of HRR-field and is specified by means of path-independent C^*-integral [8]. In our case, however, C^* is path-dependent [21] and numerical solution in whole range $0 < r < \infty$ should be applied to determine the amplitude $A(t)$. The modification of path-independent C^*-integral for coupled creep-damage theory will be discussed in forthcoming paper [22]. Simple estimation

Plane strain analysis

Eugenvalue problem. The derivation of governing equations in plane strain is similar to that in anti-plane shear. Let us denote

$$y_1 = g^{(1)}, \quad y_2 = f_{\varphi\varphi}^{(1)}, \quad y_3 = f_{r\varphi}^{(1)}, \quad y_4 = f_{rr}^{(1)},$$

and introduce the auxiliary function

$$y_5 = 4(n\lambda+1)y_3 + (y_2-y_4)' + (n-1)(y_2-y_4)y'/y,$$

where $y = (y_3^2 + (y_2-y_4)^2/4)^{1/2}$.

Equations (7b), (7c), (8b) and (9) can be written for $r \to 0$ as

$$y_1' = (\mu\, y_1 \cos\varphi - y_e^m)/\sin\varphi \qquad (16a)$$

$$y_2' = - (\lambda+\mu+2)y_3 - y_2 y_1'/y_1 \qquad (16b)$$

$$y_3' = y_2 - (\lambda+\mu+1)y_4 - y_3 y_1'/y_1 \qquad (16c)$$

$$y_4' = y_2 - \frac{(y_5 - 4(n\lambda+1)y_3)y^2 + (n-1)(y_4 - y_2)y_3 y_3'}{ny^2 - (n-1)y_3^2} \qquad (16d)$$

$$y_5' = n\lambda(n\lambda+2)(y_2 - y_4) - (n-1)y_5 y'/y \qquad (16e)$$

For plane strain conditions y_e is the linear combination of the sum $y_2 + y_4$ and y

$$y_e = (3-2\alpha-3\beta)(y_2+y_4)/2 + (\alpha+\beta)y$$

Equations (16) are the five-order system of ordinary differential equations in range $[0,\pi]$. Boundary conditions for this system follow from the traction-free conditions on the crack surface (5)

$$y_2(\pi) = 0, \quad y_3(\pi) = 0 \qquad (17a,b)$$

from the symmetry condition with respect to $\varphi = 0$

$$y_3(0) = 0, \quad y_4'(0) = 0 \qquad (17c,d)$$

and from the regularity requirement at $\varphi = 0$

$$y_1(0) = y_e^m(0)/\mu \qquad (17e)$$

Thus for plane strain we have also the eugenvalue problem for unknowns λ and $y_1(\varphi)$, i=1-5. It should be noted that boundary condition $y'_4(0) \equiv 0$ is equivalent to $y_5(0)=0$. Similarly to anti-plane shear the normalizing condition for homogeneous equations (16) and boundary conditions (17) may be chosen as

$$y(0) = 1 \qquad\qquad (17f)$$

The last initial value at $\varphi = 0$ is chosen as

$$y_2(0) = Y \qquad\qquad (17g)$$

where Y is undefined constant.

Results of solution. The solution of this eugenvalue problem is based on shooting method too. Two trial values for λ and Y have been assumed. The initial-value problem for (16) with boundary conditions (17c)-(17g) was integrated numerically by Runge-Kutta technique. Both in anti-plane shear and in plane strain there is only one solution of this eugenvalue problem. The system (16) with boundary conditions (17) for $\lambda=0$ and Y=2 can be integrated in range $[0,\pi/2]$. The analytical solution is written as

$$y_1 = \psi_0 \cos\varphi, \quad y_2 = 2\cos^2\varphi, \quad y_3 = \sin 2\varphi,$$
$$\qquad\qquad (18)$$
$$y_4 = 2\sin^2\varphi, \quad y_5 \equiv 0, \quad y \equiv 1, \quad y_e = 3-\alpha-2\beta,$$

where $\psi_0 = (3-\alpha-2\beta)^m$.

Similarly with anti-plane shear we can have the unknown y_1, y_2, y_3, y_4 and y_5 are equal zero in range $[\pi/2,\pi]$. The accepted solution satisfies to all boundary conditions (17) and continuity requirements in whole range $[0,\pi]$.

Plane stress analysis

The equilibrium equations and the damage evolution law in plane stress conditions are identical to that in plane strain, but the compatibility condition is different. For y_1, y_2, y_3, y_4 denoted in plane strain and auxiliary function

$$y_5 = 6(n\lambda+1)y_3 + (y_2-2y_4)' + (n-1)(y_2-2y_4)z'/z$$

where $z = (y_3^2 + (y_2^2 + y_4^2 - y_2 y_4)/3)^{1/2}$ the governing equations can be written as

$$y'_1 = (\mu\, y_1 \cos\varphi - y_e^m)/\sin\varphi \qquad\qquad (19a)$$

$$y'_2 = -(\lambda+\mu+2)\, y_3 - y_2 y'_1/y_1 \qquad\qquad (19b)$$

$$y'_3 = y_2 - (\lambda+\mu+1)\, y_4 - y_3 y'_1/y_1 \qquad\qquad (19c)$$

$$y_4' = \frac{y_2'}{2} - \frac{(y_5 - 6(n\lambda+1)y_3)z^2 + (n-1)(2y_4-y_2)(y_3y_3' + y_2y_2'/4)}{2(nz^2 - (n-1)(y_3^2 + y_2^2/4))} \quad (19d)$$

$$y_5' = n\lambda((2n\lambda+3)y_2 - (n\lambda+3)y_4) - (n-1)y_5 z'/z \quad (19e)$$

Equivalent function y_e has the following form

$$y_e = (2-\alpha-2\beta)(y_2+y_4)/2 + \alpha y + \beta z$$

In exact accordance with the plane strain solution let us suppose that $\lambda=0$. For this eugenvalue the system (19) with boundary conditions (17) is integrated in range $[0,\pi/2]$ and has the similar with plane strain solution

$$y_1 = \psi_0\cos\varphi, \quad y_2 = \sqrt{3}\cos^2\varphi, \quad y_3 = \sqrt{3}\sin\varphi\cos\varphi,$$

$$y_4 = \sqrt{3}\sin^2\varphi, \quad y_5 \equiv 0, \quad y \equiv \sqrt{3}/2, \quad z \equiv 1, \quad y_e \equiv \sqrt{3} + (1-\sqrt{3})\beta \quad (20)$$

where $\psi_0 = (\sqrt{3} + (1-\sqrt{3})\beta)^m$. In range $[\pi/2,\pi]$ this system has a zero solution.

DISCUSSION

At this section we discuss the results obtained above. In dimementional variables the damage and stress fields near the crack tip have the next asymptotic form

$$\psi = \mathcal{x}^m \left[\frac{C^*}{B\,I_n}\right]^{m/(n+1-m)} \left[\frac{A}{\upsilon}\right]^{(n+1)/(n+1-m)} r g^{(1)}(\varphi) \quad (21a)$$

$$\sigma_{ij}/\psi = \mathcal{x} \left[\frac{A\,C^*}{\upsilon\,B\,I_n}\right]^{1/(n+1-m)} f_{ij}^{(1)}(\varphi) \quad (21b)$$

where $f_{ij}^{(1)}(\varphi)$ and $g^{(1)}(\varphi)$ are discribed by (15), (18) or (20) for anti-plane shear, plane strain or plane stress cunsequently.

At first, it should be noted that asymptotic solution (21) is independent of time t explicitly and their implicit time dependence associates only with crack growth rate $\upsilon(t)$. Besides, the dimentionless parameter $\mathcal{x}$ and crack growth rate $\upsilon(t)$ in asymptotic solution (21) are undefined and can be sought numerically by solving the main governing equations (7)-(9) in whole range of valiables $0<r<\infty$, $0<t<\infty$ and $0<\varphi<\pi$. A complete numerical solution of this problem will be published in detail in forthcoming paper [23]. This solution allows to define the unknown value $\mathcal{E}$ and crack growth rate $\mathcal{V}(t)$ for the coupled creep-damage theory.

An important result of this investigation lies in new type of stress and damage distribution near the tip of a moving crack in the coupled creep-damage theory. The introduction of damage state variable ψ into power-law constitutive equations of creeping material permits to describe the stress redistribution and appearance near the crack tip a process zone in which the net stress is bounded and stress and continuity disappear. This is a result of damage influence on creep process taking into accout in coupled creep-damage theory. The result of stress redistribution and nullification near a smooth concentrators in coupled theory was determined by Rabotnov [15] (plate with whole) and Hayhurst [24] (notched bars). For cracked members there are only numerical finite element solutions by Hayhurst [2,14], Rieded [1] and Chaboche [25].

The asymptotic analytical solution (21) is in accordance with those numerical ones and allows to determine minor circumstances of stress and damage fields near the crack tip. The first one is linear radial dependence of stresses and damage with zero values at the crack tip. The second one is nullification of stress and continuity in circumferential direction in range $[\pi/2, \pi]$. It means that fully damaged zone (failed zone) adjoins to traction-free crack surfaces.

CONCLUSION

The nature of asymptotic stress and damage fields near a tip of a growing crack in materials described by the coupled creep-damage constitutive equations (Rabotnov-Leckie-Hayhurst's theory) is investigated. A new type of stress and damage fields near a crack tip is obtaned.

A principal result of this analysis is that there exists a process zone near a crack tip in which the net stress is bounded, stress and continuity fall to zero at crack tip as a linear functions. Just behind the crack front there appears a fully damaged (failed) zone adjoined to traction-free crack surfaces.

REFERENCES

1. Riedel, H., Recent advances in modelling creep crack growth. in Adv. Fract. Res.: Proc. 7th Int. Conf Fract. (ICF7), Pergamon Press, Oxford, 1989, V. 2, pp. 1495-523.

2. Hayhurst, D.R., Brown, P.R. and Morrison C.J., The role of continuum damage in creep crack growth. Phyl. Trans. Roy. Soc., London, 1984, A311, 131-58

3. Jansson, S., Damage, crack growth and rupture in creep. Ph. D. Thesis, Chalmers University of Technol., Göteborg, Sweeden, 1985.

4. Kubo, S., Ohji, K. and Ogura, K., An analysis of creep crack propagation on the basis of the plastic singular stress field. Eng. Fract. Mech., 1979, 11, 315-29.

5. Astafjev, V.I., Creep crack growth with taking into consideration a plastic zone near a crack tip. Prikl. Mech. i Techn. Phys., 1979, N.6, 154-8, (in Russian).

6. Riedel, H., The extension of a macroscopic crack at a elevated temperature by the growth and the coalescence of microvoids. In IUTAM Simposium on Creep at Structures, eds. A.R.S. Ponter and D.R. Hayhurst, Springer-Verlag, Berlin, 1981, pp. 504-19.

7. Cocks, A.C.F. and Ashby, M.F., The growth of dominant crack in a creeping material. Scr. Metall., 1982, 16, 109-14.

8. Riedel, H. and Rice, J.R., Tensile crack in creeping solids. ASTM STP 700, 1980, pp. 112-30.

9. Astafjev, V.I., The influence of nonstationar stress field on creep crack growth. Prikl. Mech. and Techn. Phys. 1983, N.3, 148-52, (in Russian).

10. Astafjev, V.I., Subcritical creep crack growth under variable loading. Prikl. Mech. and Techn. Phys., 1985, N.3, 152-57 (in Russian).

11. Hui, C.Y. and Riedel, H., The Asymptotic Stress and Strain Field Near the Tip of a Growing crack under Creep Conditions. Int. J. of Fract., 1981, 17, 409-25.

12. Hui, C.Y., The mechanics of self-similar crack growth in an elastic power-law creeping material. Int. J. of Solids and Struct., 1986, 22, 357-72.

13. Astafjev, V.I., A generalities of creep crack growth. In Izv. Acad. Nauk SSSR, Mech. Tv. Tela, 1986, N.1, 127-34, (in Russian).

14. Hayhurst, D.R., Dimmer, P.R. and Chernuka, M.W., Estimates of the creep rupture life time of structures using finite element method. J. of Mech. and Phys of Solids, 1975, 23, 335-55.

15. Rabotnov, Yu.N., Creep Problems in Structural Members, (English translation, ed. F.A. Leckie), North Holland, Amsterdam, 1969

16. Hayhurst, D.R., Creep rupture under multi-axial states of stress. J. of Mech. and Phys of Solids, 1972, 20, 381-90.

61

17. Leckie, F.A. and Hayhurst, Constitutive equations for creep rupture. _Acta Metall._, 1977, 25, 1059-70.

18. Kachanov, L.M., On the time to failure under creep conditions, _Izv. Acad. Nauk. SSSR, Otd. Techn. N._, 1958, N.8, 26-31. (in Russian).

19. Hutchinson, J.W., Singular behaviour at the end of tensile crack in a hardening material, _J.Mech. and Phys. of Solids_, 1968, 16, 13-31.

20. Rice, J.R. and Rosengren, G.F., plane strain deformation near a crack tip in a power-law hardening material, _J.Mech. and Phys. of Solids_, 1968, 16, 32-48.

21. Astafjev, V.I., Asymptotic of stress field near a tip of creep growing crack with damage accumulation. _Dokl. Acad. SSSR_, 1984, 279, 1327-30 (in Russian).

22. Astafjev, V.I., New path-independetn integral for coupled creep-damage theory. (Submitted for publication to Strength and Fracture).

23. Astafjev,V.I., Pastukhov, V.A. and Grigorova, T.V., Creep crack growth in damaged media. (in preparing).

24. Hayhurst, D.R., Dimmer, P.R. and Morrison, C.J., Development of continuum damage in the creep rupture of the notched bars. _Phyl. Trans. Roy. Soc._, London, 1984, A311, 103-29

25. Chaboche, J.L., Phenomenological aspects of continuum damage mechanics. In _Theor. and Appl. Mechanics_, eds. P.Germain, M.Pian and D.Gaillerie, Elsevier Applied Science Publishers, London, 1989, pp 41-56.

DISLOCATION MOVEMENT AT CRACK TIP
OF SINGLE CRYSTALS OF ICE

Yingchang Wei and John P. Dempsey
Department of Civil and Environmental Engineering
Clarkson University, Potsdam, NY 13699 USA

ABSTRACT

Etching and replicating techniques were used to study dislocation movement in the crack tip region of single crystals of ice undergoing mode I loading. Dislocation pile-ups were observed in front of the tip of an arrested cleavage crack in the prism plane. A dislocation-free-zone of length 750μm between the crack tip and the etching tracks was found for the first time in an ice crystal subjected to creep. These etching tracks corresponded to the traces of climbing dislocations and prevailed through the whole specimen, indicating that dislocation climb was the dominant mechanism for the creep of the specimen.

INTRODUCTION

The ductile versus brittle fracture behavior of a material is determined by its ability to exhibit crack tip plasticity. An important problem, therefore, is what mechanisms control the crack tip plasticity? Rice and Thomson [1] established a criterion for brittle fracture in crystals in terms of the emission of dislocations from a sharp cleavage crack. They found that when the parameter $\mu b/\gamma < 7.5 - 10$ (where μ: shear modulus, b: Burgers vector and γ: surface energy of the crystal) is approximately satisfied, dislocations will be emitted from a crack tip and the crack is blunted, while a sharp crack will be stable against blunting and brittle fracture occurs if $\mu b/\gamma > 7.5 - 10$. An alternative view on this problem was proposed by Ashby and Embury [2]. According to [2], the crack tip plasticity is independent of the stability of a crack tip against the emission of dislocations. Rather, it is controlled by the interaction between the stress field of the crack and the pre-existing dislocations near the crack tip. Briefly, at lower dislocation densities, plasticity and blunting at the crack tip may be prevented because the nearest dislocations are too far away: the crack tip stress intensification is insufficient to cause them to move and multiply. In this case, the crack may run in a brittle manner. An increase in dislocation density generates dislocations lying closer to the crack tip; a number of these dislocations interact with the crack tip field. Consequently, the near-field dislocations move and multiply, dissipating energy and blunting the crack. For the latter case, the crack will run in a less brittle manner.

Applying Rice and Thomson's criterion to ice, the value of $\mu b/\gamma$ is found to be approximately 17 for $\mu = 3.8$ GPa, $b = 0.4523$ mm and $\gamma = 0.1$ J/m^2 [3,4] — note that the value of γ has not been determined for single crystals of ice or polycrystalline ice at the temperature in question. This implies that cracking in ice may not be accompanied by crack tip blunting, as discussed first for ice by Parsons [5] and later in [6]. However, in a recent study on the fracture toughness of columnar freshwater ice [7], the observed crack-tip-opening -displacement (CTOD) indicated the existence of microplastic deformation in the vicinity of crack tip at the initiation of unstable fracture. Crack tip blunting due to creep deformation in equiaxed freshwater ice has been hypothesized in [8].

The intriguing possibility is that in some instances a crack tip in ice does not blunt and in others it does. This phenomenon may be complicated by factors such as specimen size, loading rate, temperature and ice type [6, 9]. When a sharp crack in ice is subjected to combined loading, both tensile and shear stresses can develop in the crack tip region. The competition between the tensile fracture of the bond at the crack tip and the movement of dislocations at the tip region should be governed by the crack tip field and the existing dislocations [2]. The competition between brittle and ductile processes is particularly true for ice because of its remarkably high anisotropy in plastic deformation. An ice crystal can exhibit either ductile or brittle fracture behavior depending on the orientation of the basal planes (the preferred slip plane) with respect to the load direction [3, 4]. For a favorable orientation, the dislocations near the crack tip may move easily and multiply, thereby contributing to the crack tip ductility.

In light of the above comments, more needs to be known about the behavior of dislocations in the vicinity of crack tips in ice crystals. Attempts have been made by the authors to investigate the distribution and movement of dislocations at such crack tips. This paper presents some results of this study. Mode I cleavage cracks were introduced in the $\{10\bar{1}0\}$ plane with the crack front either parallel or perpendicular to the c-axis. Etching and replicating techniques were used to reveal the basal and non-basal dislocations on the prismatic and basal planes, respectively. This investigation focused on the dislocation-free-zone and climbing movement of basal dislocations in a specimen subjected to creep.

EXPERIMENTAL PROCEDURE

Dislocation movement at each crack tip was examined in this paper by etching and replicating. These techniques have been used by a number of researchers both to observe dislocations in ice crystals [10–14] and to conduct fractographic analyses on the fracture surfaces of polycrystalline ice [12, 15]. In this study, Formvar solutions (Formvar in ethylene dichloride) of $0.5 - 6\%$ concentration were used to etch a chosen surface of an ice crystal. While etching a surface, the Formvar solution was able to replicate the etching patterns on the ice surface at the same time. After the etchant dried, a plastic replica with the details of the etching pits was left on the etched surface . The replica was peeled off, mounted on a glass slide and examined under an Olympus BHS optical microscope. The microscope with Hoffman modulation contrast proved to be useful in identifying the general characteristics of various types of dislocation pits. In addition, some replicas were prepared by vacuum deposition of a layer of gold and examined with a scanning electron microscope.

The specimens used for this study were cut from single crystals of ice grown in the Ice Mechanics Research Laboratory at Clarkson University by a wire saw along a selected crystallographic plane. The surfaces of each specimen were microtomed to acquire the desired dimensions. For the surface to be etched, care was taken to prepare (again by microtoming) the surface to a mirror-smooth finish, following the procedure recommended by Sinha [13]. Prior to introducing cracks in the specimens, the densities of basal and non-basal dislocation of the ice crystals were examined by etching the prismatic plane and the basal plane, respectively, using a 6% Formvar solution at -10° C. The non-basal dislocation density was found to be $1.2-3 \times 10^8 \text{m}^{-2}$, while the basal dislocation density varied between $0.84-8.6 \times 10^9 \text{m}^{-2}$. Note that the measurable dislocation density may be influenced by the etching conditions, such as temperature and concentration of the etchant. Therefore, the dislocation density reported here has a relative meaning only.

The characteristics of etching pits of basal and non-basal dislocations are illustrated in Fig. 1. Fig. 1a shows the typical features of the hexagonal pyramidal etch-pits on a (0001) plane, corresponding to the emergence of non-basal dislocations on this surface. The basal dislocations emerging on a ($10\bar{1}0$) plane are characterized by the centrally depressed, elongated etch-pits developed on that plane (Fig. 1b). The long axes of the central depressions are parallel to the c-axis or < 0001 > axis of the crystal. For both photos of Fig. 1, the large pits represent the grown-in dislocations while the small

(a)   (b)

Figure 1   Etching pits on (a) basal plane corresponding to non-basal dislocations
and (b) prismatic plane corresponding to basal dislocations.

pits are believed to be introduced during sawing and microtoming. The sharply pointed tips of the pits indicate the sites of the dislocation cores. All of the above features are in agreement with observations reported in the literature [10–14].

DISLOCATION MOVEMENT IN FRONT OF CRACK TIP

Dislocation Pile-up in Front of an Arrested Crack Tip

A Mode I cleavage crack was introduced into the $(10\bar{1}0)$ plane (the cleavage plane of the ice crystal) by pressing a knife against the crystal (Fig. 2). The crack front was parallel to the c-axis of the ice crystal so that the basal plane could be etched. The specimens tested in this group were fairly small with length $L = 20$ mm, width $W = 20$ mm and thickness $t = 2$ mm.

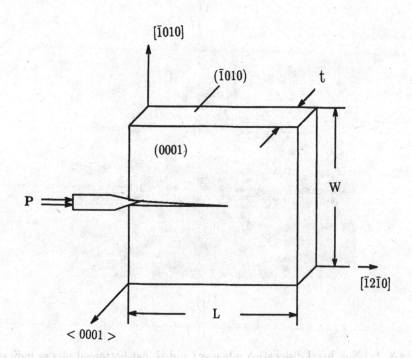

Figure 2 Orientation of crack with respect to the c-axis.

At the instant of nucleation, the crack ran very fast and stopped at a point in the crystal. Etching followed immediately to detect the dislocation behavior in the tip region of the arrested crack. Fig. 3a shows the dislocation pile-ups observed in front of an arrested crack tip etched by 6% Formvar solution at -10^0 C. For comparison, the distribution of dislocations at the tip region of a crack without loading is presented in

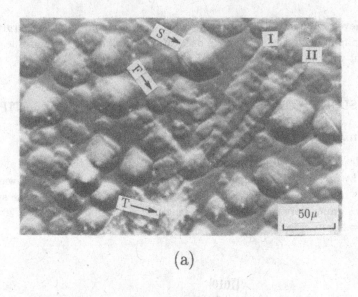

(a)

(b)

Figure 3 (a) Non-basal dislocation pile-ups I and II; flat-bottomed pits as indicated by arrow F, stationary dislocations by arrow S, and the crack tip by arrow T. (b) Dislocation pits surrounding a crack before loading.

Fig. 3b. This crack was fabricated by sharpening a wire-saw-cut notch by a hand-held razor blade (to be described later). As shown in Fig. 3b, most dislocations remain intact. By contrast, from Fig. 3a, the movement of dislocations in the specimen subjected to a wedge loading is evident and can be characterized by two types of etch-pits. One is the small sharply pointed pits aligned in the pile-ups. These dislocations were believed to be induced by the stress field of the crack tip. However, whether they emanated from the crack tip or were produced by an existing dislocation source is unknown. The angle between pile-up I and the crack is found to be 150°, indicating that the pile-up I is parallel to the direction of $< 11\bar{2}0 >$. The angle between pile-up I and pile-up II is 10°, trisecting the angle between a $< 10\bar{1}0 >$ and $< 11\bar{2}0 >$. This slip direction is not uncommon for non-basal dislocation and has been observed in crack-free ice crystals [10, 13]. The other type of etch-pits showing dislocation movement is a number of so called "flat-bottomed pits" distributed in front of the crack tip. This type of etch-pits has been previously observed in LiF crystal [16] and ice crystals [10]. Under the influence of the crack tip field, if a dislocation moves away from the position of a etching pit, deepening along the former dislocation core will stop during subsequent etching. Thus, the flat-bottomed pits are the indication of dislocation movement. However, not all the dislocations were mobiled by the crack tip stress field. Those stationary dislocations are illustrated by the large hexagonal pyramidal pits in Fig. 3a.

Dislocation-Free-Zone and Dislocation Climb in Ice Crystals Subjected To Mode I Creep Deformation

To examine the features of dislocation movement at the vicinity of a crack tip during Mode I creep deformation, tests were conducted on the specimens with the three-point bend loading configuration shown in Fig. 4. The specimen size was $L \times W \times t = 50 \times 25 \times 4$ mm. The front of a $10\bar{1}0$ crack was perpendicular the c-axis of the crystal, in the hope of being able to detect the evidence of both slip and climb of the basal dislocations. For

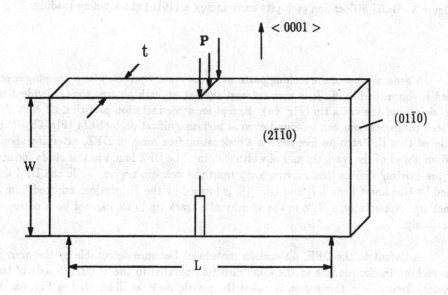

Figure 4 Orientation of crack and loading configuration for creep test.

each creep test (including those discussed earlier), a notch of length 4 mm was cut by a wire saw. The notch was then carefully sharpened by hand-held-razor blade, allowing a cleavage crack of 2-4 mm long to extend from the notch root. Fig. 5 shows the basal dislocation pits near a sharpened crack before loading. The specimen was then stressed by a constant load for 5 hours with a loading jig. As soon as the load was applied, the specimen was etched with a 0.5% Formvar solution. After unloading, the specimen was coated with a thick layer of 2.5% Formvar solution to obtain a thicker replica which could be easily peeled off.

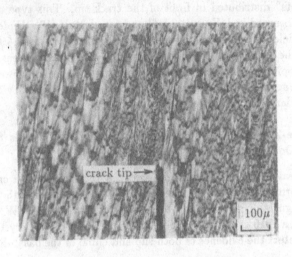

Figure 5   Basal dislocation etch-pits surrounding a $(10\bar{1}0)$ crack before loading.

A general view of the etching patterns in the crack tip region for a creeping specimen is shown in Fig. 6. It is very interesting that an etch-pit-free zone is evident in the vicinity of the crack tip (Fig. 6a). Except for some inclusion particles, no etch-pits related to dislocations can be found even at high magnification by SEM (Fig. 6b). It is believed that this etch-pit-free zone is a dislocation-free zone, or DFZ, extenting about $750\mu m$ ahead of the crack tip and of width $800\mu m$. The DFZ found in this study formed by pre-existing dislocations moving away from the near-tip region, while the DFZ defined by Ohr and Chang [17] and Ohr [18] is formed by the dislocations emitted from a crack tip. Apparently, a DFZ in the vicinity of a crack tip in ice has not been observed previously.

Adjacent to the DFZ, dislocation movement becomes detectable by the numerous etching tracks running in the same direction parallel to the $< 0001 >$ axis of the crystal. Tentatively, this region is called the plastic zone, as illustrated in Fig. 6a. It was observed using a scanning electron microscope that most tracks ran continuously throughout the width of the specimen and were consistently parallel to $< 0001 >$, as shown in Figs. 7 and 8, indicating that the plastic zone might extend across the width

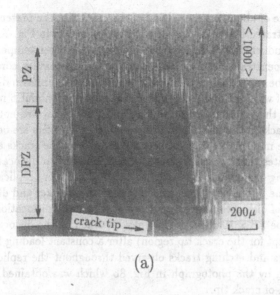

(a)

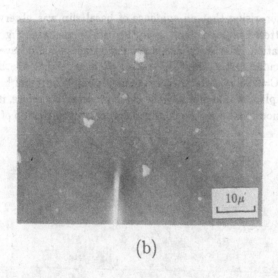

(b)

Figure 6 (a) Dislocation-free zone and etching tracks caused by dislocation climb;
(b) Crack tip viewed under high magnification (SEM photographs).

of the specimen. In the region near the DFZ, only the etching tracks can be found; no etch-pits related to the stationary dislocations were detected. These etching tracks are the indication of the climbing of basal dislocations out of their slip planes during creep, as has been demonstrated by Sinha [19]. More importantly, the climb of dislocation outside the DFZ dissipates energy and shields the crack tip from the applied stress field.

At a distance of about 2.5 mm ahead of the crack tip, the presence of stationary dislocations is illustrated by the replicated whiskers, as shown in Fig. 7. Since the etching process was conducted by a dilute 0.5% solution at a lower temperature of -15°C, the etchant could penetrate deeply along many of the dislocation cores, allowing the dislocation line to be replicated in the form of whiskers with uniform diameter. Fig. 8a shows the details of whiskers and etching tracks at a position about 6 mm ahead of the crack tip. Most of the whiskers are quite long and thin. They are not stiff enough to stand up but fall back on the surface, while some thicker whiskers are bent over the surface. Note that the majority of the whiskers are formed on the tracks and fall over the tracks. This indicates that the tracks were formed first due to dislocations climb; and that upon unloading the dislocations stopped moving and were replicated in the form of whiskers. The one to one correspondence between the whisker and dislocation makes the determination of dislocation density very easy. The basal dislocation density at this site was found to be 1.8×10^{10} m^{-2}. This may be taken as the dislocation density for the specimen (except for the crack tip region) after a constant loading for 5 hours. The features of whiskers and etching tracks observed throughout the replica are the same, as illustrated again by the photograph in Fig. 8b which was obtained from a position about 8 mm ahead of crack tip.

It is quite surprising that no evidence of basal slip was observed even though there was a shear stress component acting on the basal planes (see Fig. 4 for reference). If there were dislocation glide on basal planes, their traces should have been replicated as tracks perpendicular to the climbing tracks. It is true that the high homologous temperature ($-15°C = 0.95\ T_m$, where T_m is the melt point of ice) and high normal stress acting on the climb plane would promote the climb process. Therefore, it seems plausible to say that dislocation climb was the dominant process during creep of the specimen.

Figure 7 Etching tracks and whiskers adjacent to the DFZ and ahead of crack tip
—the crack is on the under side of the photograph (SEM photograph)

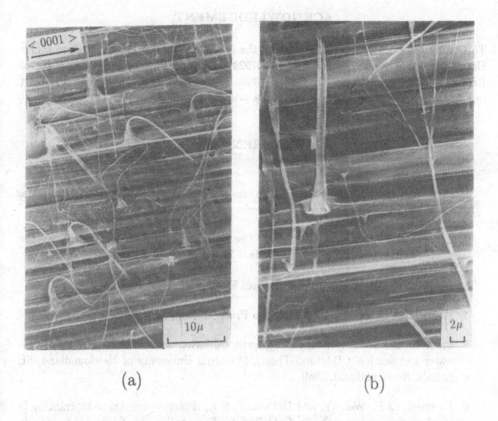

Figure 8   Details of whiskers and etching tracks (SEM photographs).

SUMMARY

Dislocation-crack interaction in single crystals of ice with Mode I loading has been studied by etching and replicating. Non-basal dislocation pile-ups were found in front of a (10$\bar{1}$0) cleavage crack. A dislocation-free zone 750 μm long and 800 μm wide was observed for the first time in the vicinity of a crack tip for a specimen subjected to creep. The DFZ was surrounded by a plastic zone characterized by the numerous etching tracks caused by extensive climbing of basal dislocations. The climbing movement of dislocations dissipated energy, shielding the crack tip from the applied stress field. While basal dislocation in the creeping crystal could be observed by a large number of whiskers, no basal slip was detected. Dislocation climb was therefore the dominant mechanism of the creep deformation under the conditions of this study.

ACKNOWLEDGEMENT

This study has been supported in part by the U.S. National Science Foundation under Grant Nos. MSM–86–18798 and MSS–90–079291 and in part by the U.S. Office of Naval Research under Grant No. N00014–90–J–1360. The assistance of Mr. William Plunkett in preparing SEM photographs is gratefully acknowledged.

REFERENCES

1. Rice, J.R. and Thomson, R., Ductile versus brittle behavior of crystals. Phil. Mag., 1974, **29**, 73–97.

2. Ashby, M.F. and Embury, J.D., The influence of dislocation density on the ductile-brittle transition in bcc metals. Scrip. Met., 1985, **19**, 557–562.

3. Michel, B., Ice Mechanics, Les Presses De L′université Laval, Québec, 1978.

4. Hobbs, P.V., Ice Physics, Clarendon Press, Oxford, 1974.

5. Parsons, B.L., Subcritical crack growth, initiation and arrest in columnar fresh-water and sea ice. Doctoral Thesis, Memorial University of Newfoundland, St. Lohn's, Newfoundland, 1989.

6. Dempsey, J.P., Wei, Y. and DeFranco, S.J., Fracture resistance to cracking in ice: intiation and growth. In Cold Regions Engineering, ed. Sodhi, D.S., ASCE, 1991, pp.579–594.

7. Wei, Y., DeFranco, S.J. and Dempsey, J.P., Crack fabrication techniques and their effects on the fracture toughness and CTOD for freshwater columnar ice. J. Glac., 1991 (in press).

8. Schulson, E.M., Hoxie, S.G. and Nixon, W.A., The tensile strength of cracked ice. Phil. Mag., 1989, **59**, 303–311.

9. Dempsey, J.P., The fracture toughness of ice. In IUTAM/IAHR Symposium on Ice-Structure Interaction, St. John's, Newfoundland, 1989 (in press).

10. Kuroiwa, D. and Hamilton, W.L., Studies of ice etching and dislocation etch pits, In Ice And Snow, ed. Kingery, W.P., The MIT Press, Cambridge, Massachusetts, 1963.

11. Muguruma, J. and Higashi, A., Observation of etch channels on the (0001) plane of ice crystal produced by nonbasal glide. J. Phys. Soc. of Japan, 1963, **18**, 1261–1269.

12. Kuroiwa, D., Surface topography of etched ice crystals observed by a scanning electron microscope. J. Glac., 1969, **8**, 475–483.

13. Sinha, N.K., Dislocations in ice as revealed by etching. Phil. Mag., 1977, **36**, 1385–1406.

14. Sinha, N.K., Observation of basal dislocation in ice by etching and replicating. J. Glac., 1978, **6**, 636–642.

15. Wei, Y. and Dempsey, J.P., Fractographic examination of fracture in S2 ice. J. Mat. Sci., 1991 (in press).

16. Gilman, J.J. and Johnston, W.G., The origin and growth of glide bands in lithium fluoride crystals. In Dislocations and Mechanical Properties of Crystals, New York: John Wiley & Sons, Inc. 1957, pp.116–163.

17. Ohr, S.M. and Chang, S.-J., Dislocation-free zone model of fracture comparison with experiments. J. Appl. Phys., 1982, **53**, 5645–5651.

18. Ohr, S.M., Dislocation-crack interaction. J. Phys. Chem. Solids, 1987, **48**, 1007–1014.

19. Sinha, N.K., Dislocation climb in ice observed by etching and replicating. J. Mat. Sci. Letters, 1987, **6**, 1406–1408.

A THEORETICAL STUDY ON THE EFFECT OF RESIDUAL STRESS ON CREEP BRITTLE CRACK GROWTH.

DAVID J. SMITH
Department of Mechanical Engineering
University of Bristol, Bristol BS8 1TR

ALAN C.F. COCKS
Department of Engineering
University of Cambridge, Cambridge CB2 1PZ

ABSTRACT

In this paper a theoretical analysis is presented for the interaction of combined mechanical and residual stresses in a brittle creeping material. At long times it is assumed that the residual stresses have completely redistributed and the near crack tip stress field is governed by C for mechanical loading alone. At short times the near tip stress field is similar to the Riedel and Rice solution, but a new analysis is given that examines the relaxation of the combined stresses to provide expressions for the crack tip field parameter C[t]. It is assumed that with the introduction of residual stress the redistribution time from the initial elastic to the creep state remains largely unchanged. The subsequent result leads to faster local relaxation rates than for mechanical loading alone. The influence of residual stresses on the rate of creep crack growth in welded steel and sintered ceramic components are assessed by combining the new expressions for C[t] with experimentally determined growth rate laws.

INTRODUCTION

At elevated temperatures structural components can fail from the time dependent propagation of pre-existing flaws which nucleate during the manufacture of the component. These cracks invariably form in regions of high residual tensile stress which develop as a result of: thermal gradients in the component; variations in the thermal and elastic properties between different phases or different crystallographic orientations; or from any localized plastic deformation introduced during processing. Typical examples can be found in: welded steel components, where defects develop in the creep brittle region of the weld heat affected zone; sintered ceramic components, which invariably microcrack when cooled to room temperature after firing; and ground components, which suffer subsurface damage during machining.

The residual stresses will relax during operation of a component at elevated temperature as the material creeps; but as these stresses relax the cracks can grow and any enhancement of crack growth due to the presence of the residual stresses can significantly influence the life of a component. At any instant in time the crack tip stress and strain-rate fields can be expressed in terms of the quantity C[t], which is related to the stress intensity factor K_I at short times and reduces to the creep J-integral C^* at long times [1]. Both theoretical and experimental studies of creep crack growth suggest that the crack growth rate is determined by the magnitude of C[t] [2,3]. Expressions for C[t] are provided by Riedel and Rice [1] for mechanical loading alone (ie no residual stress field) which take into account the redistribution of stress from the initial elastic to the steady state creep field. In the present paper we extend these results to situations where residual stresses are significant, employing approximate, yet simple, procedures which retain the essential physics of the process, and evaluate the resulting influence of the relaxing residual stress field on component life. Initially we concentrate on the response of a linear creeping material, which is relevant to the response of engineering ceramics, for which the changing value of C[t] is entirely due to the relaxation of the residual stresses. We then extend the analysis to non-linear material response where redistribution of the mechanical stress field must also be taken into account.

CRACK TIP FIELDS

In this section we consider the response of an elastic/creeping material, which deforms according to the relationship

$$\dot{\varepsilon} = \frac{\dot{\sigma}}{E} + \dot{\varepsilon}_0 \left(\frac{\sigma}{\sigma_0}\right)^n \tag{1}$$

in uniaxial tension, where σ and $\dot{\sigma}$ are the applied stress and stress-rate respectively, $\dot{\varepsilon}$ is the resulting strain-rate for a material of Young's

Modulus E which creeps at a rate $\dot{\epsilon}_0$ at a stress σ_0, and n is the creep exponent.

Riedel and Rice [1] have determined the crack tip stress and strain-rate fields for small scale creep, ie: for situations where creep deformation is only important in a small zone surrounding the crack tip, with elastic effects dominating in the rest of the body. The stresses within the creep zone are given by

$$\sigma_{ij} = \sigma_0 \left[C[t]/\dot{\epsilon}_0 \sigma_0 I_n r \right]^{1/(n+1)} \tilde{\sigma}_{ij}(\theta, n) \tag{2}$$

where I_n and $\tilde{\sigma}$ are known dimensionless functions of n, r is the radial distance from the crack tip and $C[t]/\dot{\epsilon}_0\sigma_0$ gives a measure of the amplitude of the near tip stresses in terms of the elapsed time and specimen geometry:

$$C[t] = K_I^2/(n+1)E't \tag{3}$$

where $E'=E$ under plane stress conditions and $E'=E(1-\upsilon^2)$ in plane strain, υ is Poisson's ratio and t is the elapsed time.

During steady state creep the stress field is also given by equation (1) if $C[t]$ is identified with C^*. Ainsworth et al [4] demonstrate that in this limit C^* is well approximated by

$$C^* = \dot{\epsilon}_R \sigma_R \lambda \tag{4}$$

where $\dot{\epsilon}_R$ is the uniaxial strain-rate at a reference stress given by

$$\sigma_R = \frac{P}{P_L} \sigma_y$$

where P_L is the limit load for a perfectly plastic material of yield strength σ_y and P is the applied load; and λ is a characteristic length for the component,

$$\lambda = K_I^2/\sigma_R^2 \tag{5}$$

The time for stress redistribution can be obtained approximately by determining the time at which C[t] for small scale creep equals C^*. We can now define a suitable dimensionless time

$$\tau = t/t_T$$

and $t_T = \sigma_R/(n+1) E' \dot{\varepsilon}_R$ is the time for stress redistribution, such that

$$
\left.
\begin{aligned}
C[t] &= C^*/\tau = \dot{\varepsilon}_R \sigma_R \lambda/\tau && \text{for } \tau < 1 \\
\text{and} \quad C[t] &= C^* = \dot{\varepsilon}_R \sigma_R \lambda && \text{for } \tau > 1
\end{aligned}
\right\}
\tag{6}
$$

In the limit of n = 1 it can readily be demonstrated that C[t] is given by the second of equations (6) for all τ. For small τ however the analysis of Riedel and Rice [1] is inappropriate since it is assumed that creep strains dominate within a small creep zone which develops ahead of the crack. For a linear creeping material the stresses do not change with time and creep strains do not dominate anywhere in the body until after $\tau=2$. The conditions for small-scale creep are therefore never met and the linear situation must be treated as a special case. We can, however, assign a physical interpretation to τ in this limit; when $\tau \simeq 1$ creep effects become important in determining the overall response of the body. Also, for all values of n, the creep displacement of the body equals the elastic displacement after a time $(n+1)\tau$.

COMBINED MECHANICAL AND RESIDUAL STRESSES

Cocks and Ashby [5] and Smith [6] demonstrate how the form of the above results can be obtained by examining the relaxation of stress in a uniaxial specimen subjected to an initial high stress, and subsequently constrained to creep at a rate corresponding to a much lower stress. In this section we extend the philosophy of this approach to evaluate the influence of an initial residual stress field on C[t].

Consider the situation where a cracked body is subjected to a constant primary mechanical load which results in a stress field ahead of the crack tip which we can characterize in terms of a stress intensity factor K^P. Initially the body also contains a residual stress field which provides an initial additional stress intensity K^r. The residual stresses will, however, relax with time until eventually the near tip field can be expressed in terms of C^* for mechanical loading alone.

From equations (2) and (6) it is evident that the stress and strain-rate fields ahead of the crack tip can be characterized in terms of the reference stress σ_R for mechanical loading alone. Throughout stress redistribution the material wants to creep at a rate which can be directly determined from this reference stress. In the following we seek a comparable stress for situations where the contribution from the residual stress field to the crack tip fields becomes significant. We would not expect, however, the presence of a residual stress to significantly influence the time for stress redistribution in the body and in the steady state σ_R as defined in equation (4) would completely describe the component response.

It proves convenient to represent the residual stress field by

$$\sigma^r = K^r \sqrt{\lambda}$$

where λ is given by equation (5) and K_I is interpreted at the stress intensity factor for mechanical loading alone, K^P. We assume that in the presence of a residual stress field the material still wants to creep at a rate determined from σ_R, such that

$$\dot{\varepsilon} = \frac{\dot{\sigma}}{E} + \dot{\varepsilon}_o \left(\frac{\sigma}{\sigma_o}\right)^n = \dot{\varepsilon}_R \qquad (7)$$

where $\sigma = \sigma^r + \sigma_R$. If we identify σ_o and $\dot{\varepsilon}_o$ with σ_R and $\dot{\varepsilon}_R$ respectively equation (7) becomes

$$\frac{d\Sigma}{d\tau} = (n+1)(1-\Sigma^n) \qquad (8)$$

with $\Sigma = \sigma/\sigma_R$.

In the following we present results obtained from equation (8) for the limit $n = 1$ (linear creep) as a special case and for the more general situation of $n > 1$ (non-linear creep).

Linear Creep

Integrating equation (8), and noting that $\Sigma = 1 + \alpha_K$ at $\tau = 0$, where

$$\alpha_K = \frac{\sigma^r}{\sigma_R} = \frac{K^r}{K^P}$$

yields

$$\Sigma = 1 + \alpha_K \exp(-\tau/2) \qquad (9)$$

This equation can also be expressed in terms of the stress intensity factor or creep J-integral; i.e.

$$K = K^P \{1 + \alpha_K \exp(-\tau/2)\}$$

and

$$C[t] = \eta (K^P)^2 f_1[\alpha_K, \tau] \qquad (10)$$

where $\eta = \dot{\varepsilon}_o/\sigma_o$ and $f_1[\alpha_K, \tau] = \{1 + \alpha_K \exp(-\tau/2)\}^2$

Non-Linear creep

When $n > 1$ there is no complete analytical solution to equation (8). Cocks and Ashby [5] use equations similar to (7) and (8) to evaluate the stress field for small scale creep under mechanical loading alone, using the solution for short times when $\sigma/\sigma_R \gg 1$, where here σ represents the magnitude of the stresses within the creep zone. We need not repeat a similar analysis here, but can simply note that under small scale creep conditions stress relaxation is limited to a small zone ahead of the crack tip, with the initial elastic field still being retained outside of this zone. The estimate of C[t] at short times is

$$C[t] = \dot{\varepsilon}_R \sigma_R \lambda (1 + \alpha_K)^2/\tau \qquad (11)$$

Matching this with the steady state value of C* of equation (6) to determine an approximate time for complete stress redistribution, can provide a value which is significantly greater than t_T particular for large values of α_K [7]. To provide a better description of C[t] at longer times we examine the response of equation (8) at long times when Σ is close to 1. Then

$$\Sigma = 1 + \alpha_K \exp\left[\frac{-n\tau}{n+1}\right]$$

which implies the following approximate result for C[t].

$$C[t] = \dot{\varepsilon}_R \sigma_R \lambda \left\{1 + \alpha_K \exp\left(\frac{-n\tau}{n+1}\right)\right\}^{n+1} \tag{12}$$

The amplitudes of the near tip stresses C[t] predicted by equations (11) and (12) for linear and non-linear creep for combined mechanical and residual stresses are shown in Fig.1. as functions of τ for n = 1, 2, 3 and 4 with α_K = 6 at τ = 0. The amplitudes have been normalised with respect to C* for mechanical loading alone.

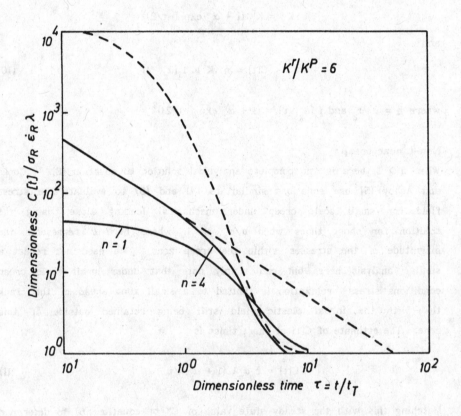

Fig. 1 Amplitude of near crack tip stress field for combined mechanical and residual stresses

For linear creep the result is unambiguous and for all times C[t] is given by equation 10. In contrast for n > 1 the analysis suggests at short times C[t] is given by equation (11) and at longer times given by equation (12). These two estimates do not match at intermediate times and it is apparent that the approximate long time solution equation (12) is an overestimate of the amplitude. The two solutions are matched approximately to give an estimate of C[t] for non-linear creep

$$C[t] = \sigma_R \dot{\varepsilon}_R \lambda \quad f_n[\alpha_K, \tau] : \quad n > 1$$

with

$$f_n[\alpha_K, \tau] = \frac{(1+\alpha)^2 \tau^{-1} \left\{1 + \alpha_K \exp\left(\frac{-n\tau}{n+1}\right)\right\}^{n+1}}{(1+\alpha)^2 \tau^{-1} + \left\{1 + \alpha_K \exp\left(\frac{-n\tau}{n+1}\right)\right\}^{n+1}}$$

(13)

CREEP CRACK GROWTH

In this section we consider the effect of residual stress on the process of creep crack growth in sintered ceramic and welded steel components. Ceramics often have creep exponents between 1 and 2 [8], while steels have creep exponents for practical operating conditions of 2 and greater [9]. There are many similarities between the crack growth laws developed for ceramic and metallic materials. In particular Thouless [10], Cocks and Ashby [2], Riedel [11] and Nikbin, Smith and Webster [3] suggest that creep crack growth rates can be described by

$$\frac{da}{dt} = \dot{a} = B(\Delta a)^{1-\phi} \{C[t]\}^\phi$$

(14)

where B and ϕ are material constants, a is the crack length and Δa is the increment of crack growth from the original crack length a_o. The governing parameter is C[t], which for linear creep, is ηK^2 with the crack growth rate proportional to $K^{2\phi}$. This represents an expression often used to describe creep crack growth in ceramics, where ϕ is in the range 0.5 to 1. For power law creeping materials such as steels there is a wide range of experimental evidence [3] that suggests C[t], equation 6, is the governing parameter with ϕ less than or equal to 1. An important feature of equation (14) is that the growth rate is not only a function of the load parameter C[t] but also the increment of crack growth, which in the early stages of crack growth is significant. Clearly when $\phi = 1$ this effect does not occur.

It proves convenient to express equation (14) in the following dimensionless form

$$\dot{\bar{a}} = \frac{d\bar{a}}{d\tau} = \bar{B} \, (\Delta \bar{a})^{1-\phi} \left\{ f \, [\alpha_K, \tau, n] \right\}^{\phi} \tag{15}$$

where

$$\bar{B} = Ba_o^{\,1-\phi} \, t_T \left(\sigma_R \dot{\epsilon}_R \lambda \right)^{\phi}$$

and $f[\alpha_K, \tau, n]$ is given by f_1 for $n = 1$ and f_n for $n > 1$.

To provide some insight into the effect of residual stress on the propagation of a crack we consider the following simple problem.

An infinite body contains a through-thickness crack of initial length a_o which lies along the x-axis of a linear co-ordinate system, whose origin lies at the centre of the crack. The body is subjected to a remote uniaxial stress σ^P normal to the crack, and contains a residual stress field:

$$\sigma^r = \sigma_o^r \cos\left(\frac{2\pi x}{L} \right)$$

along the line of the crack which includes a characteristic distance L. The stress intensity factor K^r for this field is well approximated by

$$K^r = \sigma_o^r \sqrt{\pi a} \; g \, [a/L]$$

with

$$g[a/L] = \sqrt{2} \; \cos\left(\frac{2\pi a}{L} - \frac{\pi}{4} \right) \Big/ \left(\frac{2a\pi^2}{L} + 1 \right)^{1/2} \tag{16}$$

The variation of dimensionless $K^r/\sigma_o^r \sqrt{\pi a}$ {= g[a/L]} is shown in Figure 2 for characteristic lengths, L = 0.2, 1 and 5. When L is small the residual

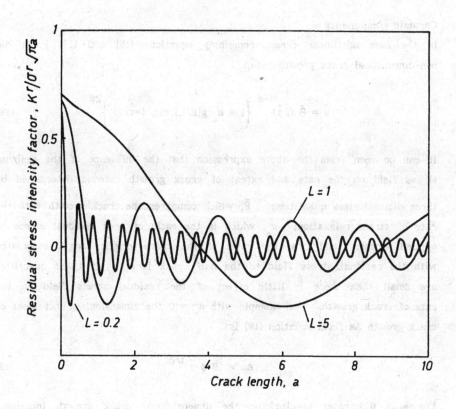

Fig. 2 Variation of stress intensity factor for residual stress for different characteristic lengths, L

stress varies rapidly with increasing crack length. When L is large relative to the crack length K^r, for small increments of crack growth, remains relatively unchanged.

Noting that $K^p = \sigma^p \sqrt{\pi a}$, an expression for α_K is

$$\alpha_K = \alpha_o \, g[a/L] \tag{17}$$

with $\alpha_o = \sigma_o^r / \sigma_o^p$.

Ceramic components

In the case of linear creep, combining equations (15) and (17) gives the non–dimensional crack growth rate;

$$\dot{\bar{a}} = \bar{B} \, (\Delta\bar{a})^{1-\phi} \left\{ 1 + \alpha_o \, g[a/L] \, \exp{(-\tau/2)} \right\}^{2\phi} \tag{18}$$

It can be seen from the above expression that the influence of the residual stress field on the rate and extent of crack growth rate is determined by three dimensionless quantities; $\bar{B}$, which compares the crack growth with the rate of stress relaxation; α_o, which is the ratio of peak residual stress to applied stress; and L/a, the ratio of the characteristic length associated with the residual stress field to the half crack length. If these quantities are small then there is little effect of the residual stress field on the rate of crack growth. For example with $\alpha_o = 0$ the dimensionless increment of crack growth $\Delta\bar{a}$ from equation (18) is

$$\Delta\bar{a} = (\bar{B} \, \phi \, \tau)^{1/\phi} \tag{19}$$

Figure 3 illustrates results for the dimensionless crack growth increment $\Delta\bar{a}/\bar{B}$ according to equation (19) for $\phi = 1$. Also shown are results obtained by numerically integrating equation (18) for $n = 1$, $\phi = 1$ and with $L/a_o = 0.2$, 1 and 5, corresponding to the sinusoidally varying residual stress intensities shown in Fig.2. If the total $C[t]$ becomes zero or negative crack growth is arrested and the crack will not start propagating again until the residual stress has relaxed sufficiently to provide a positive net value of $C[t]$. In the case of $L/a_o = 0.2$ crack growth is initially rapid over a small increment, but as $C[t]$ becomes negative crack growth slows down with increasing time; however, the residual stress field gradually relaxes allowing $C[t]$ to increase and the crack to grow again. Eventually when the influence of the residual stress diminishes with time and crack length, the crack will grow at a rate according to the mechanical loading alone. This is evident in Fig.3 for $L/a_o = 0.2$ at longer times where the crack growth increment becomes approximately parallel to the case for $\alpha_o = 0$.

For larger values of L relative to a_o, the initial increment of rapid crack growth becomes larger. This is followed by larger periods of time where cracking will remain essentially dormant, until the residual stress relaxes

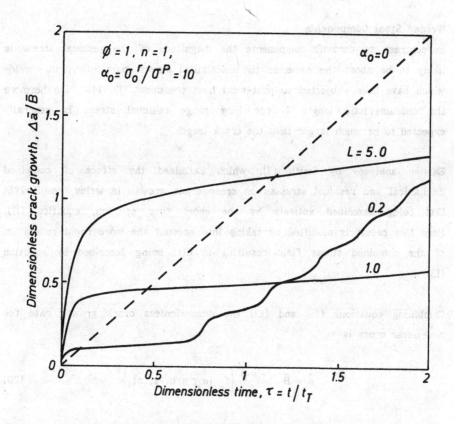

Fig. 3 Creep crack growth in sintered ceramics containing residual stresses with different characteristic lengths

sufficiently to allow C[t] to increase.

The type of response predicted in Fig 3 for $L/a_o < 1$ is typical of what might occur in sintered components where cracks formed on cooling are several grains in length and any residual stress resulting from thermal and elastic mismatch varies over a length scale comparable to the grain size [12]. If, however, the crack is adjacent to a large inhomogeneity or if the residual stress field results from grinding of the surface then the length scale associated with the residual stress field can be much larger than the grain and crack sizes, and the results for $L/a_o = 5$ are more relevant to the description of component response.

Welded Steel Components

In contrast to ceramic components the magnitude of the residual stress is likely to be about the same as the mechanical stress, particularly in welds which have been subjected to post-weld heat treatment [13, 14]. Furthermore the characteristic length L for long range residual stress is generally expected to be much larger than the crack length.

Earlier analyses by Smith [7], which examined the effect of combined mechanical and residual stresses on creep crack growth in welds, resulted in C[t] being described entirely by the short time solution, equation (11). Here this result is modified by taking into account the more rapid relaxation of the combined stress field resulting in C[t] being described by equation (13).

Combining equations (15) and (13) the dimensionless crack growth rate for non-linear creep is

$$\dot{\bar{a}} = \bar{B} \, (\Delta\bar{a})^{1-\phi} \left\{ f_n \, [\alpha_o, \, a/L, \, \tau, \, n] \right\}^{\phi} \tag{20}$$

As for linear creep the influence of the residual stress field on creep crack growth is determined by the dimensionless quantities $\bar{B}$, α_o and a/L. Fig.4 illustrates results obtained by numerical integration of equation 20 for n = 4, ϕ = 1, L/a_o = 10.0 and α_o = 1, 2 and 3. In each case the crack grows rapidly initially, but then the rate decreases as the residual stress relaxes and the crack grows in the compressive part of the residual stress field. Similar to residual stresses in ceramic components eventually the net C[t] will become positive and crack growth will recommence.

CONCLUSIONS

It has been shown that for combined residual and mechanical stresses there are faster relaxation rates of near crack tip stress fields. In materials exhibiting linear creep, such as sintered ceramics, the near crack tip stress fields can be described entirely as a combination of the stress intensity factors for mechanical and residual stresses. For non-linear creeping materials an approximate expression is obtained which at short times matches the Riedel and Rice solution.

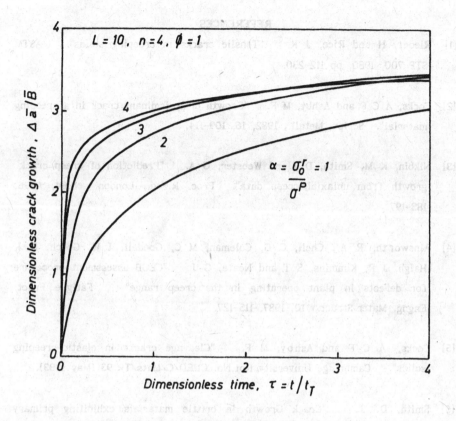

Fig. 4 Creep crack growth for different levels of residual stress representative of welded components

The results are also expressed in terms of the creep fracture mechanics parameter C[t], and incorporated into a creep crack growth law applicable to both ceramic and welded steel components. Assuming a sinuoidally varying residual stress it has been shown that, depending on the ratio of characteristic length of the varying residual stress to the initial crack length, and also the magnitude of the residual stress, there can be either initial rapid crack growth followed by a period of negligible growth, or vice-versa.

REFERENCES

[1] Riedel, H and Rice, J R. "Tensile cracks in creeping solids". ASTM STP 700, 1980, pp 112-230.

[2] Cocks, A C F and Ashby, M F. "Growth of a dominant crack in a creeping material". Script. Metall., 1982, 16, 109-114.

[3] Nikbin, K M, Smith, D J and Webster, G A. "Prediction of creep crack growth from uniaxial creep data". Proc. R.Soc., London. 1984, A396, 183-197.

[4] Ainsworth, R A., Chell, G G, Coleman, M C, Goodall, I W, Gooch, D J, Haigh, J R, Kimmins, S T and Neate, G J. "CEGB assessment procedure for defects in plant operating in the creep range". Fatigue Fract. Engng. Mater.Struct. 10, 1987, 115-127.

[5] Cocks, A C F and Ashby, M F. "Cleavage cracks in elastic-creeping solids". Cambridge University Rpt.No. CUED/C/Mats/TR.93 (May 1983).

[6] Smith, D J. "Crack Growth in brittle materials exhibiting primary creep". Eng. Fract. Mechs., 37,1, 1990, 27-41.

[7] Smith, D J. "Predicting creep failure in high temperature cracked welded components contining residual stresses". In: High Temperature Crack Growth,, I.Mech.E., London, 1987, 41-50.

[8] Cannon, R W and Langdon, T G. "Review: Creep of ceramics, Part I Mechanical Characteristics". Jnl. of Mats. Science, 18, 1983, 1-50.

[9] Goodall, I W. "Multiaxial data requirements for structural integrity assessments in creep". In: Techniques for Multiaxial Creep Testing, Elsevier Applied Science, 1986, 1-27.

[10] Thouless, M D. "Modelling creep-crack growth processes in ceramic materials." In: Mechanics of Creep Brittle Materials, 1. Elsevier Applied Science, 1989, 50-62.

[11] Riedel, H.   "A continuum damage approach to creep crack growth". In: Fundamentals of Deformation and Fracture, Proceedings of the Eshelby Memorial Symposium, CUP, 1984, 293-309.

[12] Ortiz, M and Molinari, A.   "Microstructural thermal stresses in ceramic materials". J.Mech,.Phys. Solids, 36, 4, 1988, 385-400.

[13] Fidler, R.   "Relaxation of Residual Stresses in a CrMoV/2CrMo Steam pipe weld under service operation". CEGB Report R/M/R298 January 1980.

[14] Smith, D J and Webster, G A.   "Case studies in the prediction of creep failure in cracked components operating at elevated temperatures".   In: Creep and Fracture of Engineering Materials and Structures, Inst. of Metals, London, 1987, 549-562.

CREEP CRACK GROWTH PRIOR TO STRESS REDISTRIBUTION

P J BUDDEN, D W DEAN* AND G E TURNER*
Nuclear Electric plc, Berkeley Nuclear Laboratories,
Berkeley, Gloucestershire, GL13 9PB, UK
* Nuclear Electric plc, Barnett Way, Barnwood,
Gloucestershire, GL4 7RS, UK.

ABSTRACT

Crack growth in an elastic-creeping body prior to redistribution of stress under constant applied load is considered. The cases of isothermal creep and also a step change in temperature are studied. Theoretical estimates of C(t) and hence crack growth are obtained. Finite-element analysis of a centre-cracked plate subject to a step temperature change is described and the results compared with the theory. For isothermal creep, it is shown that the crack growth rate is bounded by the sum of short-time and steady state approximations. When the temperature changes, it is shown to be conservative to neglect prior redistribution due to creep strain when estimating subsequent crack growth.

INTRODUCTION

Integrity assessments of high temperature plant require consideration of failure by creep mechanisms. In particular, defects can occur in plant due to the manufacturing or welding processes or they can arise in service due to the operating conditions. A procedure for assessing the integrity of cracked structures under steady creep loading is given by Ainsworth et al [1] and forms part of the R5 Procedure [2] developed for use within the UK electricity generating industry. In R5, reference stress methods are used to estimate C*; C* [3] characterises the steady-state stress and strain-rate fields ahead of a stationary crack in Mode I for a power-law creeping material. Estimates of the time to initiate a growing crack, and of subsequent crack growth, in [1, 2] were based on C*; a reference stress formulation enabled generalisation to arbitrary creep laws.

During the transition period between initial elastic loading and steady-state response, the singularity fields are governed by $C(t)$; $C(t) \rightarrow C^*$ as time $t \rightarrow \infty$ and $C(t) \sim K^2/E'(n+1)t$ as $t \rightarrow 0$ [4] when creep strain rate $\dot{\varepsilon}_c$ satisfies

$$\dot{\varepsilon}_c = B\sigma^n \qquad (1)$$

with B, n constants, K the stress intensity factor and $E' = E$, $E/(1 - \nu^2)$ in plane stress, plane strain, respectively. Estimation formulae for $C(t)$ were given in [5, 6]; $C(t) \simeq C^*$ for $t \geq t_{red}$, where t_{red} is defined by equality of creep and elastic strain at the reference stress and $t_{red} = K^2/E'C^*$ when eqn. (1) holds. Then crack growth rate can be correlated with $C(t)$:

$$\dot{a} = A' C(t)^q \qquad (2)$$

where A' and $q < 1$ are constants.

For materials operating at temperatures such that practical assessment times exceed t_{red}, crack growth can be simply estimated by scaling the equivalent total steady state growth [7, 8].

For structures creeping rapidly, crack growth may occur early in life. Conversely, for structures operating at temperatures where creep rates are low, t_{red} can be large. In either case, estimates of crack growth at times $t < t_{red}$ are required, taking proper account of any enhancement of instantaneous growth rates.

In this paper, bounds to the crack growth expressions in [7] are obtained and generalised to arbitrary creep deformation and crack growth laws. The effect of a step change in temperature at time t_1 is then considered; the cases $t_1 < t_{red}$ and $t_1 \geq t_{red}$ are distinguished. Finite-element validation of the estimates is described.

CRACK GROWTH WHEN $t < t_{red}$

For an elastic, creeping material which satisfies eqn. (1), the stress and strain-rate fields are singular at the crack tip; in particular the stress field is given by

$$\sigma_{ij}(r, \varphi, t) \sim (C(t)/BI_n r)^{1/n+1} \tilde{\sigma}_{ij}(\varphi,n) \text{ as } r \to 0 \tag{3}$$

where (r,φ) are polar co-ordinates centred at the crack tip and I_n, $\tilde{\sigma}_{ij}$ are dimensionless functions [11]. Ainsworth and Budden [5] derived the estimate

$$C(t) = C*(1 + \tau)^{n+1}/\{(1 + \tau)^{n+1} - 1\} \tag{4}$$

where $\tau = t/t_{red} \equiv E'C*t/K^2$ is dimensionless time. Then $C(t) \to C*$ as $t \to \infty$ and $C(t) \sim K^2/E'(n+1)t$ as $t \to 0$.

If crack growth rate à is given by eqn. (2) then Ainsworth and Budden [7] showed that, for $q = n/n+1$ and for crack growth $\Delta a \ll$ crack size, a:

$$\Delta a \simeq A'C*^{n/n+1} \quad t\{1 + t_{red}/t\} \tag{5}$$

when $t \geq t_{red}$. This was generalised [8] to

$$\Delta a \simeq A'C*^q t \{1 + \varepsilon_e(\sigma_{ref})/\varepsilon_c(\sigma_{ref},t)\} \tag{6}$$

for general creep deformation and crack growth laws. Here $\sigma_{ref} \equiv P\sigma_y/P_L(a,\sigma_y)$ is the reference stress corresponding to load P and limit load P_L assuming rigid-plastic behaviour with yield stress σ_y and $\varepsilon_e(\sigma_{ref}) = \sigma_{ref}/E$. Equations (5) and (6) assume crack growth starts at $t = 0$ but are not appropriate to $t/t_{red} \ll 1$ since they do not produce zero crack growth in the limit $t \to 0$.

Crack Growth Estimates

From eqns. (2) and (4), defining $f \equiv à/A'C*^q$ and $g \equiv f - (n+1)^{-n/n+1}$ $\tau^{-n/n+1}$ it is possible to show that the function g satisfies $0 \leq g(\tau,n) \leq g(1, n) \leq 1$ for all τ, n when $q = n/n+1$ and $\tau \leq 1$. An upper bound to f is then

$$f(\tau,n) = (n+1)^{-n/n+1} \tau^{-n/n+1} + 1 \tag{7}$$

which can be generalised to $q \neq n/n+1$ as

$$f(\tau,n) = (n+1)^{-q} \tau^{-q} + 1. \tag{8}$$

Note that eqns. (7) and (8) are simply the sum of short-time and steady-state approximations to $\dot{a}/A'C*^q$.

Often creep data are not given in the form of Norton's law, eqn. (1). Then an estimate of f which is independent of n is required. From eqns (7) - (8),

$$(n+1)^{-q} = (n+1)^{1/n+1} (1 - q) \leq 1.445 (1 - q)$$

when $q = n/n+1$. Then eqn. (8) can be approximated by

$$f = 1.445 (1 - q) \tau^{-q} + 1 \tag{9}$$

for $\tau \leq 1$ and for general creep crack growth and deformation laws, with $\tau \equiv \varepsilon_c/\varepsilon_e$ in general.

Note that $\dot{a} \to \infty$ as $t \to 0$. It is then preferable to estimate Δa directly for the initial timestep in a numerical procedure.

Estimates of Δa when $t \ll t_{red}$

If $\Delta a \ll a$ it follows, from eqn. (8) for example, that

$$\Delta a \simeq A'C*^q t_{red}\{(n+1)^{-q} (1-q)^{-1} \tau^{1-q} + \tau\}.$$

If $\tau \ll 1$ then the second term in brackets is negligible and then

$$\Delta a \simeq A'(K^2/E'(n+1))^q t^{1-q}/(1-q)$$

since $t_{red} = K^2/E'C*$. Similarly it follows from eqn. (9) for general creep laws that, when $\tau \ll 1$:

$$\Delta a \simeq 1.445 A' (K^2/E')^q t^{1-q}.$$

STEP TEMPERATURE CHANGES

A step change in temperature at time t_1 is considered in this section, corresponding to constants $B = B_1$ and B_2 in eqn. (1) for $t < t_1$ and $t > t_1$, respectively. The stress index n in eqn. (1) and the external load are both assumed constant. Then $B_1 > B_2$ or $B_1 < B_2$ for a fall or rise in temperature, respectively.

Equation (1) can be re-written as

$$d\epsilon_c/dT = B_1 \sigma^n$$

where modified time $T = t$ for $t < t_1$ and $T = t_1 + (B_2/B_1)(t - t_1)$ for $t > t_1$. At any 'real' time t, σ_{ij} is continuous (in the absence of thermal stresses) and satisfies eqn. (3) with the appropriate B. Hence $C(t)/B$ is continuous at $t = t_1$, that is

$$C(t_1^+)/C(t_1^-) = B_2/B_1 = C_2^*/C_1^* \tag{10}$$

where the subscripts on C* correspond to values appropriate to the two temperatures.

Approximation to C(t)

Following [5] it is argued that the J-integral is well approximated by a linear function of the modified time T, that is

$$J(T) = K^2/E' + C_1^*T. \tag{11}$$

J is then a bi-linear function of t, with slopes C_1^* and C_2^* for $t < t_1$ and $t > t_1$, respectively. Finite-element validation of this assumption is given in the following section. The differential equation connecting J and C derived in [5] then holds here for all t. For $t < t_1$, the solution is eqn. (4). For $t > t_1$, solution of the differential equation gives

$$C(t) = \frac{C_2^* J^{n+1}(t)}{J^{n+1}(t) + J^{n+1}(t_1) \{C_2^*/C(t_1^+) - 1\}} \geq C_2^* \tag{12}$$

since $C_2^*/C(t_1^+) = C_1^*/C(t_1^-) \leqq 1$ from eqns. (4) and (10). From eqn. (12), $C(t)$ falls monotonically from $C(t_1^+)$ at $t \to t_1^+$ to C_2^* as $t \to \infty$. At $t = t_1$ there is a step discontinuity given by eqn. (10). Qualitatively, if $t_1 \geqq t_{red}^{(1)} = K^2/E'C_1^*$, then $C(t_1^-) \simeq C_1^*$ and hence $C(t_1^+) \simeq C_2^*$, that is $C(t) \simeq C_2^*$ for all $t > t_1$. This means that as redistribution is essentially complete at $t = t_1$, $C(t)$, and hence $\dot{a}$, jump instantaneously to new steady-state values. Conversely, if $t_1 < t_{red}^{(1)}$ then further redistribution occurs; $C(t)$ continues to drop and additional enhancement of crack growth rate ensues. Finite-element validation of these conclusions is addressed in the next section.

Effect on Crack Growth when $t_1 < t_{red}$

It can be shown, by equating creep strain accumulated under the two conditions (at the same σ_{ref}) with the elastic strain, that the redistribution time t_{red} satisfies

$$t_{red}/t_1 = (C_1^*/C_2^*)(t_{red}^{(1)}/t_1 - 1) + 1 \geqq 1.$$

Then two cases are apparent: (i) $t_1 < t < t_{red}$ and (ii) $t_1 < t_{red} < t$. Case (ii) is a trivial extension of case (i) since steady state growth rates apply for times exceeding t_{red}.

For case (i), total crack growth $\Delta a \equiv \Delta a_1 + \Delta a_2$, where Δa_1 follows from the previous section as the growth at constant temperature between times 0 and t_1, and Δa_2 is the subsequent growth from time t_1 to the assessment time, t. Then, assuming $\Delta a \ll a$ as before:

$$\Delta a_2 = A' \int_{t_1}^{t} C(\bar{t})^q \, d\bar{t}$$

where, following substitution for J in eqn. (12) and some simplification

$$\frac{C(t)}{C_2^*} = (\alpha + \bar{t}/t_{red}^{(2)})^{n+1}/\{(\alpha + \bar{t}/t_{red}^{(2)})^{n+1} - 1\} \tag{13}$$

with $\alpha \equiv 1 + t_1/t_{red}^{(1)} \geqq 1$ and $\bar{t} \equiv t - t_1$. Hence $C(t) \leqq C(t)|_{\alpha = 1}$; that is, $C(t)$ is bounded by the value it would attain if the load were first applied at time t_1. It is then conservative to regard the two isothermal periods as independent, that is to ignore prior accumulation of creep strain when assessing crack growth in any such period.

FINITE-ELEMENT ANALYSIS

Finite-element, elastic-creep analyses of a centre-cracked plate (CCP) with crack length, 2a, to plate width, 2w, ratio a/w = 0.25 were performed using BERSAFE [12]. Norton law creep with n = 5 and a constant remote stress normal to the crack plane were assumed under plane strain conditions. The finite-element mesh, representing $\frac{1}{4}$ of the CCP contained 777 nodes and 242 quadratic isoparametric elements with a highly refined crack tip region (see Fig. 1). Four analyses were performed, corresponding to B_2/B_1 = 5, 0.2 and $t_1/t_{red}^{(1)}$ = 0.1, 4. C(t), C* and J(t) were taken as the mean of values computed from eight contours about the crack tip, each contour consisting of piecewise-linear arcs at distances between 0.025a and 0.8a from the crack tip. The minimum contour radius was five times the crack tip element radius.

Typical results are plotted in Figs. 2-3 in terms of graphs of $C(t)/C_1^*$ and J(t)/J(0) against normalised time $\tau = t/t_{red}^{(1)} = E'C_1^*t/K^2$. The estimation formulae, eqns. (4), (11) and (13), are also plotted for comparison. It can be seen that agreement is good, in particular that the predicted change in C(t) by a factor of five (in these cases) at $t = t_1$ is accurately reproduced. Similarly, when $t_1 > t_{red}^{(1)}$, the computed results show that $C(t) \simeq C_2^*$ for $t > t_1$.

CONCLUSIONS

Crack growth prior to stress redistribution in an elastic-creeping material has been considered. The applied load was assumed constant and the cases of isothermal creep and also a step change in temperature were analysed. In particular it was found that:

(1) Under isothermal conditions, the crack growth rate is bounded by the sum of short-time and steady-state approximations.

(2)   Following a step change in temperature prior to stress redistribution, a further period of enhanced growth ensues. For calculations
of crack growth it is conservative to assume that the isothermal
periods are independent; that is, to neglect creep strain accumulated
prior to any such period.

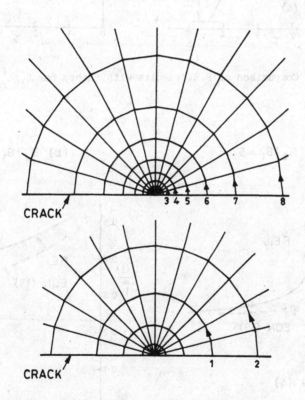

Figure 1. Finite-element mesh crack-tip refinement.
The integration contours are indicated.

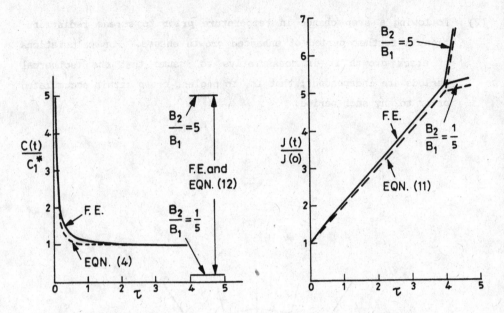

Figure 2. Comparison of F.E. results with theory for $t_1/t_{red}^{(1)} = 4$.

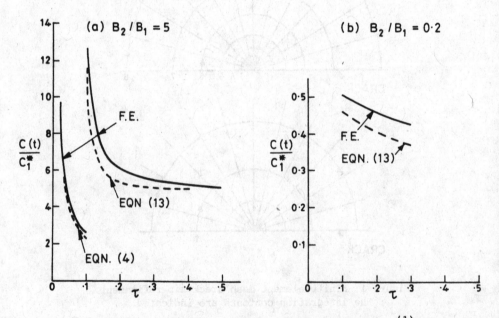

Figure 3. Comparison of F.E. results with theory for $t_1/t_{red}^{(1)} = 0.1$.

ACKNOWLEDGEMENT

This paper is published with the permission of Nuclear Electric plc.

REFERENCES

1. Ainsworth, R.A., Chell, G.G., Coleman, M.C., Goodall, I.W., Gooch, D.J., Haigh, J.R., Kimmins, S.T. and Neate, G.J., CEGB Assessment Procedure for Defects in Plant Operating in the Creep Range. Fatigue Fract. Eng. Mater. Struct. 1987, **10**, 115-127.

2. Goodall, I.W. and Ainsworth, R.A., An Assessment Procedure for the High Temperature Response of Structures. In Proceedings IV IUTAM Symposium on Creep in Structures, 1990, Krakow, Poland.

3. Landes, J.D. and Begley, J.A., A Fracture Mechanics Approach to Creep Crack Growth. In Mechanics of Crack Growth, ASTM STP 590, 1976, pp 128-148.

4. Riedel, H. and Rice, J.R., Tensile Cracks in Creeping Solids. In Fracture Mechanics: Twelfth Conference, ASTM STP 700, 1980, pp 112-130.

5. Ainsworth, R.A. and Budden, P.J., Crack Tip Fields under Non-Steady Creep Conditions - I Estimates of the Amplitude of the Fields. Fatigue Fract. Engng. Mater. Struct., 1990, **13**, 263-276.

6. Ehlers, R. and Riedel, H., A finite element analysis of creep deformation in a specimen containing a macroscopic crack. In Proc. 5th Intnl. Conf. on Fracture, ed. D Francois, Vol. 2, Pergamon, Oxford, 1981, pp 691-698.

7. Ainsworth, R.A. and Budden, P.J., Crack Tip Fields under Non-Steady Creep Conditions - II Estimates of Associated Crack Growth. Fatigue Fract. Engng. Mater. Struct., 1990, **13**, 277-285.

8. Ainsworth, R.A., Stress Redistribution Effects on Creep Crack Growth. In Proceedings Mechanics of Creep Brittle Materials, Leicester, Elsevier Appld. Science, 1988.

9. Rice, J.R. and Rosengren, G.F., Plane Strain Deformation near a Crack Tip in a Power-Law Hardening Material. J. Mech. Phys. Solids, 1968, **16**, 1-12.

10. Hutchinson, J.W., Singular Behaviour at the End of a Tensile Crack in a Hardening Material. J. Mech. Phys. Solids, 1968, **16**, 13-31.

11. Shih, C.F., Tables of Hutchinson-Rice-Rosengren Singular Field Quantities. Brown Univ. Report MRL E-147, Providence, R.I., 1983.

12. Hellen, T.K., The BERSAFE System, BERSAFE User's Guide Volume 1, Nuclear Electric Manual, 1991.

CREEP-FATIGUE CRACK GROWTH IN WELDED FUSION BOUNDARIES

R H PRIEST, D N GLADWIN AND D A MILLER
Technology Division
Nuclear Electric plc
Bedminster Down, Bridgwater Road, Bristol, Avon, BS13 8AN

ABSTRACT

The creep-fatigue crack growth response of welded fusion boundaries is examined. Displacement-controlled creep-fatigue tests have been performed at 565°C on two types of weld. The tests incorporated hold periods of up to 190 hours. A comparison is made between the response of the welded fusion boundaries in both an Incoweld A "cold" weld and a conventional 2¼Cr1Mo weld laid in ½CrMoV steel parent. It is shown that faster crack growth rates are obtained for the Incoweld A weld. Furthermore, dwell time dependent cyclic rates were obtained for the Incoweld A material. Mean and upper bound equations are given which describe the creep-fatigue crack growth response of both materials in terms of the relevant fracture mechanics parameters.

INTRODUCTION

CrMoV creep resisting steels are widely used for high temperature components in UK power plant operating typically at 540°C or 565°C. The weld repair of these items is costly and time consuming, involving high pre-heat (200°C) and post weld heat treatment (705°C) and often elaborate steelwork support. Major expenditure is incurred on weld repairs during plant overhauls. On those occasions when repairs are required during unplanned outages, additional costs may arise from the need to replace lost generating capacity with less efficient plant. Rapid "cold" weld repair techniques which dispose of the need for heat treatment therefore have clear financial benefits. The use of high nickel electrodes for fabricating "cold" welds has been investigated in recent years, the integrity of such welds needing to be assessed against possible fracture mechanisms.

Operational experience has shown that high nickel "cold" welds are particularly vulnerable to cracking by thermal fatigue [1-3] due to the mismatch of thermal expansion coefficients at the fusion line; such a mismatch is unavoidable with dissimilar welds. Of the numerous cold weld repairs that have been carried out in the UK on CrMoV components,

ENiCrFe-3 electrodes (e.g. Inconel 182) have generally been used. However, it is not clear that this is the optimum electrode for all applications since Inconel 182 produces a high strength weld metal with a higher coefficient of expansion than the parent material. The use of an alternative weld metal, also of high strength but with a coefficient of expansion which more closely matches that of the parent is clearly beneficial in terms of thermal fatigue resistance; one such weld metal is Incoweld A (ENiCrFe-2).

In order to be assured of the long term integrity of components welded with Incoweld A, a measure of the response to thermally-induced creep-fatigue deformation is required, particularly for weld fusion boundary zones. This paper therefore describes the results of a series of creep-fatigue tests performed to examine the crack growth response of fusion boundaries associated with Incoweld A "cold" welds. These properties are compared with similar data obtained on fusion boundaries associated with conventional 2¼Cr1Mo welds, as used during the fabrication of welded joints in CrMoV components.

EXPERIMENTAL

Material

Test welds were made to simulate, as far as possible, typical welds on power plant. At the same time, care was taken to produce a test block which would facilitate subsequent test specimen manufacture. The test blocks were 400 mm long and made from ½CrMoV steel. Details of the welding configuration used are shown in Figure 1. For both 2¼Cr1Mo and Incoweld A welds, stringer bead deposition was employed on a surface inclined at 20° to the vertical, an orientation typical of actual repairs. This geometry was adopted so that the HAZ associated with the initial stages of the weld would eventually lie in the plane of the starter notch in the machined specimens. This would be the material sampled by the growing cracks during testing, see below.

Specimens

All testing was performed using plane sided, 25 mm thick standard compact tension specimens. The orientation of the specimens is indicated in Figure 2. Prior to creep-fatigue testing, specimens were macro-etched to show up the fusion boundary and then spark notched. The notches were aligned with the fusion boundaries and machined to a depth as close as practicable to the boundaries. The orientation of the starter notch was designed to ensure crack propagation along the fusion line of the weld.

Mechanical Testing

To simulate the thermally-induced cycling experienced during service, isothermal displacement-controlled creep-fatigue tests were performed. The testing configuration has been described previously [4]. Tests were performed at 565°C ± 2°C. Crack length measurements were made during the tests using a dcpd technique with a current of 40A. The pd results were used to linearly interpolate crack lengths between the initial and final values measured fractographically following the tests.

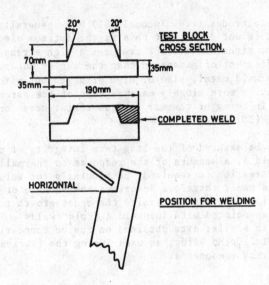

FIG.1. TEST BLOCK CONFIGURATION.

Incoweld A: Fully reversed continuous cycling tests were performed at frequencies of 5×10^{-2} to 1.4×10^{-4} Hz. In addition, tests with hold periods of 4, 22, 94 and 190h at peak tensile displacement were performed; the cycle time for these fully reversed dwell tests was 2h.

$2\frac{1}{4}$Cr1Mo Weld: Fully reversed continuous cycling tests were performed at a frequency of 1.4×10^{-4} Hz. In addition, tests with hold periods of 22, 94 and 190h at peak tensile displacement were performed, the cycle time again being 2h.

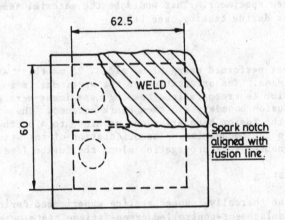

FIG.2. ORIENTATION OF 25mm THICK COMPACT TENSION SPECIMENS.

Metallography/Fractography

Following the completion of each test, specimens were sectioned in half longitudinally. One of the halves was cleaved open at liquid nitrogen temperature. Fracture surfaces were viewed both optically and using a scanning electron microscope. Optical examination enabled initial crack lengths and the extent of crack growth to be determined from the fracture surfaces. Scanning electron microscopy enabled the detailed mechanism of crack growth to be confirmed.

A section of the remaining specimen half which contained the major crack was mounted and polished metallographically by conventional means to 1 μm. A polish-etch sequence was then carried out, and the resulting specimens viewed in an optical microscope; specimens were etched in 2% Nital. The metallographic specimens were used to identify crack propagation paths.

ANALYSIS

In general, the creep-fatigue tests exhibited crack growth during both the hold periods and the cyclic excursions. Total crack growth rates, da/dN, were therefore described by the equation:

$$da/dN = (da/dN)_f + (da/dN)_c \qquad (1)$$

where $(da/dN)_f$ represents the cyclic contribution and $(da/dN)_c$ the creep contribution to growth.

Consistent with earlier approaches [4,5] and Nuclear Electric's R5 procedure [6], cyclic growth rate contributions were described by:

$$(da/dN)_f = C\Delta K_{eff}^l \qquad (2)$$

where $\Delta K_{eff} = (1 + R/5)\Delta K_{total}$ for $-2.5 < R < 0$. The coefficients C and l are found by experiment. The expression for the total stress intensity factor range, ΔK_{tot}, has been given previously [4]; ΔK_{eff} is the stress intensity factor range for which cracks are open and R is the ratio of minimum to maximum load in a cycle [4-6].

Creep crack growth rates during the hold periods were described by:

$$da/dt = AC*^q \qquad (3)$$

where A and q are found by experiment. The amount of creep crack growth occurring during a hold period of duration t_h hours is:

$$(da/dN)_c = \int_0^{t_h} AC*^q \, dt \qquad (4)$$

The equation describing C* has been given previously [4].

RESULTS

Incoweld A

Cyclic crack growth rates: Crack growth rates resulting from the continuous cycling tests at different frequencies are plotted as a function of ΔK_{eff} in Figure 3. Although relatively few data were obtained, a trend of increasing cyclic crack growth rates with decreasing frequency is discernible.

Cyclic contributions for the hold period tests are also shown as a function of ΔK_{eff} in Figure 3. The data show that the inclusion of a 4h hold period at peak displacement has resulted in an increase in crack growth rates. Indeed, despite the scatter in the cyclic contribution data, the dataset as a whole can be interpreted as exhibiting increasing crack growth rates with increasing hold period duration. It is noted, however, that occasional large increases in crack growth increments are apparent, even for hold times as low as 4h. This is likely to be due to particularly rapid crack growth through brittle areas of microstructure, see below.

In view of the variability in measured crack growth rates, it is prudent to describe an upper bound to the cyclic growth contributions. An upper bound to the cyclic crack growth database can be described by:

$$(da/dN)_f = 7 \times 10^{-9} \Delta K_{eff}^{3.6} \quad m.cycle^{-1} \qquad (5)$$

where ΔK_{eff} is measured in MPa$\sqrt{m}$. It is noteworthy that this is two orders of magnitude faster than the upper bound fatigue crack growth equation of Skelton [7], which has been found to bound cyclic crack growth rate data from a range of ferritic steels.

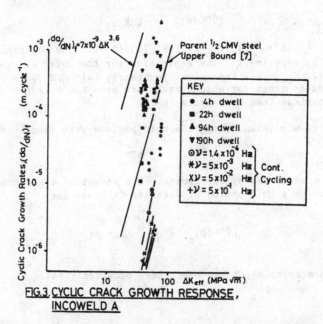

FIG.3. CYCLIC CRACK GROWTH RESPONSE, INCOWELD A

Creep crack growth rates: Crack growth rates measured during the constant displacement hold periods are plotted as a function of experimentally measured C* in Figure 4. As with the cyclic data, significant scatter is evident in the creep crack growth rates measured. However, unlike the cyclic data, there is no apparent trend towards increasing crack growth rates with increasing hold time duration.

The data shown in Figure 4 are compared with the mean and upper bound crack growth rates previously described for ½CrMoV steel [8], given by:

$$\frac{da}{dt} = 0.006 \ C*^{0.8} \quad mh^{-1} \tag{6}$$

and

$$\frac{da}{dt} = 0.06 \ C*^{0.8} \quad mh^{-1} \tag{7}$$

respectively, where C* is given in units of $MPamh^{-1}$. These equations can be seen to reasonably describe the Incoweld A fusion boundary results.

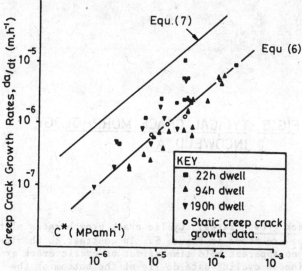

FIG.4. CREEP CRACK GROWTH RESPONSE, INCOWELD. A

Metallography: From the optical metallography, a range of crack growth patterns was discernible. For all tests, initial crack tips were located in the fusion zone area. In the majority of cases, subsequent creep-fatigue crack propagation tended to follow the profile of the fusion boundary, with the crack tending to reside in the HAZ adjacent to the fusion line. Depending on the exact details of when testing was stopped, for a given test crack tips were seen to stop in the HAZ, parent or indeed on the boundary itself. In general, propagation alternated between the HAZ and the fusion line; Figure 5 shows the typical morphology, in this case from a 22h dwell test.

Some cavitation was discernible local to the crack tip. Moreover, creep damage was also observed in the HAZ adjacent to the fusion boundary, though this was not associated directly with creep damage from the crack tip. Similar features were observed in all other tests except the 4h dwell test, in which little evidence of any intergranular creep damage was seen.

FIG.5. TYPICAL CRACK MORPHOLOGY, INCOWELD A.

X 12.8

2¼Cr1Mo Weld

Cyclic crack growth rates: Cyclic crack growth rates plotted as a function of ΔK_{eff} are shown in Figure 6. In contrast to the Incoweld A data, there was no apparent hold time effect on cyclic crack growth rates, though the continuous cycling data do lie at the bottom of the scatterband. Also shown on Figure 6 is the upper bound cyclic rate equation for ½CrMoV steel [7]. Although this gives a reasonable representation of the mean response, a significant proportion of the measured data lie above the ½CrMoV steel upper bound line. An upper bound to the 2¼Cr1Mo weld fusion boundary data is given by:

$$(da/dN)_f = 2 \times 10^{-9} \Delta K_{eff}^{3.6} \quad m.cycle^{-1} \qquad (8)$$

where ΔK_{eff} is in MPa√m. This is more than an order of magnitude higher than the ½CrMoV steel upperbound, but lower than the comparable upper bound for Incoweld A, equation (5).

Creep crack growth rates: Few creep crack growth rate data have been gathered; these are shown in Figure 7. Comparison with Figure 4 shows that the 2¼Cr1Mo weld fusion boundary crack growth rates are similar to those obtained on the Incoweld A, if anything lying towards the lower end of the Incoweld A scatterband. As with the Incoweld A data, the mean equation derived for ½CrMoV steel adequately describes the 2¼Cr1Mo weld data, Figure 7, though the upper bound appears to be too conservative. In this context it is worth noting that data on 2¼Cr1Mo weld metal, HAZ and fusion boundary have been obtained by Liaw et al [9] and Saxena et al [10].

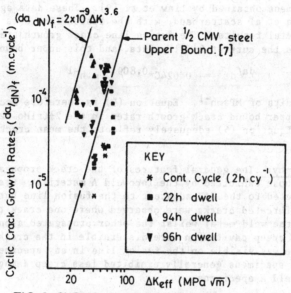

$(da/dN)_f = 2 \times 10^{-9} \Delta K^{3.6}$

Parent ½ CMV steel Upper Bound. [7]

Cyclic Crack Growth Rates, $(da/dN)_f$ (m.cycle^{-1})

KEY
* Cont. Cycle (2h.cy^{-1})
■ 22h dwell
▲ 94h dwell
▼ 96h dwell

20   100 ΔK_{eff} (MPa √m)

FIG. 6. CYCLIC CRACK GROWTH RESPONSE, 2¼ Cr 1 Mo WELD.

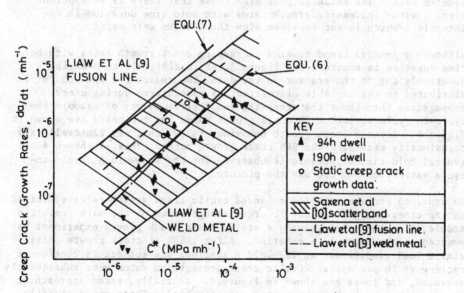

EQU.(7)

LIAW ET AL [9] FUSION LINE.

EQU.(6)

Creep Crack Growth Rates, da/dt (mh^{-1})

KEY
▲ 94h dwell
▼ 190h dwell
○ Static creep crack growth data.
▨ Saxena et al [10] scatterband
--- Liaw et al[9] fusion line.
-·- Liaw et al[9] weld metal.

LIAW ET AL [9] WELD METAL.

C^* (MPa mh^{-1})

10^{-6} 10^{-5} 10^{-4} 10^{-3}

FIG. 7. CREEP CRACK GROWTH RESPONSE, 2¼ Cr 1Mo WELD.

A comparison between the current data and the scatterband derived by Saxena et al [10] for weld/HAZ/fusion boundary is shown in Figure 7. It can be seen that the current fusion boundary crack growth data for 2¼Cr1Mo weld metal lie well within the scatterband produced by Saxena et al. Also shown in Figure 7 are the best estimate trend lines for fusion line and weld metal specimens obtained by Liaw et al [9]. These data again fall within the Saxena et al scatterband, with the fusion line data falling above the weld metal results. The fusion line crack growth trend curve adequately bounds the current 2¼Cr1Mo data, and this upper bound is described by:

$$\mathrm{da}/\mathrm{dt} = 0.024C*^{0.809} \quad \mathrm{m.h}^{-1} \tag{9}$$

where C* is in units of MPamh^{-1}. Equation (9) is therefore recommended for describing upper bound crack growth rates in the 2¼Cr1Mo welded fusion boundary zone. Equation (6) adequately reflects the mean crack growth rates.

Metallography: The general features of the crack propagation paths are similar to those exhibited by the Incoweld A material. Cracks were generally restricted to the HAZ adjacent to the fusion line but, occasionally, bifurcated cracks were observed where one crack tip propagated into the weld metal whilst the other propagated along the HAZ. Small amounts of creep cavitation were discernible in the crack tip regions, but none was visible on the fusion line in any specimen. Thus, the 2¼Cr1Mo weld specimens generally exhibited less creep damage than comparable Incoweld A specimens.

DISCUSSION

Data have been presented on the creep-fatigue crack growth response of fusion boundaries associated with both an Incoweld A "cold" weld and a 2¼Cr1Mo weld. The evidence presented shows that there is an apparent trend towards increasing growth rates with hold time duration in the Incoweld A which is not apparent with the 2¼Cr1Mo weld metal.

Although a general trend towards increasing crack growth rates with hold time duration is apparent for Figure 3, it is difficult to describe accurately due to the scatter in the data. The reason for the scatter is attributed to the variable microstructure encountered during crack propagation throughout the creep-fatigue tests. By way of example the cycle-to-cycle variation in crack growth rates for two tests are shown in Figure 8. Erratic crack growth behaviour is evident, the 22h dwell data occasionally exceeding the 94h crack growth rates. Thus, although a general hold time dependence is observed for the Incoweld A, cycle-to-cycle variations can confuse the picture.

In order to confirm that the enhanced cyclic crack growth rates exhibited during creep-fatigue are due to creep damage ahead of the main crack tip accelerating the crack growth, a static creep crack growth experiment was mounted on the Incoweld A material. After 760h of crack growth under static load conditions, an Incoweld A specimen was subject to continuous cycling at 2h per cycle. Cyclic crack propagation rates were subsequently measured, and these are shown in Figure 9. Initially, rates approaching the upper bound crack growth rates were obtained. These rates steadily fell to approach the baseline continuous cycling rate as cycles progressed. This is in line with expectation.

It is noteworthy that such a test technique is of benefit in defining upper bound cyclic rates. In addition, prior creep crack growth rates can also be used to define the hold period contributions during creep-fatigue. Thus, the two components of creep-fatigue crack growth can be obtained for a given ΔK_{eff} from one specimen.

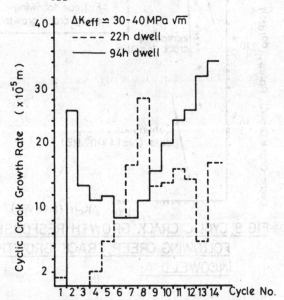

FIG. 8. VARIATION IN CYCLIC CRACK GROWTH RATE AS A FUNCTION OF CYCLE NUMBER.

Arguably the most important observation is that creep-fatigue crack growth rates along the fusion zone associated with Incoweld A are high compared with $2\frac{1}{4}$Cr1Mo welds. Thus, the decision to employ "cold" weld repair techniques is a balanced one. The cost saving of such a repair has to be offset against the faster crack growth rates expected in service for the Incoweld A. In this context, two points are noteworthy. First, the crack growth rates in the Incoweld A samples are only slightly faster than in the $2\frac{1}{4}$Cr1Mo weld samples, ie a factor of ~3. Secondly, no information has been obtained on the comparative time to crack initiation in $2\frac{1}{4}$Cr1Mo weld and Incoweld A fusion zones. All these factors, in addition to the expected remaining life of the power plant component to be repaired, must clearly be considered when choosing a "cold" weld repair technique over a conventional repair.

CONCLUSIONS

1. A range of creep-fatigue crack growth tests has been performed which examines the welded fusion boundary in an Incoweld A "cold" weld and a $2\frac{1}{4}$Cr1Mo weldment at 565°C.

2. Crack growth rates have been described in terms of the fracture mechanics parameters ΔK_{eff} and C^*.

3. Crack growth rates in the Incoweld A samples were higher than in the $2\frac{1}{4}$Cr1Mo weld specimens.

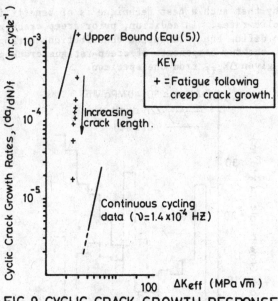

FIG.9. CYCLIC CRACK GROWTH RESPONSE
FOLLOWING CREEP CRACK GROWTH,
INCOWELD.A.

4. Increasing hold periods resulted in increasing crack growth rates in
    the Incoweld A specimens. No such effect was discernible for the
    2¼Cr1Mo weld specimens.

5. The Incoweld A samples showed more of a tendency for creep damage
    formation than the 2¼Cr1Mo weld test pieces.

6. Scatter in the data has been ascribed to the varying microstructure
    sampled by the propagating crack.

REFERENCES

1. Zemzin V N, Titiner Z K, Shron R E, Bukin Yu A, Krechet L E and
    Bereznev A T. Reliability of welds to repair defects in cast
    components made from CrMoV steel. Energomashinostroenie, 1978
    No. 12, p 25.

2. Damman C N, Loney C and Randolph D. TVA's experience with casings.
    EPRI Workshop Proceedings: Life Assessment and Repair of Steam
    Turbine Casings, California, 1985, June 5.

3. Harris P and Jones W K C. A cold welding technique for repairs to
    creep resisting steel casings. CEGB Report RD/M/R318, 1981.

4. Gladwin D N, Miller D A and Priest R H. Examination of fatigue and
    creep-fatigue crack growth behaviour of aged type 347 stainless
    steel weld metal at 650°C. Mat.Sci. and Tech., 1989, 5.

5. Skelton R P, Beech S M, Holdsworth S R, Neate G J, Miller D A and Priest R H. Round robin tests on creep-fatigue crack growth in a ferritic steel at 550°C. To be published. 1991.

6. Goodall I W (Ed), R5: An assessment procedure for the high temperature response of structures, Vol. 5: Creep-fatigue crack growth. CEGB Report TPRD/B/1007/R87, 1987.

7. Skelton R P. Cyclic crack growth and closure effects in low alloy ferritic steels during creep-fatigue at 550°C. High Temp. Tech., 1989, 7, pp 115-128.

8. Priest R H, Gladwin D N and Miller D A. The high temperature crack growth response of a ½CrMoV steel. Proc. 4th Int. Conf. Creep and Fracture of Engng. Mats. and Structs., 1990, pp 765-778.

9. Liaw P K, Saxena A and Schaefer J. Estimating remaining life of elevated temperature steam pipes - Part 1, materials properties. Eng.Frac.Mech., 1989, 32, 5, pp 675-708.

10. Saxena A, Han J and Banerji K. Creep crack growth behaviour in power plant boiler and steam pipe steels. J.Press.Ves.Tech., 1987.

ACKNOWLEDGEMENT

The assistance of Dr S J Brett, National Power plc, in procuring the welded test blocks is gratefully acknowledged.

This paper is published with the permission of Nuclear Electric plc.

MICROSCOPIC MODELS AND MACROSCOPIC CONSTITUTIVE LAWS FOR HIGH TEMPERATURE CREEP AND CREEP FRACTURE OF METALLIC AND CERAMIC MATERIALS

B WILSHIRE

IRC in 'Materials for High Performance Applications'
Department of Materials Engineering, University College Swansea SA2 8PP

ABSTRACT

The model-based θ Projection Concept offers materials constitutive relationships which allow data obtained from short-term high-precision constant-stress creep curves to be extrapolated to predict long-term creep properties with impressive accuracy. The practical aspects of this approach to creep data prediction are discussed for creep brittle and creep ductile materials, with special attention focused on microstructurally-unstable alloys. The factors affecting trends in creep ductility for a variety of materials are then considered in relation to long-term creep failure prediction using the θ methodology.

INTRODUCTION

For almost half a century, most theoretical and practical approaches to high temperature creep and creep fracture have been based on analyses of the variations in secondary creep rate ($\dot{\varepsilon}_s$) and rupture life (t_f) with stress (σ), grain size (d) and temperature (T), using power law equations of the form

$$1/t_f \propto \dot{\varepsilon}_s = A\sigma^n (1/d)^m \exp-(Q_c/RT) \qquad (1)$$

Unfortunately, when this type of relationship is used to describe creep and creep fracture behaviour, the stress exponent (n), the grain size exponent (m) and the activation energy for creep (Q_c) are themselves functions of stress and temperature. This problem of "variable constants" is usually overcome by assuming that different mechanisms of creep and creep fracture, each associated with different values of n, m and Q_c, become dominant in different stress-temperature regimes. Thus, the gradual decrease in stress exponent from n = 4 or more at high stresses to n = 1 at low stresses has commonly been interpreted in terms of a transition from dislocation to diffusional creep processes. However, the few experimental observations widely quoted as proof that diffusional creep processes become dominant at low stresses are inconclusive (1). Indeed, unambiguous evidence exists to show that diffusional creep processes are irrelevant at temperature of practical importance.

Diffusional creep processes can make only an insignificant contribution to the overall creep rate of single crystals. Yet, in every study carried out with metallic and ceramic materials, single crystals have been found to display creep rates comparable with those observed for polycrystalline samples at both high and low stresses (2-6). Dislocation creep processes must therefore be dominant at all stress levels.

In addition to disproving the widely-held view that diffusional creep processes become dominant at low stresses, comparisons of the creep behaviour of single crystals and polycrystalline samples also establish that even the concept of a "steady state" period should be abandoned. Hence, normal creep curves are recorded for polycrystalline aluminium of standard purity, with the tertiary stage and eventual fracture caused by intergranular damage development. In contrast, when constant-stress creep tests are carried out with single crystals and polycrystalline samples of superpurity aluminium under comparable conditions, no evidence of intergranular crack formation is discernible and the creep curves are characterized by a continuously decaying creep rate, even after creep strains in excess of 0.4 (7). Similar behaviour patterns have been reported for single and polycrystalline magnesia (8). These results show that, in the absence of damage development or other processes which can lead to a tertiary period of accelerating creep rate, only a continuously decaying primary curve is displayed. Thus, the idea of a "steady state" creep rate should be discarded, ie for normal creep curves, all that can reasonably be defined is a minimum rate which merely appears to be constant when the decaying creep rate during the primary stage is offset by the gradual acceleration in rate associated with the tertiary stage.

The idea that normal creep curves can be envisaged as the sum of a decaying primary and an accelerating tertiary stage is an essential feature of a radical alternative to traditional power-law approaches, termed the θ Projection Concept (7, 9). Detailed micromodelling of the dislocation processes governing creep behaviour and the damage and degradation processes which can cause tertiary creep leads (7) to a quantitative description of the accumulation of creep strain (ε) with time (t) for normal creep curves as

$$\varepsilon = \theta_1\left(1 - e^{-\theta_2 t}\right) + \theta_3\left(e^{\theta_4 t} - 1\right) \tag{2}$$

where θ_1 and θ_3 scale the primary and tertiary stages with respect to strain, while θ_2 and θ_4 are rate parameters which govern the curvatures of the primary and tertiary components respectively. Thus, unlike traditional "steady-state" approaches which totally ignore the primary and tertiary stages, the θ methodology defines the shape of individual creep curves and the systematic changes in creep curve shape with stress and temperature, ie equation 2 provides a basis for describing the fact that, at a fixed creep temperature, the primary stage generally becomes less pronounced and the tertiary stage more dominant with increasing test duration. The theoretical and practical advantages of the θ Projection Concept can then be demonstrated by discussing the rationalization of creep and creep fracture behaviour afforded by quantifying creep curve shape.

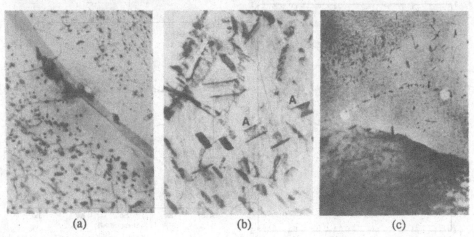

(a)   (b)   (c)

Figure 1. Microstructures developed in $\frac{1}{2}Cr\frac{1}{2}Mo\frac{1}{4}V$ steel showing (a) coarsening of VC particles and denuded zone formation after 35 000h. (x 21000), (b) H-type carbide formation (x 21000) and (c) grain boundary migration isolating voids present after 60 000h. (x 7000).

CREEP OF MICROSTRUCTURALLY UNSTABLE ALLOYS

Using the θ methodology to analyse the shapes of creep curves recorded in tests of comparatively short duration allows the effects of gradual changes in microstructure during long-term creep exposure to be predicted. The validity of this statement can be demonstrated by results obtained for $\frac{1}{2}Cr\frac{1}{2}Mo\frac{1}{4}V$ ferritic steel. With this material, the fine vanadium carbide dispersion present initially can change markedly during creep, as the VC precipitates coarsen, molybdenum carbides develop and grain boundary migration occurs as denuded zones form, Figure 1 (10). The extent to which these microstructural changes take place depends on the test duration and temperature. In very short-term tests, microstructural instability can be ignored. Under these conditions, tertiary creep and fracture are caused by grain boundary cavity and crack development. However, it should be possible to monitor the gradual loss of creep strength as changes in carbide dispersion become progressively more important as the test duration or temperature increases, because variations in the acceleration in tertiary creep rate are quantifiable through the stress and temperature dependences of θ_4. The θ variations for $\frac{1}{2}Cr\frac{1}{2}Mo\frac{1}{4}V$ steel are shown in Figure 2. These results can be described quantitatively as

$$\theta_1 = G_1 \exp H_1 (\sigma/\sigma_y)$$
$$\theta_2 = G_2 \exp -[(Q_2 - H_2\sigma)/RT]$$
$$\theta_3 = G_3 \exp H_3 (\sigma/\sigma_y) \qquad\qquad (3)$$
$$\theta_4 = G_4 \exp -[(Q_4 - H_4\sigma)/RT]$$

where G_i and H_i (with i = 1, 2, 3, 4) are constants for the material, σ_y is the rapid yield stress at the creep temperature, and Q_2 and Q_4 are the activation energies associated with the rate parameters θ_2 and θ_4 respectively. The magnitude of Q_2 then provides evidence as to the dominant creep mechanism, while Q_4 indicates the processes affecting tertiary behaviour.

The observation that $Q_2 = 224$ kJmol^{-1} is consistent with the fact that creep occurs by the generation and movement of dislocations controlled by lattice self diffusion in the ferrite matrix. In contrast, Q_4 is stress dependent, proving that more than one process contributes to the acceleration in tertiary creep rate, with each process characterized by a different activation energy, ie the relative importance of tertiary processes such as intergranular damage accumulation and microstructural instability vary with stress and temperature. Although gradual changes in microstructure progressively modify the creep behaviour as the test duration increases towards 100 000 hours and more, the results in Figure 3 provide evidence that dislocation creep processes remain dominant even at very low stresses.

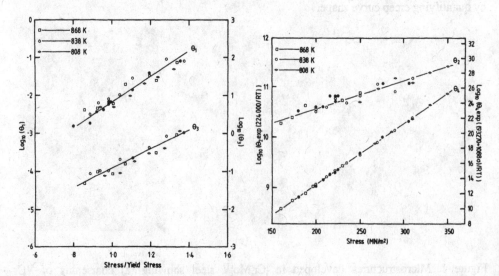

Figure 2. Rationalization of θ data for $\frac{1}{2}Cr\frac{1}{2}Mo\frac{1}{4}V$ steel.

The linearity of the stress/$\log\theta_i$ plots in Figure 2 suggests that the θ relationships allow interpolation and reasonable extrapolation of creep data, ie equations 2 and 3 permit creep curves to be constructed for a wide range of stresses over the temperature ranges investigated. Since creep curves can be constructed, any creep strain or creep rate parameter can be calculated easily. Figure 3 then shows the stress/minimum creep rate relationships at 838K for $\frac{1}{2}Cr\frac{1}{2}Mo\frac{1}{4}V$ steel. The accuracy with which the extrapolated curve matches the measured long-term data emphasizes the impressive predictive capability of the θ methodology, even for microstructurally unstable alloys. Moreover, the data in Figure 3 shows that n decreases from ~10 at high stresses to around unity at low stresses. Yet, even in the n ~ 1 regime, there is ample evidence of dislocation generation and movement within the grains and within denuded zones at grain boundaries, Figure 4. Since similar dislocation arrangements were also found under conditions giving n ~ 4 or more (10), it must be concluded that dislocation creep processes are dominant over the entire range of conditions represented in Figure 3.

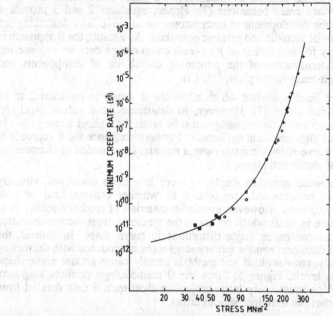

Figure 3. The predicted stress/minimum creep rate curve (solid line) compared with actual long-term data (o ●) for the same batch of $\frac{1}{2}Cr\frac{1}{2}Mo\frac{1}{4}V$ steel.

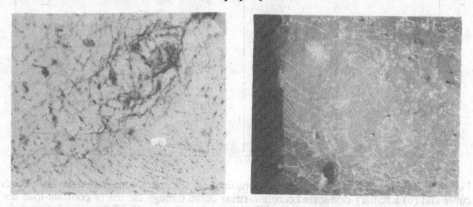

Figure 4. Dislocations present (a) in the grains and (b) in the denuded zones of $\frac{1}{2}Cr\frac{1}{2}Mo\frac{1}{4}V$ steel after creep exposures exceeding 50 000h. at 838K (x 24000).

PRACTICAL ASPECTS OF CREEP DATA PREDICTION

The results in Figure 3 show that the θ methodology allows creep curves with a maximum duration of only about 1,000 hours to be analysed (7, 9) to predict exactly the creep behaviour determined independently (11) under stress-temperature conditions giving creep lives in excess of 100,000 hours. Moreover, since this impressive predictive capability has been demonstrated for a material which can exhibit marked changes in carbide dispersion during creep exposure, the θ relationships are obviously capable of quantifying the long-term effects of continuous changes in microstructure. It must also be emphasized that, since there is conclusive evidence to prove that dislocation creep processes are dominant at high and low stresses at temperatures of practical relevance, there can be no valid objection to the θ predictions on the assumption that the dominant creep mechanism changes as the test duration increases.

The θ relationships can be derived by modelling the micro-mechanisms known to determine primary and tertiary creep behaviour (7). Hence, equations 2 and 3 provide a sound theoretical basis for the development of macroscopic constitutive laws describing the detailed creep characteristics of metallic and ceramic materials. As a result, the θ approach is now being used increasingly for acquisition of long-term creep design data for engineering materials and even for determination of the remaining useful life of components and structures in operational high-temperature plant, (12, 13).

Computer codes have been published which allow the θ values in equation 2 to be derived for any normal creep curve (7). However, to determine these values precisely, several hundred accurate creep strain/time readings must be taken throughout a creep test. In turn, this involves the use of high-precision equipment. Furthermore, since the θ approach is based on analysis of creep curve shapes, constant-stress machines are needed to characterize the creep behaviour of most metals and alloys.

With metallic and ceramic materials displaying very low creep ductilities, virtually identical creep curves will be obtained, irrespective of whether constant-load or constant-stress conditions are employed. However, even with materials of modest ductility (5 to 10%), the continuous increase in stress which occurs as the specimen cross-section decreases with increasing creep strain can cause major distortions in curve shape. In general, the primary stage becomes less dominant and the tertiary stage more pronounced with decreasing applied stress. Consequently, constant-load test methods usually cause greater curve shape distortions at higher stress levels, Figure 5. Since the θ methodology predicts long-term behaviour by analysis of high-stress creep tests, projecting short-term θ data derived from constant-load test programmes can introduce serious errors.

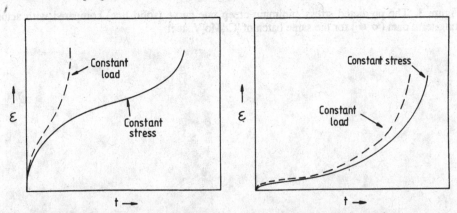

Figure 5. Schematic illustration of the distortion of (a) a primary-dominanted constant-stress curve and (b) a tertiary-dominated constant-stress curve through the use of constant-load test procedures. As the stress dependence of the creep rate increases, the differences between the constant-stress and the constant-load curves also increases.

Even for materials displaying substantial ductilities at all stress levels, the gradual transition in creep curve shape generally observed with increasing test duration means that almost identical creep curves will be recorded at very low stresses, independent of whether constant-load or constant-stress machines are used, Figure 5b. As a result, the vast body of long-term constant-load data available for many commercial alloys is perfectly valid. Thus, in the case of the results shown for $\frac{1}{2}$Cr$\frac{1}{2}$Mo$\frac{1}{4}$V steel in Figure 3, long-term data obtained using constant-load equipment can legitimately be used to validate the predictions based on θ analysis of short-term constant-stress creep curves. Yet, while constant-load test methods are perfectly adequate for long-term data acquisition, the cost differences between constant-load and constant-stress machines are minimal. The continued use of constant-load facilities can therefore be justified only for long-term data acquisition or for studies of creep-brittle materials.

Even when high-precision constant-stress equipment is used for data acquisition, it is important to recognize that, as with any forecasting procedure, the accuracy of the predictions must inevitably improve as the extent of the extrapolation decreases. Moreover, since forecasts are normally made on the basis of "surprise-free" continuity of trends, the shorter the extent of the extrapolation, the lower is the risk of unforeseen events influencing the predictions. In seeking to predict long-term creep properties, projections should therefore be based on analyses of comprehensive data sets obtained over wide ranges of stress at several different temperatures.

The need for a carefully selected test matrix can be illustrated by results available for an oxide-dispersion-strengthened 13% chromium ferritic steel, a fuel pin cladding material developed for fast breeder reactor applications (14, 15). When this material was tested in the as-solution-treated condition, while normal creep curves were displayed under most test conditions studied, abnormal curve shapes were found when recrystallization occurred in relatively long-term tests at the highest test temperatures investigated, as shown in Figure 6. Analyses of the normal curves observed for the majority of the test conditions investigated would not predict the occurrence of a sudden event, such as recrystallization. However, by completing a substantial test matrix, with the inclusion of tests of reasonable duration at temperatures above those normally encountered during service, the stress-temperature combinations resulting in recrystallization can be predicted over broad ranges of conditions.

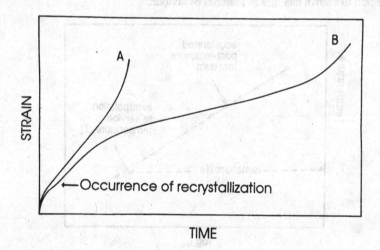

Figure 6. Creep curves observed for an ODS ferritic steel, showing the abnormal creep curve shape resulting from recrystallization at low stresses and high temperatures. Curve A was obtained at 180 MNm^{-2} and 973K, giving t_f = 23ks and ε_f = 0.18, while curve B was recorded at 165 MNm^{-2} and 973K, giving t_f = 929ks and ε_f = 0.15.

The results obtained for the ODS ferritic steel and for $\frac{1}{2}Cr\frac{1}{2}Mo\frac{1}{4}V$ ferritic steel illustrate the importance of distinguishing between different forms of microstructural instability in relation to the prediction of creep behaviour. With $\frac{1}{2}Cr\frac{1}{2}Mo\frac{1}{4}V$ steel, the progressive changes in carbide dispersion as the test duration increases (Figure 1) are quantified through the systematic trends in the θ relationships derived from short-term creep curves (Figure 2), so that trend extrapolation exactly forecasts the effects of continued microstructure evolution during long-term exposure (Figure 3). In contrast, creep curves recorded under conditions of grain size stability contain no information about the occurrence of recrystallization under different test conditions (Figure 6). Only by completing a comprehensive test matrix can the stress-temperature conditions giving rise to recrystallization be predicted. For this type of reason, the θ methodology has never been presented as a substitute for long-term test programmes but only as the most reliable method currently available for reducing the scale and costs of the programmes now required for characterization of the long-term properties of newly-developed engineering materials.

In recommending the completion of creep tests over a broad spectrum of conditions, it is essential to check that the "material character" does not vary over the temperature ranges selected. For instance, with many steels and other precipitation-hardened alloys, the precipitate type and dispersion can differ at different temperatures. Clearly, creep data obtained in one temperature regime cannot be used to predict the behaviour patterns in another regime where a different type of precipitate dispersion is formed prior to creep testing. Hence, no attempt should be made to extrapolate the θ data obtained for $\frac{1}{2}Cr\frac{1}{2}Mo\frac{1}{4}V$ steel to temperatures significantly outside the ranges specified in Figure 3, because the precipitate type and dispersion prior to creep testing is different above and below about 900K. For this reason, conventional temperature-accelerated post-exposure test methods for estimating the remanent creep life of service components must often be regarded with suspicion, because the temperature ranges selected may extend across regimes where different precipitate types are stable, Figure 7.

It is also important not to misinterpret the statement that dislocation processes are dominant at all temperatures of practical relevance. Thus, different dislocation creep mechanisms operate at temperatures above and below about $0.4T_m$, so that data obtained at high temperatures cannot predict low-temperature logarithmic creep properties. Even with the highly successful θ methodology, extrapolation must therefore be performed only with sensible regard to known changes in material behaviour.

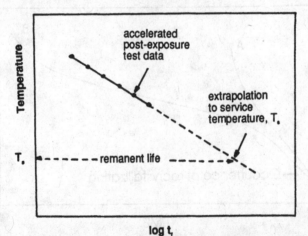

Figure 7. Schematic representation of the estimation of remanent creep life using stress-rupture data obtained over a range of temperatures at the service stress. With low alloy steels, the upper temperate limit can be around 1000K. This need for very high test temperatures is avoided when the remanent life is determined using the θ methodology (12, 13).

PREDICTION OF CREEP FRACTURE BEHAVIOUR

With the θ relationships able to describe a normal creep strain/time curve accurately, the time to fracture can be specified conveniently as the time to reach a limiting creep strain (ε_t). This separation of the curve shape and the occurrence of failure allows straight-forward classification of different types of creep fracture behaviour. Thus, at one extreme, a developing microcrack may propagate rapidly to terminate a decaying primary curve, so that no significant tertiary component is observed. This allows creep brittle materials to be defined (16) through the expression

$$\varepsilon_t/\varepsilon_m = 1 \tag{4}$$

where ε_t is the creep ductility and ε_m is essentially the strain to the minimum creep rate. Many materials show negligible tertiary stages. Indeed, this type of behaviour is particularly common with brittle ceramics, even under compressive creep loading. With only a primary stage displayed, a modified form of equation 2 must be used to describe the truncated creep curve (8). Essentially, the exponential tertiary term in equation 2 is expanded by means of its Taylor series as

$$\varepsilon = \theta_1\left(1 - e^{-\theta_2 t}\right) + \theta_3(\theta_4 + (\theta_4 t)^2/2 + ...) \tag{5}$$

and, since $\theta_4 t \ll 1$, higher order terms in the second bracket can be ignored, yielding

$$\varepsilon = \theta_1\left(1 - e^{-\theta_2 t}\right) + \theta_3\theta_4 t \tag{6}$$

Equation 6 has been shown to offer a theoretically sound and practically convenient basis for analysis of the creep and creep fracture behaviour of ceramics and other creep brittle materials, (8, 17, 18). A creep ductile material can then be defined as a material which is able to accommodate tertiary deformation without forming a microcrack which propagates rapidly, with adequate ductility obtained when $\varepsilon_t/\varepsilon_m > 10$ (16).

For materials displaying creep ductile behaviour, the θ methodology allows estimates to be made of long-term stress-rupture properties (7, 9). The procedures involved can be outlined by reference to the data available for $\frac{1}{2}$Cr$\frac{1}{2}$Mo$\frac{1}{4}$V steel. The short-term creep tests carried out to obtain the θ values in Figure 2 were continued to fracture, with the creep ductilities recorded for each test presented in Figure 8. In order to predict long-term stress-rupture behaviour, it is then necessary to be able to use the short-term values in Figure 8 to estimate the creep ductilities expected in tests of long duration.

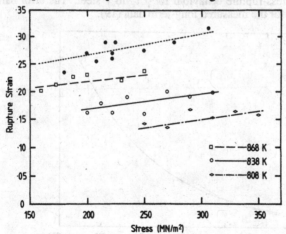

Figure 8. The stress dependence of the rupture strains for two batches of $\frac{1}{2}$Cr$\frac{1}{2}$Mo$\frac{1}{4}$V steel (shown as open and closed symbols). At 838K, the strains at which necking and void formation become apparent were similar for both batches, suggesting that creep ductility depends strongly on the ease of crack link-up and propagation.

The data in Figure 8 shows that the creep ductility appears to decrease linearly at each creep temperature. Linear projection of these results then provides estimates of the failure strains at low stresses. For each creep temperature, using the θ relationships in Figure 2 to construct a full creep curve at some low stress level, the time to failure can be calculated as the time to reach the estimated failure strain at this stress. The results predicted using this procedure are shown in Figure 9, together with the published long-term stress-rupture properties (19). Clearly, the extrapolated logσ/logt$_r$ curves are in excellent agreement with the measured data. However, this quality of fit may be fortuitous. Specifically, while the information in Figure 3 suggests that prediction of long-term curve shapes is reasonable, it is important to assess whether linear projection of short-term creep ductility measurements is valid. In the case of ductile materials, there are grounds for confidence. Figure 8 contains data for two batches of ½Cr½Mo¼V steel. The θ parameters governing creep curve shape were identical for both batches, but one batch displayed ductilities of 25 to 30% while the other was characterized by ductilities between 15 and 20% over the stress ranges investigated at 838K. Yet, since the primary strains are negligible and the creep strain accumulates rapidly only late during the tertiary stage in tests lasting more than 1000 hours, these differences in ductility do not significantly alter the stress-rupture predictions in Figure 9.

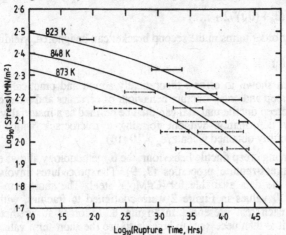

Figure 9. Predicted stress-rupture behaviour for ½Cr½Mo¼V steel. The error bars represent the scatter bands reported for the measured long-term data (19).

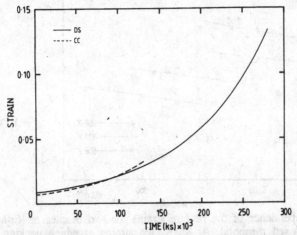

Figure 10. High temperature creep curves for conventionally cast (dashed line) and directionally solidified (solid line) Mar M002 at a stress of 250 MNm⁻².

For materials displaying adequate creep ductilities at all stress levels, appreciable variations in the actual magnitude of the rupture strain estimates for tests of long duration do not therefore appear to cause serious errors in the predicted creep lives. In contrast, significant changes in rupture life can occur when the creep ductility decreases from a high to a low value, or vice versa. This situation can be illustrated by considering the behaviour patterns shown in Figure 10 for the nickel-base superalloy, Mar M002.

The θ parameters governing the creep curve shape were identical for conventionally cast Mar M002 (CC) having an equiaxed grain structure and for directionally solidified material (DS) having a very large grain aspect ratio. However, the creep ductility of the CC alloy is around 2% whereas that of the DS material is about 15% under the test conditions used to obtain the creep curves presented in Figure 10. As with the two batches of $\frac{1}{2}Cr\frac{1}{2}Mo\frac{1}{4}V$ steel displaying differences in creep ductility (Figure 8), voiding becomes apparent at similar low creep strains for both the CC and DS materials, but the difficulties of crack link-up and propagation in the DS alloy results in considerably greater ductility. Since the creep curve shapes were identical for the two forms of Mar M002, terminating the curves at the appropriate ductilities leads to the time to fracture for the DS material being considerably greater than that for the CC alloy.

A transition from a reasonably high ductility to a low ductility is frequently observed in tests of increasing duration. For instance, Figure 11a shows the rapid decrease in the magnitude of the rupture strains with decreasing stress reported for the aluminium alloy, 2124-T851 (20). This material is one of a series of newly-developed high-strength aluminium alloys being evaluated as an eventual replacement for Hid RR58 for high temperature airframe applications. Inspection of the data in Figure 11a reveals that, with the creep ductility decreasing rapidly with decreasing applied stress at temperatures from 373 to 463K, linear extrapolation of these trends would lead to a zero creep ductility (and therefore a zero rupture life) at stress levels known to give long times to failure. Clearly, in seeking to predict long-term creep and creep rupture properties for this type of material by analysis of short-term data, more realistic methods of estimating the stress dependence of the rupture ductilities are required.

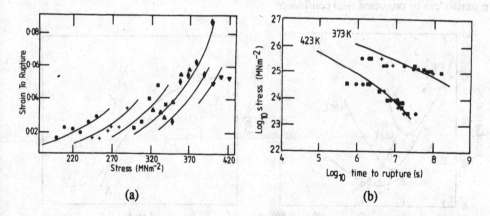

Figure 11.   (a) The stress dependence of the creep ductilities recorded for alloy 2124-T851 at 373K (▼), 391K (♦), 409K (▲), 427K (■), 445K (+) and 463K (●).
(b) The stress dependence of the rupture life for alloy 2124-T851 at 373 and 423K. The solid lines predicted from the θ relationships can be compared with the measured long-term values obtained independently for three batches of plate (●+■). The predictions, based on θ analysis of constant-stress creep curves, overestimate the measured short-term constant-load data but correlate well with the low stress data, as would be expected from Figure 5.

Several methods of extrapolating the creep ductility data in Figure 11a were then investigated, with satisfactory results achieved by assuming that the rupture strain decreases exponentially to a minimum value of ~1%, a value consistent with the information derived from a limited number of extended tests taken to completion. As evident from Figure 11b, the resulting creep life predictions obtained using the θ methodology agreed well with actual long-term stress-rupture data provided independently by RAE, Farnborough (20).

While decreases in creep ductility are often observed with decreasing applied stress, evidence is available which suggests that this trend can be reversed in tests of very long duration. For example, with ½Cr½Mo¼V steel, the isolation of developing voids as grain boundary migration occurs (Figure 1c) should lead to increased ductilities during prolonged creep exposure. This type of behaviour is illustrated in Figure 12a which shows that the creep ductility of 0.5Mo steel decreases almost linearly to a minimum value and then increases as the test duration increases further. Moreover, the time to the minimum value decreases with increasing temperature, demonstrating once again that valuable information on behaviour trends can be derived from tests carried out over wide ranges of stress at temperatures above the normal operating ranges for the material. It should also be noted that, if the changes in creep ductility shown in Figure 12a are of general applicability during high temperature creep of precipitation-hardened alloys, the downward trend of the logσ/logt_f projections in Figure 9 could change, such that the rupture lives at very low stresses could be greater than expected, as indicated schematically in Figure 12b.

CONCLUSIONS

Results are available for a wide range of metallic and ceramic materials to illustrate the advantages of using the θ methodology to quantify creep curve shape, with the time to fracture determined as the time to reach a limiting creep strain. With the proven ability of the θ relationships to predict creep behaviour accurately, attention can now be focused on developing a comprehensive understanding of the factors governing creep ductility. When this is achieved, the long-term creep and creep fracture properties of metallic and ceramic materials can be predicted with confidence.

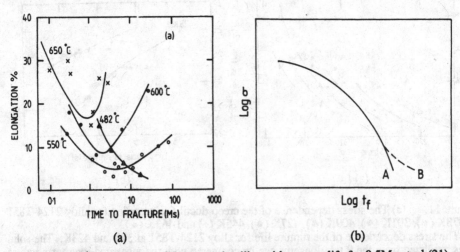

Figure 12. (a) The variation of the creep ductility with rupture life for 0.5Mo steel (21).
(b) If the rupture strain for a material decreases to a low value and then increases once again as the test duration increases further, this increase in ductility at low stresses could affect the projected downward slope of logσ/logt_f plots (shown as line A) to give longer creep lives (shown as line B).

REFERENCES

1. Wilshire, B., Proc. 4th Int. Conf. on Creep and Fracture of Engineering Materials and Structures, ed. B. Wilshire and R.W. Evans, The Institute of Metals, London, 1990, 1.

2. Harper, J.G. and Dorn, J.E., Acta Metall., 1957, 5, 654.

3. Barrett, C.R., Muehleisen, E.C. and Nix, W.D., Mater. Sci. Eng., 1972, 10, 33.

4. Mohamed, F.A., Murty, K.L. and Morris, J.W., Met. Trans., 1973, 4, 935.

5. Malakondaiah, G. and Rama Rao, P., Scripta Metall., 1979, 13, 1187.

6. Dixon-Stubbs, P.J. and Wilshire, B., Philos. Mag., 1982, 45A, 519.

7. Evans, R.W. and Wilshire, B., Creep of Metals and Alloys, The Institute of Metals, London, 1985.

8. Evans, R.W., Scharning, P.J. and Wilshire, B., Creep Behaviour of Crystalline Solids, ed. B. Wilshire and R.W. Evans, Pineridge Press, Swansea, 1985, 201.

9. Evans, R.W., Parker, J.D. and Wilshire, B., Recent Advances in Creep and Fracture of Engineering Materials and Structures, ed. B. Wilshire and D.R.J. Owen, Pineridge Press, Swansea, 1982, 135.

10. Williams, K.R. and Wilshire, B., Mater. Sci. Eng., 1981, 47, 151.

11. Browne, R.J., Cane, B.J., Parker, J.D. and Walters, D.J., Int. Conf. on Creep and Fracture of Engineering Materials and Structures, ed. B. Wilshire and D.R.J. Owen, Pineridge Press, Swansea, 1981, 645.

12. Wilshire, B. and Evans, R.W., Life Assessment and Life Extension of Power Plant Components, ed. T.V. Narayanam, ASME, New York, 1989, 217.

13. Wilshire, B., Applied Stress Analysis, ed. T.H. Hyde and E. Ollerton, Elsevier Applied Science, London, 1990, 180.

14. Huet, J.J. and Le Compte, C., Ferritic Steels for High Temperature Applications, A.S.M. New York, 1983, 210.

15. Huet, J.J., Coheur, L., De Bremaeker, A., De Wilde, L., Gedopt, J.B., Hendrix, W. and Vandermeulen, W., Nucl. Tech., 1985, 70, 215.

16. Goodall, I.W., Leckie, F.A., Ponter, A.R.S. and Townley, C.H.A., J. Eng. Mat. Tech., 1979, 101, 137.

17. Evans, R.W., Murakami, T. and Wilshire, B., Brit. Ceram., 1988, 87, 54.

18. Wilshire, B., Science of Ceramics 14, ed. D. Taylor, The Institute of Ceramics, Stoke-on-Trent, 1988, 61.

19. Johnson, R.F., May, M.J., Trueman, R.J. and Mickleraith, J., Proc. Conf. on High Temperature Properties of Steels, ISI, London, 1967, 229.

20. Evans, R.W., Fadlalla, A.A., Wilshire, B., Butt, R.I. and Wilson, R.N., Proc. 4th Int. Conf. on Creep and Fracture of Engineering Materials and Structures, ed. B. Wilshire and R.W. Evans, The Institute of Metals, London, 1990, 1009.

21. Glen, J., J. Iron Steel Inst., 1955, 179, 320.

CHARACTERISTIC TEMPERATURE OF DEFORMATION OF MATERIALS AND COLD BRITTLENESS OF BCC METALS AND CERAMICS

Yu.V. MILMAN

Institute for Problems of Materials Science, 252180 Kiev, USSR

ABSTRACT

The physical mechanisms of deformation and fracture of materials in the temperature regions of cold, warm and hot deformation are considered. It is shown that the characteristic deformation temperature is the natural boundary between the regions of cold and warm deformation while the recrystallization temperature separates the warm and hot deformation regions. The cold brittleness phenomenon is observed only in the cold deformation region, and the ductile-brittle transition temperature may be decreased by the structural factors. The mechanisms of deformation and fracture of BCC metals and ceramics are discussed.

INTRODUCTION

The concept of the characteristic deformation temperature, t^*, has been developed by the author in collaboration with V.I.Trefilov [1-3]. The temperature t^* has been determined as the temperature below which the motion of dislocations is controlled by the Peierls-Nabarro stress. Correspondingly a sharp growth of the flow stress is observed when the temperature drops below t^*. The experimental and theoretical results obtained in the last decade show that the temperature t^* is the fundamental characteristic of a variety of crystalline materials - covalent crystals, diamond, refractory compounds (carbides, oxides, borides etc.), semiconducting crystals $A^{III}B^V$ and $A^{II}B^{VI}$, BCC metals etc. Using the notion of the characteristic deformation temperature we can divide the whole temperature interval from 0 K to the melting point into three temperature regions, namely the regions of hot, warm and cold deformation. The boundary temperature between hot and warm deformation is well-known — it is the dynamical recrystallization temperature, t_r.

This classification allows one to formulate clearly the typical features of the deformation and fracture in each temperature interval. Using these concepts we can also establish a considerable analogy between the mechanisms of deformation for different crystals, for example for BCC metals and for covalent crystals of Si and Ge.

CHARACTERISTIC DEFORMATION TEMPERATURE

When comparing the influence of temperature on the physical properties of various materials it is common to use a concept of a homologous temperature, $t = T/T_m$, where T_m is the melting temperature and T is the test temperature, both in degrees K. Utilization of the homologous temperature allows one to study various physical properties of different materials under comparable conditions, i.e. when their homologous temperatures are equal. This is particularly important for the self-diffusion phenomenon and for a large complex of properties connected with the self-diffusion.

However, when studying the effect of temperature on the mechanical properties of crystals the application of the homologous temperature does not allow to compare properties of crystals with different types of the interatomic bonding and of the crystalline structure. For example the covalent crystals with a diamond-type structure (Si, Ge etc.) are ductile only at temperatures close to the melting point while numerous metals maintain their ductility at extremely low temperatures close to the absolute zero.

Such a situation can be explained by the fact that mechanical properties are structure sensitive characteristics mainly determined by the resistance to the motion of dislocations in a crystal and by the rate of relaxation processes, and these factors depend substantially on the type of bonding and of the crystalline structure. For this reason, the aspiration for determining some general regularities of the temperature influence on the structural state and on mechanical properties seems to be natural. This problem is solved to a considerable extent by the use of the concept of the characteristic deformation temperature [1-3].

Under continuously increasing external load the elastic and plastic deformations occur in a solid. The effect of temperature on the resistance to elastic deformation may be characterized by the temperature dependence of the Young's modulus which has been studied in detail for most common materials. In the absence of phase transitions the Young's modulus, E, decreases monotonously as the temperature increases. A decrease in the Young's modulus does not usually exceed 30-40% within a wide temperature range. On the other hand the value of E is proportional to the melting temperature as the Leibfried number is nearly constant for different materials. This might indicate that the comparison of elastic properties should be carried out at the same homologous temperatures. At the same time the resistance to plastic deformation which is usually characterised by the value of the flow stress has a considerably more pronounced temperature dependence than the elastic properties. The onset of a sharp growth of the flow stress for decreasing deformation temperature is observed in various materials at significantly different homologous temperatures. This onset temperature acquires great importance as a reference point of the temperature scale and as a point which should be avoided in any practical application.

In pure undoped crystals a sharp growth of the flow stress and hardness is observed at the certain temperature, t^*, at which the friction stress of the lattice, i.e. the stress of thermally activated overcoming of the Peierls-Nabarro barriers, becomes significant (Fig. 1). We found that for different materials the value of the characteristic deformation temperature in a homologous temperature scale, t^*, could be approximated as:

$$t^* = T^* / T_m \approx 0.22\, \alpha^{1/2} \tag{1}$$

where $\alpha = U/kT_m$, U is the activation energy of the dislocation motion, k is the Boltzmann constant. Values of α and t^* for different covalent and partially covalent crystals are given in Table 1.

TABLE 1
Characteristic temperature of deformation of crystalline materials

Materials	U (eV)	T_m (K)	$\alpha = U/kT_m$	$t^* = T^*/T_m$	T^* (^oC)	t_T^o
Ge	1.6	1210	15.3	0.83	720	0.89
Si	2.2	1683	15.1	0.82	1100	0.84
InSb	0.7	800	10.1	0.67	260	–
Al_2O_3	1.9	2303	9.7	0.65	1250	–
ZrC	3.12	3800	9.5	0.65	2185	–
TiC	2.55	3520	8.4	0.61	1890	0.51 - 0.60
WC	1.5	3163	5.49	0.49	1500	0.52 - 0.55
$NbC_{0.75}$	1.8	3673	5.7	0.48	1500	0.53
Fe_3C	0.78	1830	4.9	0.46	570	–
TiB_2	1.13	3220	4.06	0.42	1070	0.51 - 0.53
HfB_2	0.8	3200	2.9	0.35	860	–
ZrB_2	0.6	3313	2.09	0.29	700	0.47
Be	0.40	1556	2.98	0.35	275	–
Cr	0.20	2200	1.05	0.21	170	0.3 - 0.4
Mo	0.19	2880	0.77	0.17	220	0.3 - 0.4
W	0.49	3670	1.54	0.25	640	0.3 - 0.4
V	0.18	2225	0.94	0.19	150	0.3 - 0.4
Nb	0.24	2740	0.85	0.18	220	0.3 - 0.4
Ta	0.30	3270	1.06	0.20	380	0.3 - 0.4
Fe	–	1810	–	0.19	80	0.3 - 0.4

In some crystals, e.g. in metals with close-packed structures, the Peierls-Nabarro stress is negligible and there is no sharp increase of the flow stress for decreasing temperature. Therefore it can be conditionally considered that the whole temperature interval is above the characteristic temperature of deformation. It has been demonstrated that the α and t^* parameters characterize the degree of covalence of the interatomic bond from the view-point of mechanical properties of crystals and they can be used to classify crystals according to this criterion. The crystals with purely covalent bonds have the highest values of $\alpha \approx 15$ and $t^* \approx 0.85$ (Table 1). In the carbides of the group IVA transition metals (ZrC, TiC) the values of α and t* parameters are higher than in NbC and WC which conforms to the modern concepts of the electronic structure of carbides. In the BCC metals $\alpha \approx 1$ and $t^* \approx 0.2$.

The dislocation structure formed in the process of plastic deformation is significantly changed in the proximity of t^*. If the deformation is performed at the temperature $t < t^*$ than the uniform random distribution of dislocations prevails while in the case of $t > t^*$ a cell dislocation structure is formed after a certain critical degree of deformation, ε_c, which is given as $\varepsilon_c \sim \sigma^2 d$, where σ is the thermal component of the flow stress and d is the grain size [1]. From this expression it is evident that the value of ε_c grows sharply below t^* and the formation of a cell dislocation structure becomes difficult.

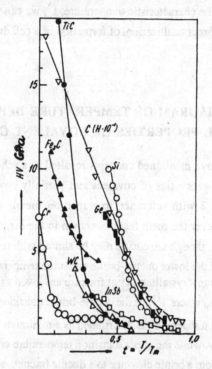

Figure 1. The temperature dependence of hardness for covalent crystals.

It is known that the formation of a cell structure changes the deformation mechanism [1,3]. This change at temperatures close to t^* is also caused by the fact that at and above this temperature the velocities of screw and edge dislocations practically coincide while at lower temperatures the mobility of screw component is essentially lower than that of edge ones.

The risk of occurrence of the cold brittleness in crystals which are plastic at high temperatures is revealed only at $t < t^*$. The ductile-brittle transition temperature corresponding to the onset of the macroscopic plasticity can be made lower than t^* due to structural factors but the nature of fracture below t^* remains either brittle or quasibrittle and only above t^* the risk of brittle fracture disappears completely and the fracture becomes ductile.

It has been demonstrated that annealing of deformed crystals at a temperature below the characteristic temperature, t^*, does not change the dislocation structure essentially [2]. On the other hand annealing at $t > t^*$ results in the intensive polygonization and recrystallization. For all covalent and partially covalent crystals the condition $t_r^o > t^*$ is satisfied as can be seen from Table 1 (t_r^o is the temperature of the onset of recrystallization). The physical reason for this phenomenon is a sharp reduction of the dislocation mobility at $t < t^*$ which makes the processes of polygonization, annihilation of dislocations, and polygonizational creation of the recrystallization nuclei impossible.

Thus when we know the characteristic temperature, t^*, we can predict the temperature intervals of the beginning of recrystallization, of formation of a cell dislocation structure, and of the cold brittleness.

GENERALIZED DIAGRAM OF TEMPERATURE DEPENDENCE OF MECHANICAL PROPERTIES OF COVALENT CRYSTALS

The development of the above mentioned concepts resulted in a scheme of the temperature dependence of mechanical properties of covalent and partially covalent crystals [2]. The diagram is shown in Fig. 2 with reference to transition metals carbides in single and polycrystalline states. However the main features shown in the diagram are common for all crystals under consideration, though the curves may be shifted with respect to the temperature axis. The temperature t^* and the lower ductile-brittle transition temperatures of polycrystals for intercrystalline (T^{li}_b) and transcrystalline (T^{lt}_b) fracture are given as the reference points. In this case we have $T^{ls}_b \approx T^{lt}_b$, where T^{ls}_b is the ductile-brittle transition temperature of a single crystal. The lower transition temperature is determined as a minimum temperature of presence of the macroscopic plasticity, while the upper transition temperature corresponds to the change of the fracture mechanism from a brittle cleavage to a ductile fracture, so its value is close to t^*.

The main characteristics of strength and plasticity including the fracture toughness, K_{IC}, are given in Fig. 2 which allows to trace a change of mechanical properties below and close to t^*. It can be seen that in the brittle region the single crystals appear to be stronger than the polycrystals, while at intermediate temperatures an opposite situation is observed. Typical features at this plot are the peaks in the temperature dependence of the fracture stress within the region of ductile-brittle transition temperatures.

COLD, WARM AND HOT PLASTIC DEFORMATION

The nature of a deformation process is connected with the structural and substructural changes in the course of deformation, i.e. with the changes in a grain structure and with the formation of a cell structure. According to different kinds of those changes usually we distinguish between hot, warm and cold deformation. The hot deformation should be carried out above the temperature of dynamical recrystallization, t_r, which results in the creation of an equiaxial granular structure free of internal stresses and with a low dislocation density. Unfortunately there is no precise concept of the boundary between warm and cold deformations. Taking into account our previous considerations it seems reasonable to use the characteristic deformation temperature, t^*, as such transition point. Temperature intervals for these three regions are illustrated in Fig. 3. The value of t_r is plotted by the dashed line since its precise position depends on the presence of impurities and on deformation conditions.

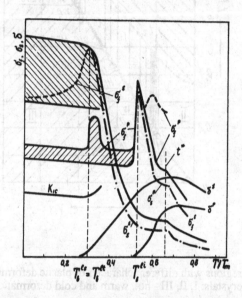

Figure 2. The generalized diagram of the temperature dependence of mechanical properties of crystals with the high values of the Peierls-Nabarro stress; σ_s - flow stress, σ_f - fracture stress, δ - plasticity, indices s and p correspond to single crystals and polycrystals, respectively.

Structural features specific for each type of deformation are summarized in Table 2. The hot deformation results in the formation of equiaxial grains with a low dislocation density and a low level of internal stresses. The warm deformation gives a non-equiaxial granular structure and a fragmented or cellular intragranular structure. The level of internal stresses and strain hardening increases but the fracture is generally still ductile. After cold deformation the granular structure is non-equiaxial like after warm deformation though the formation of a cell structure is difficult. Even after very high degree of deformation the dislocation forest is the main substructural element, and only for strains exceeding 90% it is possible to observe the formation of cells. It is difficult to achieve such deformation since due to an extremely high strain hardening the ductility of materials is insufficient and fracture is either brittle or quasibrittle. For a number of materials cold deformation might be connected with the presence of deformation twins.

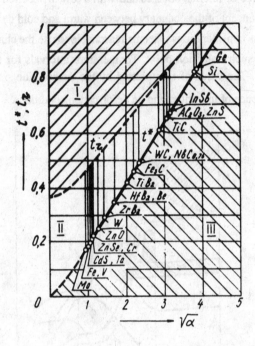

Figure 3. Temperature regions with different character of plastic deformation for covalent and partially covalent crystals: I, II, III - hot, warm and cold deformations, respectively.

TABLE 2

Effect of deformation on structure and substructure

Type and temperature of deformation	Granular structure	Dislocation substructure
hot, $\quad t > t_r$	equiaxial	low density of dislocations, $\rho \approx 10^6$ cm^{-2}
warm, $t^* < t < t_r$	non-equiaxial	fragments and dislocation cells*
cold, $\quad t < t^*$	non-equiaxial	high density of randomly distributed dislocations, $\rho > 10^9 - 10^{10}$ cm^{-2}

* In the upper region of the temperature interval fragments or cells are either equiaxial or weakly non-equiaxial; in the lower part of the interval they are strongly non-equiaxial.

STRUCTURAL SENSIBILITY OF MECHANICAL PROPERTIES IN VARIOUS TYPES OF ALLOYS

In order to study the effect of structural factors on mechanical properties of refractory metals we divide all the alloys considered into three main groups (Table 3) and consider them in turn.

TABLE 3
Types of refractory metals alloys

No	Alloy type	Examples (weight %)	Structural factors determining mechanical properties
I	Pure metals, low-alloyed metals and alloys with stable second phase particles	Cr + 0.5% Y ЦМ-2A (Mo + 0.07-0.3% Ti + 0.08-0.25% Zr, <0.004% C)	Grain size and dislocation cell (or fragment) size; density of forest dislocations; dispersion hardening by second phase particles; normal dependence of mechanical properties on structural factors
II	Alloys with unstable second phase particles	Cr alloyed with V, B, Y (B ≈ 0.05%)	Factors of the group I + dissolution of second phase particles under deformation strengthening; anomalous dependence of mechanical properties on structural factors is possible
III	High alloys	Systems Mo-Re, W-Re, Cr-Fe	Change of electronic structure and of the character of dislocation structure, lowering of the stacking fault energy and of the Peierls stress, processes of solid solution dissociation

The influence of structural factors has been studied most extensively for the group I alloys. The main characteristic for improving the mechanical properties, i.e. for increasing strength and lowering the ductile-brittle transition temperature, is the effective grain size, d_{eff}. For the recrystallized state d_{eff} is the grain size, after warm deformation one has to use the cell size. For this group of alloys the Petch's relation for the yield stress is satisfied:

$$\sigma_s = \sigma_0 + K_y \, d_{eff}^{-1/2} \qquad (2)$$

and for the ductile-brittle transition temperature we have:

$$1 / T_b = A - B \lg d_{eff} \qquad (3)$$

Here σ_0 characterizes the resistance to the dislocation motion in a grain (or in a single crystal), and K_y characterizes the difficulty of the glide transfer through a grain boundary. Refinement of metals from interstitial impurities facilitates such transfer thus lowering the value of K_y. The value of σ_0 may be increased significantly by alloying or dispersion hardening and has a strong temperature dependence below t^*. The temperature dependence of the second term in Eq. (2) is significantly weaker.

Parameters A and B in Eq. (3) are material constants, and $B \approx 1.5k / U$. The more general and detailed expression for the dependence of T_b on structural factors is given in [1]. We shall call below the the group I alloys with Eqs. (2) and (3) being satisfied the alloys with normal dependence of mechanical properties on structural factors.

The presence of the dispersed second phase particles unstable under thermomechanical treatment is typical for the group II alloys. Particles dissociation during plastic deformation causes supersaturation of the solid solution and intensive impurity segregation on dislocations and other structural defects. The deformation strengthening and embrittlement may be so essential that this effect influences mechanical properties to a greater extent than the decrease of the grain size and cell formation.

The special place is given to the group III alloys - high alloyed chromium, molybdenum and tungsten. For the VIA group metals the conditions of a resonance covalent bond are satisfied as there is a deep minimum in the electronic density of states at the Fermi level. This kind of electronic structure explains the structural features and mechanical properties specific for these metals - high stacking fault energy and easy cross slip, high Peierls-Nabarro stress and sharp temperature dependence of the flow stress at $t < t^*$, low temperature brittleness, low solubility of interstitials and tendency to their segregation on crystal lattice defects.

Alloying of the VIA group metals with rhenium and some other elements changes significantly their electronic structure and thus increases their low temperature plasticity together with improving their strength characteristics. The observed growth of the density of states at the Fermi level is accompanied by lowering of the Peierls-Nabarro stress, decrease of the stacking fault energy, increase of the activation volume and of the interstitial solubility, and it even activates another deformation mechanism - twinning.

Finally I shall make some remarks on mechanical properties of ceramics. The above mentioned results give evidence of the analogy between deformation and fracture mechanisms for BCC metals and for covalent crystals including ceramics. The difference is that in BCC metals the ductile-brittle transition temperature T_b may be lowered below the room temperature by changing structural parameters, while in covalent crystals and ceramics where the value of t^* is high one always has T_b which is significantly higher than the room temperature. Therefore in order to optimise properties of ceramics we should create the structure which provides high fracture toughness, K_{IC}, in the region of cold deformation and brittle fracture.

Two main reasons could be mentioned which lead to the delay of crack formation and thus increase fracture toughness and strength. They are the effect of screening of the crack tip by compressive stress in the material and the interaction of a crack tip with the elements of structure (second phase particles, tough layers etc.). Screening can be achieved due to the phase transformation with the volume increase or due to the formation of a structure with microcracks. For the fibre reinforced ceramic materials interaction of a crack with fibres accounts for the main part of the energy release, and it is so important that may result in "drawing" fibres out of matrix.

REFERENCES

1. Trefilov, V.I., Milman, Yu.V. and Firstov, S.A., *Physical Bases of Strength of Refractory Metals*, Naukova Dumka, Kiev, 1975, Chapter 4.

2. Trefilov, V.I., Milman, Yu.V. and Grigoriev, O.N., Deformation and rupture of crystals with covalent interatomic bonds. *Crystal Growth and Charact.*, 1988, **16**, 225-77.

3. Trefilov, V.I. and Milman, Yu.V., Physical basis of thermomechanical treatment of refractory metals. In *Proceedings of the Plansee Seminar*, 1981, Part 1, 107-31.

ON THE LINKING-UP OF MICROCRACKS IN CREEPING POLYCRYSTALS WITH GRAIN BOUNDARY CAVITATION AND SLIDING

ERIK VAN DER GIESSEN[†] and VIGGO TVERGAARD[‡]

[†]Delft University of Technology, P.O. Box 5033, NL-2600 GA DELFT, The Netherlands
[‡]The Technical University of Denmark, Dep. of Solid Mechanics, DK-2800 LYNGBY, Denmark

ABSTRACT

The development of intergranular creep fracture in metals at elevated temperatures is studied in terms of a planar model analysis. Free grain boundary sliding and grain boundary cavitation is accounted for at all grain boundary facets, including continuous cavity nucleation and growth of cavities by diffusion and creep, while the grains deform by power law creep. The creep rupture process is modelled from the early stage on, where a few cavities have formed, until final failure by link-up of small microcracks resulting from cavity coalescence.

INTRODUCTION

The main mechanism for creep rupture in polycrystalline materials at high temperatures is the nucleation and growth of microscopic cavities at the grain boundaries. Coalescence of these cavities leads to microcracks on grain boundary facets and the final intergranular fracture occurs as nearby grain-size microcracks link-up to form a macroscopic crack.

Experiments have shown that cavitation occurs predominantly on grain boundary facets that are normal to the direction of the maximum macroscopic principal tensile stress [1]. Accordingly, micromechanical analyses of creep failure have focused on the nucleation and growth processes of cavities on isolated, or well-separated, transverse grain boundary facets (e.g. [1,2,3]). The time to microcrack formation by cavity coalescence on transverse facets, as obtained from such analyses, has been widely used as an estimate of the rupture lifetime, although final failure actually occurs somewhat later when the microcracks link up to form a macroscopic crack. Such estimates also neglect the interaction between neighbouring cavitating facets at the last stages of the rupture process. An analysis of the opposite limiting situation where all transverse facets are uniformly cavitated as in [4] accounts for interaction, but is expected to strongly underestimate the actual lifetime.

Evidently, the process of linking-up of microcracks, and the formation and subsequent growth of the macroscopic crack is an extremely complex phenomenon. In order to improve the insight in this process, with the ultimate goal of providing more accurate estimates of the rupture lifetime, we perform a numerical study of this process here in terms of a planar cell model of a polycrystal. In this model all transverse facets are assumed to be cavitated, but the initial cavitation states and/or the cavity nucleation rates are taken to be different on different facets. In addition, grain boundary sliding as well as cavitation are taken into account on intermediate grain boundary facets inclined to the maximum principal stress direction, which are most likely the key mechanisms for the linking-up of microcracks. These latter aspects have been investigated previously by the authors [5,6] using an axisymmetric model and it has been found that these failure mechanisms are already active at an early stage of the process.

The basic features of the present plane strain model were studied in a preliminary investigation [7], with focus on cases of reasonably well-separated cavitated facets. In the present work, we extend the study to consider the final stages of creep rupture where cavitated facets are closely spaced. In particular, we focus on the interaction between the cavitation processes on neighbouring transverse facets for different diffusion and cavity nucleation rates, and for various applied stress states. Some of the numerical computations are continued beyond cavity coalescence to study in detail the formation of open microcracks and their linking-up.

THE POLYCRYSTAL MODEL

For the purpose of this study, we wish to consider the situation where there is a high density of cavitation in the polycrystalline material. Therefore, we consider the 2D model polycrystal shown in Fig. 1, consisting of a periodic array of hexagonal grains, where cavitation may take place on every grain boundary facet but possibly with different rates. Furthermore, we assume free sliding on all grain boundaries. The polycrystal is taken to be subjected to a macroscopic stress state specified by principal stresses σ_1 and σ_2 under plane strain conditions. In this study, we limit ourselves to situations where cavitation on half

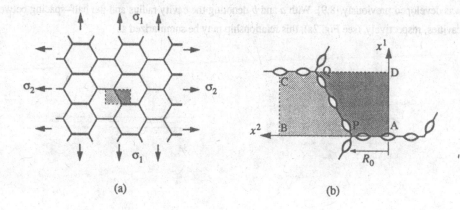

(a)   (b)

Figure 1. The 2D polycrystal model (a). The unit cell analysed (b).

of the grain boundary facets (emphasized in Fig. 1a) that are transverse to the maximum macroscopic tensile stress σ_1 is identical, and similarly for the remaining transverse facets and inclined facets, so that the unit cell as shown in Fig. 1b appears. When it is assumed that there is no cavitation along QC (and PQ), the model is expected to represent situations earlier in the process where the cavitating transverse facets are still reasonably well-separated, so that there is likely to be little interaction between those facets.

This plane strain model is an alternative to the axisymmetric model studied in [3,5,6], where the x^1-axis was taken as the axis of symmetry and the truncated cone APQD represented half a central grain. This axisymmetric model was designed so as to take into account the constraints imposed on a grain by its surrounding grains for situations where cavitating facets are well-separated. The present plane strain model gives a less realistic description of the central grain, but a more realistic description of the neighbouring grains in cases with closely spaced cavitating facets, and therefore is better suited for the present investigation. The predictions of the plane strain model for well-separated cavitating facets have been briefly discussed in [7], thus providing a link with the axisymmetric model. Anderson and Rice [4] considered an actual 3D polycrystal model, but their analysis is confined to situations where all transverse facets are uniformly cavitating. Here however, nonuniform cavitation is taken into account with possibly different cavitation rates along the three groups of cavitating facets AP, PQ and QC, respectively.

In addition to elastic deformations, the grains are taken to deform by dislocation creep governed by the power law expression

$$\dot{\varepsilon}_e^C = \dot{\varepsilon}_0 (\sigma_e/\sigma_0)^n \tag{1}$$

for the effective creep rate $\dot{\varepsilon}_e^C$ in terms of the effective Mises stress σ_e. Here, n is the creep exponent, and σ_0 and $\dot{\varepsilon}_0$ are reference stress and strain-rate parameters. Finite strains are accounted for in the analyses, but in all cases to be presented here the materials fail in a creep brittle manner with small failure strains.

The grain boundary cavities grow by diffusion as well as by creep of the grains. On the basis of detailed micromechanical studies by several authors, an approximate expression for the volumetric growth rate $\dot{V}$ of a quasi-equilibrium spherical–caps shaped cavity at a transverse grain boundary facet was developed previously [8,9]. With a and b denoting the cavity radius and the half–spacing between cavities, respectively, (see Fig. 2a), this relationship may be summarized as

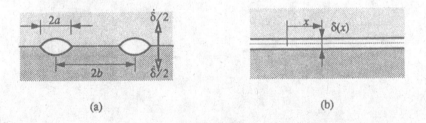

(a)   (b)

Figure 2. (a) Equally spaced cavities on a grain boundary; (b) Discrete cavity distribution smeared out over a grain boundary layer of thickness δ.

$$\dot{V} = \dot{V}_1 + \dot{V}_2 , \quad \text{for } a/L \leq 10 , \quad f = \max\left[\left(\frac{a}{b}\right)^2 , \left(\frac{a}{a+1.5L}\right)^2 \right] , \tag{2}$$

where

$$\dot{V}_1 = 4\pi \mathcal{D} \frac{\sigma_n - (1-f)\sigma_s}{\ln(\frac{1}{f}) - \frac{1}{2}(3-f)(1-f)} , \tag{3}$$

$$\dot{V}_2 = \begin{cases} \pm 2\pi \dot{\varepsilon}_e^C a^3 h(\psi) \left[\alpha_n \left| \frac{\sigma_m}{\sigma_e} \right| + \beta_n \right]^n , & \pm \frac{\sigma_m}{\sigma_e} > 1 \\[2ex] 2\pi \dot{\varepsilon}_e^C a^3 h(\psi) \left[\alpha_n + \beta_n \right]^n \frac{\sigma_m}{\sigma_e} , & \left| \frac{\sigma_m}{\sigma_e} \right| < 1 \end{cases} \tag{4}$$

Here, σ_n is the average normal stress on the grain boundary facet, while σ_m and σ_e are the average mean stress and effective stress, respectively, in the vicinity of the cavity; the sintering stress σ_s is neglected. Furthermore, $\mathcal{D} = D_B\delta_B\Omega/kT$ is the grain boundary diffusion parameter ($D_B\delta_B$ is the boundary diffusivity, Ω is atomic volume, T is absolute temperature and k is Boltzmann's constant) and the constants are given by $\alpha_n = 3/2n$, $\beta_n = (n-1)(n+0.4319)/n^2$, $h(\psi) = \left[(1+\cos\psi)^{-1} - \frac{1}{2}\cos\psi \right]/\sin\psi$, with the cavity tip angle ψ chosen here equal to $75°$. The diffusive path length

$$L = [\mathcal{D}(\sigma_e/\dot{\varepsilon}_e^C)]^{1/3} , \tag{5}$$

introduced in [10], determines whether coupled cavity growth occurs: for a/L smaller than say 0.1, cavity growth is dominated by diffusion, while local creep deformations play an increasing role for larger values of a/L. Ultimately, with $\dot{V}$ according to (2), the change of the cavity radius a is found as $\dot{a} = \dot{V}/(4\pi a^2 h(\psi))$. It is noted that cavity growth on the inclined grain boundaries like PQ will be affected by grain boundary sliding, which may give rise to non-equilibrium void shapes [1]. Assuming relatively low sliding rates, however, these effects are small enough so that the same growth relations (2)–(4) can be applied on these boundary facets.

The nucleation of new cavities, which usually takes place continually during the creep process, is still not very well understood and existing theoretical models have so far been only partly successful in explaining the experimental observations [1]. Following earlier work [3,6,7], we use a nucleation law which was motivated by the work of Dyson [11] and which is expressed by the evolution relation

$$\dot{N} = F_n(\sigma_n/\Sigma_0)^2\dot{\varepsilon}_e^C \tag{6}$$

for the number of cavities N per unit undeformed grain boundary area. Here, σ_n and $\dot{\varepsilon}_e^C$ have the same interpretation as in the cavity growth model discussed above, F_n is a material constant and Σ_0 is a normalization constant. This expression is in keeping with many experiments indicating that the cavity density is proportional to the effective strain, while on the other hand it is reminiscent of the stress dependence observed macroscopically in [12]; but it should be noted that previous computations with this nucleation law have not been able to predict this macroscopic stress dependence [6].

Cavity coalescence takes place by failure of the ligament between cavities and is taken to occur here when $a/b = 0.9$. Since cavitation is usually nonuniform over a grain boundary facet, cavity coalescence will occur first at a specific location and then propagate over the rest of the facet, until a facet sized open microcrack has appeared. In this study, loss of integrity of the unit cell depicted in Fig. 1 would make an operational failure criterion; but, this is not necessarily equivalent with open microcracks on all grain boundary facets. For the present planar model, the occurrence of open microcracks on all transverse facets like AP and QC already implies failure: with grain boundaries sliding freely, equilibrium can no longer be maintained and the material falls apart immediately.

METHOD OF ANALYSIS

Assuming that the cavities remain small as compared with the grain size, we employ a 'smeared out' model of a cavitated grain boundary facet. Thus, a facet containing a discrete distribution of cavities is replaced in the analysis with a grain boundary layer to which continuous distributions $a(x)$ and $b(x)$ of the cavitation parameters a and b is attributed (Fig. 2b). The average separation between the two adjacent grains due to cavitation defines a thickness δ_c of this grain boundary layer, $\delta_c = V/(\pi b^2)$; when cavitation is nonuniform along the facet, this thickness δ_c varies continuously along the layer. Thus, the growth rate of the layer thickness is given by

$$\dot{\delta}_c = \frac{\dot{V}}{\pi b^2} - \frac{2V}{\pi b^2} \frac{\dot{b}}{b},$$

(7)

and is governed by the growth rate expressions (2)–(4) but depends also on the rate of change of the cavity spacing, which is primarily due to the nucleation of new cavities. Continuous cavity nucleation is taken into account in an approximate manner, just as in previous work [3,6,7]. Thus, the main effect of nucleation is expressed in terms of a reduction of the cavity spacing $2b$, leading to the following expression for the ration $\dot{b}/b$, to be substituted into (7):

$$\frac{\dot{b}}{b} = \frac{1}{2}\dot{\epsilon}_I - \frac{1}{2}\frac{\dot{N}}{N}.$$

Here, $\dot{N}$ is given locally by the nucleation law (6), and $\dot{\epsilon}_I$ is the principal logarithmic strain-rate in the plane of the grain boundary. It is numerically convenient to add a fictitious layer of linear elastic springs to the grain boundary layer with a normal stiffness k_n which is high enough to keep the difference between δ_c and the actual thickness δ small. Thus, the normal facet stress is governed by the constitutive equation $\dot{\sigma}_n = k_n(\dot{\delta} - \dot{\delta}_c)$ for the grain boundary layer. When microcracking has occurred on (part of) the facet, k_n is set to 0.

The analyses are carried out numerically, using a finite element model of the grains APQD and QPBC with quadrilateral elements consisting of four "crossed" triangular elements. The mesh used (see Fig. 3a) is rather crude, but it was found in [7] that the time to develop a full facet microcrack in a typical case was reduced by only 2% when using a finer mesh with twice the number of elements. Special purpose finite elements are used to model the grain boundary layers along the cavitating facets. These elements are designed to allow for cavitation and thickening of the grain boundary layer, according to (7),

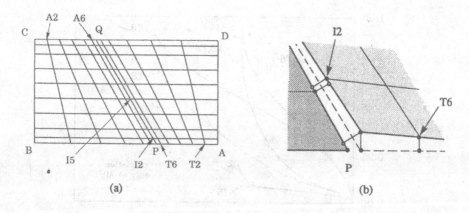

Figure 3. (a) Finite element mesh used; (b) Magnification of region near triple point P, showing grain boundary nodes (•) and phantom nodes (◦) on the grain boundary layers on AP and PQ.

as well as for free sliding. Due to symmetry, only half of the grain boundary layer needs to be accounted for along the transverse facets AP and QC, with due account of absence of grain boundary sliding. Further details concerning the computational procedure may be found in [5].

Uniform displacement boundary conditions normal to the boundary ABCD of the region shown in Fig. 1b are prescribed, such that the average true stress σ_1 and σ_2 retain specified constant values.

RESULTS

In the present study, we focus on a material with a Poisson's ratio $v = 0.3$ and a creep exponent $n = 5$, while the parameter $\dot{\varepsilon}_0/\sigma_0^n$ is used to define the reference time $t_R = \sigma_e/(E\dot{\varepsilon}_e^C)$. Here, E is Young's modulus, and σ_e and $\dot{\varepsilon}_e^C$ are taken as the macroscopic applied effective stress and the corresponding creep rate according to (1), respectively. This macroscopic effective stress is approximated by the expression $\sigma_e = \frac{1}{2}\sqrt{3}|\sigma_1 - \sigma_2|$ for pure plane strain creep, while the macroscopic principal stresses are prescribed such that $\sigma_e/E = 0.5 \times 10^{-3}$ for all cases analysed here; the macroscopic mean stress σ_m is approximated by $\sigma_m = \frac{1}{2}(\sigma_1 + \sigma_2)$. Hence, the reference time is identical to that in the previous studies [3,5,6,7]. Whenever cavitation is taken to occur at a grain boundary facet, a low initial cavity density N_I is assumed as specified by $N_I = 1/A_I$ with $A_I = \pi R_0^2$, while the initial cavity radius a_I is taken as $a_I/R_0 = 0.01$ ($2R_0$ is the length of a facet in the 2D model polycrystal, Fig. 1b). Several nucleation rates are considered by taking F_n to range from 0 to a value of $F_n = 100/A_I$ which has been used also in the earlier studies; the parameter Σ_0 is chosen equal to the macroscopic effective stress. A maximum density $N_{max} = 100/A_I$ is prescribed in all cases to be presented here, but was never reached prior to failure. The grain boundary diffusion parameter $\mathcal{D}$ is specified in terms of the length scale L through Eq. (5) measured relative to the initial cavity radius; in this study, we will concentrate on taking $(a/L)_I = 0.025$, but we will also discuss some results for $(a/L)_I = 0.1$.

140

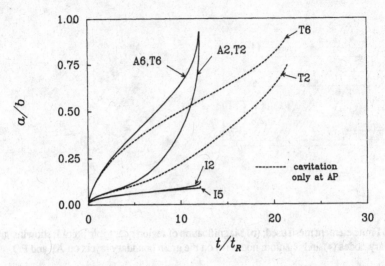

Figure 4. Development of damage for $\sigma_2 = 0.5\sigma_1$ and $(a/L)_I = 0.025$ with $F_n = 100/A_I$ on all grain boundary facets. The case where there is cavitation only at AP is shown for comparison.

Figure 4 shows the evolution of the damage parameter a/b at various locations (see Fig. 3) for a case with $(a/L)_I = 0.025$ where the nucleation rate parameter F_n is identical on all grain boundary facets, $F_n = 100/A_I$, and the transverse macroscopic stress is half the tensile stress, $\sigma_2 = 0.5\sigma_1$. Due to symmetry of the problem in this case, cavitation progresses in an identical way on AP and QC, while we observe little increase of cavitation along the inclined boundary. Microcracking initiates at the triple points P and Q at $t/t_R \approx 12$, and in this case swiftly proceeds until full facet microcracks are created and immediate failure occurs as described above. For comparison, results are also shown for the same parameters but with cavitation assumed only on AP, corresponding to a situation where half of all transverse facets are cavitating. This case presumably gives a good representation of a situation where cavitating facets are still well-separated, as studied in our previous work [3,5,6,7]. Compared with this case, cavitation on all facets enhances cavity nucleation and growth drastically, leading to a reduction of the time to first coalescence by a factor of about 2. This is primarily due to the fact that, since $(a/L)_I = 0.025$ corresponds to a high diffusion rate relative to the creep rate, cavitation in the case of well-separated transverse facets is strongly constrained by the creep deformations in the x^1-direction of the surrounding grains (see e.g. [1,2,3]) and the material fails in a creep brittle manner (the macroscopic strain in x^1-direction $\varepsilon_1 = 0.08$ just before failure). This creep constraint completely disappears in the 2D model when cavitation is identical on all transverse facets, as in the case analysed here.

To further illustrate this point, the case of Fig. 4 has been repeated for $(a/L)_I = 0.1$ (no results shown). It has been found in previous studies (e.g. [3,7]) that cavity growth is no longer completely dominated by diffusion, and that there is much less constraint by creep deformations. Indeed, cavitation on all transverse facets is found to be only slightly enhanced as compared with the case of well-separated facets, and the time to first cavity coalescence is reduced by about 15%.

The interaction between such closely spaced cavitating facets of the type AP and of the type QC

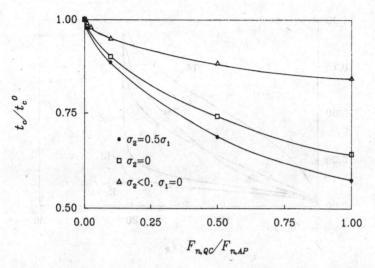

Figure 5. Time to first cavity coalescence for $(a/L)_I = 0.025$ with $F_{n, AP} = F_{n, PQ} = 100/A_I$ on the facets AP and PQ, for different values $F_{n, QC}$.

obviously depends on the nucleation rate on QC relative to that on AP, and is most pronounced for the case shown in Fig. 4. Therefore, the analysis has been repeated for various other values of F_n on QC, but with otherwise identical parameters as in Fig. 4. The time t_c to first cavity coalescence (typically near P on the central transverse facet) is plotted in Fig. 5 relative to the time to coalescence t_c^0 when $F_{n, QC} = 0$. As this interaction may be stress dependent, three different macroscopic plane strain stress states are studied, namely biaxial tension ($\sigma_2 = 0.5\sigma_1$) as in Fig. 4, uniaxial tension ($\sigma_2 = 0$) and lateral compression ($\sigma_1 = 0$, $\sigma_2 < 0$). For the compressive stress state, cavitation on AP and QC completely relies on grain boundary sliding along PQ (see e.g. also [13]); cavitation on PQ is excluded in the analyses for this stress state, since the compressive facet stresses would close existing cavities. It is seen that the interaction between the nearby cavitating facets AP and QC depends sensitively on the applied stresses. For the stress states considered here, the triaxiality σ_m/σ_e takes values $\sqrt{3}$ for biaxial tension, $1/\sqrt{3}$ for uniaxial tension and $-1/\sqrt{3}$ for lateral compression, and the results show a decreasing interaction effect with decreasing triaxiality.

The case with equal cavitation on all transverse facets analysed in Fig. 4 clearly corresponds to an extreme situation in which failure occurs immediately upon developing a full facet microcrack on the central facet AP. In Fig. 6, we now consider a case in more detail where cavitation does occur on all facets (with identical initial cavitation states), but with very slow nucleation of new cavities on QC, as specified by $F_{n, QC} = 0.025F_{n, AP}$, and for otherwise identical parameters as in Fig. 4. It is seen that, even though considerable cavity growth occurs on QC, there is only a small interaction between QC and AP, so that the evolution of damage on AP is virtually unaffected in comparison with the situation where only AP is cavitating (see Fig. 4), except for just before coalescence. In this case, failure is not equivalent with the attainment of a full facet microcrack on AP, but awaits the formation of a microcrack along QC before immediate sliding off can take place. It is seen that when the facet AP has fractured, the cavitation proc-

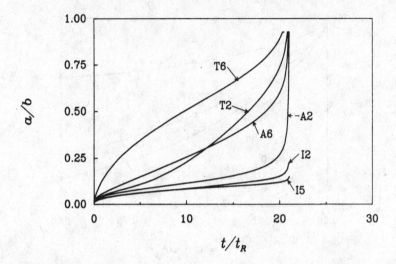

Figure 6. Development of damage for $\sigma_2 = 0.5\sigma_1$ and $(a/L)_I = 0.025$ with cavitation on all facets, but with $F_n = 100/A_I$ on AP and PQ, while $F_n = 2.5/A_I$ on QC.

esses on both QC and PQ are accelerated strongly and a full facet microcrack on QC is created very short-ly afterwards in this case. There is considerable damage developing on PQ after failure of AP, especially near the triple grain junction at P, but the subsequent time to failure of QC is too short to allow for cavity coalescence on PQ. This computation has been repeated for other nucleation rates on QC and other stress states (as in Fig. 5) and it was found that final rupture occurred prior to any coalescence on PQ in all cases.

To give an indication of the time span of the final stages of the creep rupture process, Fig. 7 shows the time to final failure, t_f, relative to the time to first coalescence, t_c, for various ratios of the nucleation

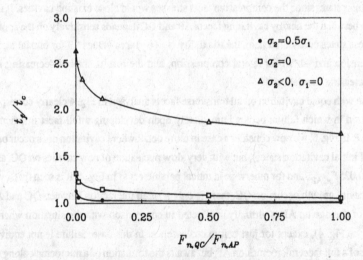

Figure 7. Creep rupture times relative to time to first cavity coalescence for $(a/L)_I = 0.025$ with $F_{n,\,AP} = F_{n,\,PQ} = 100/A_I$ on the facets AP and PQ, for different values $F_{n,\,QC}$.

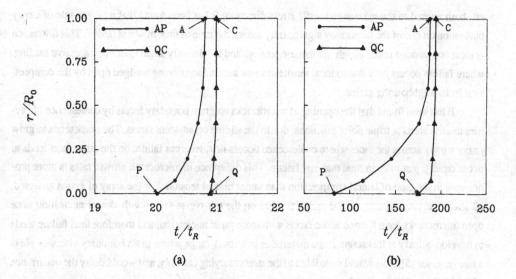

Figure 8. Microcracking along AP and QC for $(a/L)_I = 0.025$, $F_{n, AP} = F_{n, PQ} = 100/A_I$ and $F_{n, QC} = 0.025 F_{n, AP}$ with $\sigma_2 = 0.5\sigma_1$ (a) or $\sigma_1 = 0$, $\sigma_2 < 0$ (b).

rate parameters F_n on AP and on QC. We see that in situations where nucleation on facets like QC that are adjacent to the central facet AP is relatively slow, the time to first coalescence may significantly underestimate the time to actual failure; but, this effect is markedly dependent on the applied stress state.

Figure 8 finally shows the growth of microcracks along the transverse facets for a material with $F_{n, QC} = 0.025 F_{n, AP}$ subjected to either biaxial tension or lateral compression. The length of the microcracks on AP and QC is measured from the triple points P and Q, respectively (see Fig. 3). These results emphasize that under lateral compression it takes a relatively long time to create a full microcrack along AP, while failure of the remaining transverse facet takes place promptly. Another interesting difference between these two cases is that under biaxial tension coalescence on QC awaits failure of the central facet, whereas under lateral compression microcracking along QC already starts prior to failure of AP; this is probably associated with the intense grain boundary sliding in the case of compression.

DISCUSSION

The planar array of hexagonal grains with free grain boundary sliding has the special property that the material separates immediately, if open microcracks have formed at all grain boundary facets normal to the maximum tensile stress. In a real 3D array of grains this final separation requires creep accommodation that takes some extra time [4]. However, in the more realistic case where microcracks do not form simultaneously on all these facets, studies based on the planar model are expected to give a reasonably realistic picture of the final link-up process.

In the present investigation the focus has been on the interaction between two closely spaced facets, both normal to the maximum tensile stress direction. It has been found that a slower rate of cavity nucleation on one of the facets may significantly reduce the time to final material failure. This difference is most pronounced under a high hydrostatic tension, and is relatively small under compressive loading, where failure occurs only due to local tensile stresses across facets being wedged open by the compression from neighbouring grains.

It has been found that the opening of microcracks on grain boundary facets by coalescence of cavities usually starts at triple point junctions, due to the sliding on adjacent facets. The crack tends to grow rather slowly across the facet where coalescence occurs first, whereas failure on the other facet tends to occur rapidly, just prior to final material failure. This difference in microcrack growth rates is more pronounced in the case of lateral compression than under biaxial tension. For the array of grains analysed, all stress carrying capacity of the material relies on the transverse facets with slower nucleation once open microcracks have formed at the facets with more rapid nucleation, and therefore final failure tends to develop quickly at this stage. If the material is in a stress range where grain boundary viscosity plays a role (e. g. see [5]), this would contribute to the stress carrying capacity, and would delay the occurrence of final failure after the first formation of open microcracks.

The cases discussed here in detail are for a material with relatively fast grain boundary diffusion, corresponding to a large value of L. Hence, cavity nucleation and growth are severly constrained by creep of the grains, and rupture occurs at small strains. For smaller values of L, corresponding to slower diffusion and/or higher stress levels, the creep constraint is less active and the interaction between nearby transverse facets is less than found in the above. On the other hand, preliminary results [7] for smaller values of L indicate that the difference between the time to first coalescence and formation of a full facet microcrack may be even more pronounced than observed here, with a strong tendency to wedge shaped microcracks.

A more realistic representation of interaction effects in the final failure process could be obtained by considering a larger array of grains, with different nucleation rates on a number of different transverse facets. Then, the first formation of an open microcrack would not immediately focus all remaining stress carrying capacity on one other type of facet, but failure would grow by a gradually increased number of failed facets. Also in such more complex studies the effect of grain boundary viscosity could be important.

Acknowledgement – The work of EvdG was made possible by a fellowship of the Royal Netherlands Academy of Arts and Sciences.

REFERENCES

1. Argon, A.S., Mechanisms and Mechanics of Fracture in Creeping Alloys. In Recent Advances in Creep and Fracture of Engineering Materials and Structures, eds. B. Wilshire and D.R.J. Owen, Pineridge Press, Swansea, 1982, pp. 1–52.

2. Cocks, A.C.F. and Ashby, M.F., On Creep Fracture by Void Growth. Progr. Mater. Sci., 1982, 27, 189–244.

3. Tvergaard, V., Effect of grain boundary sliding on creep constrained diffusive cavitation. J. Mech. Phys. Solids, 1985, 33, 447–469.

4. Anderson, P.M. and Rice, J.R., Constrained creep cavitation of grain boundary facets. Acta Metall., 1985, 33, 409–422.

5. Van der Giessen, E. and Tvergaard, V., A creep rupture model accounting for cavitation at sliding grain boundaries. Int. J. Fracture (in press).

6. Van der Giessen, E. and Tvergaard, V., On cavity nucleation effects at sliding grain boundaries in creeping polycrystals. In Proc. of Fourth Int. Conf. on 'Creep and Fracture of Engineering Materials and Structures', eds. B. Wilshire and R.W. Evans, Elsevier Appl. Sci., Swansea, 1990, pp. 169–178.

7. Van der Giessen, E. and Tvergaard, V., On Microcracking due to Cavitation and Grain Boundary Sliding in Creeping Polycrystals. In Proc. of Fourth IUTAM Symposium 'Creep in Structures', Springer Verlag (in press).

8. Sham, T.–L. and Needleman, A., Effects of triaxial stressing on creep cavitation of grain boundaries. Acta Metall., 1983, 31, 919–926.

9. Tvergaard, V., On the creep constrained diffusive cavitation of grain boundary facets. J. Mech. Phys. Solids, 1984, 32, 373–393.

10.Needleman, A. and Rice, J.R., Plastic creep flow effects in the diffusive cavitation of grain boundaries. Acta Metall., 1980, 28, 1315–1332.

11.Dyson, B.F., Continuous cavity nucleation and creep fracture. Scripta Metall., 1983, 17, 31–37.

12.Dyson, B.F. and McLean, D., Creep of Nimonic 80A in torsion and tension. Metal Science, 1977, 2, 37–45.

13.Chan, K.S., Lankford, J. and Page, R.A., Viscous cavity growth in ceramics under compressive loads. Acta Metall., 1984, 32, 1907–1914.

LONG-TERM CREEP DUCTILITY MINIMA IN 12%CrMoV STEEL

R Timmins and P F Aplin
ERA Technology Ltd
Cleeve Road, Leatherhead, Surrey KT22 7SA

ABSTRACT

Minima in the rupture ductility versus long-term rupture life relationship for 12CrMoV steel in the temperature range 525 to 675°C have been investigated by reference to previously published long-term creep data together with some new metallographic information. The observed minima are similar to those exhibited by most other creep-resistant ferritic steels. The ductility behaviour has been modelled in terms of creep cavitation theory. The theory predicts a diffusional cavity growth controlled decrease in rupture ductility with decreasing stress at creep lives that are less than those at which the minimum occurs and a nucleation-controlled increase in rupture ductility at longer times. Insufficient materials property data are currently available to allow a rigorous quantitative verification of the model for the present materials. However certain quantitative predictions are in good agreement with the experimental data and metallographic observations. The further work that is underway to develop the model's capability for very long-term ductility behaviour prediction of creep-resistant ferritic steels in general, is highlighted.

NOTATION

A	:	Fraction of damaged grain boundary facets
d	:	Grain size
f	:	Area fraction of cavities on a grain boundary facet
k	:	Boltzmann's constant
K	:	A parameter depending on the nature of the creep damage mechanism
N	:	Cavity density
n	:	Stress exponent
r	:	Cavity radius
T	:	Temperature
D_b	:	Grain boundary diffusion coefficient
Q_b	:	Activation energy for grain boundary diffusion
Q_v	:	Activation energy for volume diffusion
Q_c	:	Activation energy for creep
$\dot{\epsilon}$	:	Global creep rate
$\dot{\epsilon}_o$	:	Temperature dependent strain rate at stress σ_o
ϵ_f	:	Rupture strain

λ	:	Cavity half spacing
σ	:	Applied tensile stress
σ_0	:	Normalising stress
ω	:	Grain boundary thickness
Ω	:	Atomic volume

INTRODUCTION

12CrMoV steels are widely used in high temperature power plant for applications such as steam pipelines, turbine casings, blading, bolts and tubes. This is because they have superior rupture strength and corrosion resistance compared with the low alloy ferritic steels, yet are cheaper to fabricate and are more resistant to thermal-shock than austenitic alloys.

In common with many creep-resistant ferritic materials 12CrMoV steel exhibits a minimum in the rupture life versus ductility relationship beyond which rupture ductility increases with increasing creep time. A knowledge of the time at which this minimum occurs, and the value of minimum ductility is very important for the design and life assessment of components operating for long times in the creep range.

The present paper addresses the occurrence of rupture ductility minima in 12CrMoV steel by reference to previously published long-term ductility data on this material (1). Additional metallographic data are presented and a model is developed for describing the observed behaviour. The findings are relevant to creep-resistant ferritic steels in general.

OBSERVED DUCTILITY BEHAVIOUR

A previous analysis (1) has shown the following general features in terms of long term rupture ductility behaviour of 12CrMoV steels (see Fig.1):

i) low ductility levels occurred at increasingly shorter times as the temperature increased between 525°C and 675°C.

ii) a ductility minimum was observed at temperatures within the range 575°C to 675°C and the lower the test temperature then:

148

a) the longer was the time to reach the ductility minimum; and

b) the lower was its value.

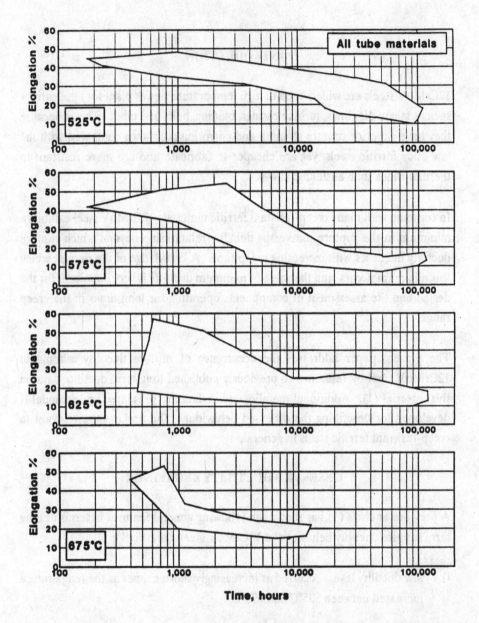

Fig.1: Trends in rupture elongation of 12CrMoV tube steels at 525, 575, 625 and 675°C

These aspects are better illustrated in Fig.2 in which a typical set of ductility data for one cast is plotted against stress at various temperatures. The plot clearly indicates that the value of the temperature dependent ductility minimum decreases with increasing stress (or decreasing time to fracture). The features are consistent with ductility trends exhibited by ½Mo, 1CrMo, ½Mo¼V, (2) and 1Cr1Mo¼V steels (3) and other low alloy ferritic steels (4). An important implication for the present materials is that the value of ductility will fall significantly below 10% at 525°C as the test duration increases beyond 100000 h. With reference to Fig.2, a tentative extrapolation of the loci of ductility minima suggests that the minimum value of rupture elongation at 525°C will be approximately 7% and will occur at a stress of approximately 170 MPa (this corresponds to a rupture life of approximately 200000 h).

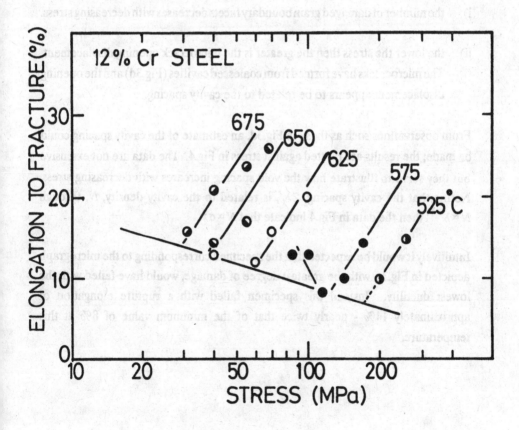

Fig.2: Plot of rupture strain versus stress at 525, 575, 625 and 675°C illustrating rupture ductility minima

METALLOGRAPHIC OBSERVATIONS

For the test data given in Fig.1, metallographic examination of ruptured test-pieces revealed a number of different modes of fracture ranging from high ductility necking failure, favoured at high stresses, to low ductility cavitation failure favoured at low stress. This general behaviour was observed across the whole temperature range (1).

Micrographs of test-pieces corresponding to the data set in Fig.2 which failed at a series of different durations at 575°C are shown in Fig.3. An interesting observation is that the extent of cavitational damage **decreases** with **decreasing** stress in the **low stress regime.** Specifically the extent of damage decreases in two ways:-

i)   the number of damaged grain boundary facets decreases with decreasing stress.

ii)   the lower the stress then the greater is the microcrack opening displacement. The microcracks have formed from coalesced cavities (Fig.3d) and the opening displacement appears to be related to the cavity spacing.

From observations such as those in Fig.3d, an estimate of the cavity spacing could be made; the results are plotted against stress in Fig.4. The data are not extensive but they serve to illustrate how the void spacing increases with decreasing stress. Noting that the cavity spacing, 2λ, is related to the cavity density, N, through $N \propto \lambda^{-\frac{1}{2}}$, then the data in Fig.4 indicate that $N \propto \sigma^3$.

Intuitively it would be expected that the specimen corresponding to the micrograph depicted in Fig.3c, with the greatest degree of damage, would have failed with the lowest ductility. Instead this specimen failed with a rupture elongation of approximately 14% - nearly twice that of the minimum value of 8% at this temperature.

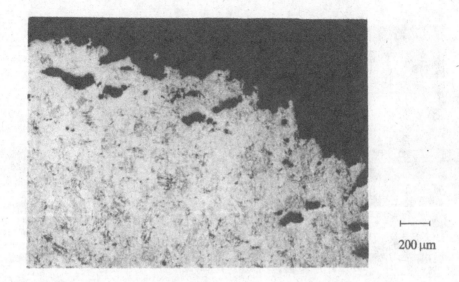

200 μm

a)   σ = 85 MPa, t$_f$ = 116353 h, ϵ_f = 13%

200 μm

b)   σ = 116 MPa, t$_f$ = 41558 h, ϵ_f = 8%

Fig.3:   Micrographs of fractured creep specimens at 575°C

152

c)   σ = 170 MPa, t_f = 4717 h, ε_f = 14%

d)   σ = 170 MPa, t_f = 4717 h, ε_f = 14%

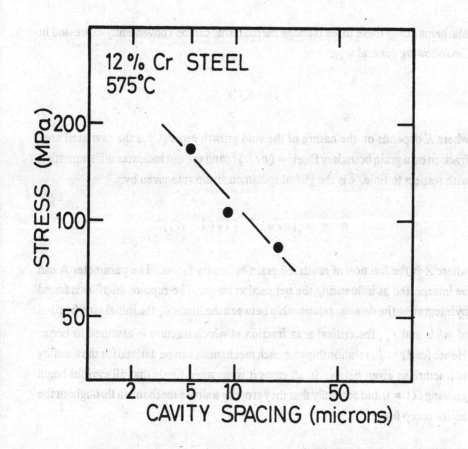

Fig.4:   Plot showing relationship between cavity spacing and applied stress

RUPTURE DUCTILITY MODEL

A model for describing the observed ductility behaviour is detailed below. Three cavity growth mechanisms are considered:

i)   Continuum or 'plastic hole' growth (5)

ii)   Diffusional cavity growth (6)

iii) Constrained cavity growth (7)

Mathematically these three damage mechanisms can be conveniently expressed in the following general way:

$$\dot{f} = K\dot{\epsilon}$$

where K depends on the nature of the void growth process, f is the cavitated area fraction on a grain boundary facet, $= (r/\lambda)^2$, and the dot indicates differentiation with respect to time. $\dot{\epsilon}$ is the global specimen strain rate given by:

$$\dot{\epsilon} = \dot{\epsilon}_o[\sigma/\sigma_o]^n \ [1/(1-A)]^n$$

where A is the fraction of cavitated grain boundary facets. The parameter A can be interpreted as influencing the net section stress. The rupture ductility is found by integrating the damage relationship between the limits f_i, the initial area fraction of voids and f_c, the critical area fraction at which fracture is assumed to occur. Hence for $f_i \ll f_c$ relationships for each mechanism can be derived for the ductility at fracture as given below. In all cases it is assumed firstly that all cavities begin growing at t = 0 and secondly that they grow by a single mechanism throughout the entire creep life.

For void growth by power law creep:

$$K = f$$

which leads to:

$$\epsilon_f = ln[1/f_i] \tag{1}$$

When cavities grow by unconstrained grain boundary diffusion then:

$$K = \frac{2D_b w \Omega \sigma}{f^{\frac{1}{2}} kT \lambda^3} \ \frac{1}{\dot{\epsilon}_o} \left[\frac{\sigma_o}{\sigma}\right]^n (1-A)^n$$

giving:

$$\epsilon_f = \frac{kT\lambda^3}{3D_b w\Omega\sigma} \dot{\epsilon}_o\left(\frac{\sigma}{\sigma_o}\right)^n \frac{1}{(1-A)^n} f_c^{3/2} \qquad [2]$$

It should be noted that equation [2] is only an approximation: in particular it ignores the strain due to the dilation of cavities and any plastic hole growth. Nevertheless the relationship is an adequate representation for the present purpose.

When the specimen strain rate is less than that due to void growth, cavity growth becomes constrained and:

$$K = \frac{1}{f^{1/2}} \frac{d}{2\lambda}$$

leading to:

$$\epsilon_f = \frac{4}{3} \frac{\lambda}{d} f_c^{3/2} \qquad [3]$$

PREDICTIONS OF THE MODEL

The important observations from equations [1], [2] and [3] are:-

i) Rupture ductility is independent of stress and temperature under continuum and constrained growth conditions ie:- independent of the creep rate. It follows that it is also independent of the proximity of cavitated facets (which is proportional to A). This aspect is supported through a more detailed analysis of the constrained cavity growth process (8).

ii) The elongation to fracture under constrained conditions is the minimum. This is because it corresponds to the strain due solely to dilation through cavity growth in contrast to the situation pertaining to diffusional growth (see v).

iii) The predicted ductility under continuum void growth conditions is very large. In practice other effects such as necking intervene and the actual elongation to failure will be much lower.

iv) Under diffusional growth conditions ductility depends on the global creep rate and will increase monotonically with increasing stress. The effect of temperature enters through the temperature dependent terms:-

$$\epsilon_f \;\; \alpha \;\; \frac{T\dot{\epsilon}_o}{D_b} \;\; \alpha \;\; \frac{T\exp(-Q_c/RT)}{\exp(-Q_b/RT)}$$

Because Q_c is generally greater than Q_b then ductility increases with temperature.

v) Under diffusional growth conditions the creep rate increases faster with stress and temperature than the cavity growth rate, and the amount of strain produced through specimen extension is therefore unrelated to the cavity spacing. λ enters equation [2] because it is related to the time for cavity growth to reach the critical area fraction for fracture on a single grain boundary facet. As mentioned above equation [2] is only an approximation. A more detailed analysis shows that λ enters into the ductility equation in a more complicated way (6), although in general terms ductility does increase as λ increases.

APPLICATION OF MODEL TO PRESENT DATA

The above expressions can be conveniently presented in schematic form as shown in Fig.5, which indicates how the three mechanisms leading to failure interact in $\epsilon_f - \sigma$ space at two different temperatures. On the basis that the cavity spacing increases as the stress decreases (as observed experimentally), then the minimum rupture elongation increases with increasing temperature. This occurs as the diffusional cavity growth line intersects the constrained growth mechanism line at increasingly higher values of rupture ductility. The fact that ductility increases with decreasing stress at stresses lower than where the minimum occurs can be explained on the basis of a reduced incidence of cavity nucleation, which is consistent with the observation of an increased cavity spacing with decreasing stress. The overall behaviour is represented by the bold lines in Fig.5. A comparison with the data set in Fig.2 indicates that the observed ductility behaviour is well explained qualitatively. At the present time a rigorous quantitative comparison with experimental data is made difficult by a lack of materials property data for use in equations [1] to [3]. Furthermore a fully quantitative analysis would have to take into account that cavities often nucleate continuously during creep and that the cavity growth mechanism will change as the cavity grows, thus requiring an iterative approach to evaluate the fracture strain.

Nevertheless the limited data that are available enable certain quantitative predictions of the model to be verified. Firstly, equation [2] predicts, using values of $2\lambda = 9\,\mu m$, $d = 65\,\mu m$ and $f_c = 1$, a rupture elongation of 9.2%, which compares favourably with the experimentally measured elongation of 8%. Secondly the slope of the lines in Fig.2 corresponding to the diffusional cavity growth regime is proportional to $n - 1$. Limited creep data for these materials indicates that n lies between 5-10. The slopes in Fig.2 are however far less than this (slope ~ 2). This may be reconciled by noting that the slope will decrease if $\lambda \alpha 1/\sigma^m$ as observed experimentally with m between 2 and 3. Finally the ductility increase with temperature at a given stress in the diffusional regime is proportional to the difference, ΔQ, between the activation energy for creep and that for grain boundary diffusion. Analysis of the data in Fig.2 indicates the value of ΔQ to be around $80\,kJmol^{-1}$. Assuming that $Q_c \sim Q_v \sim 250\,kJmol^{-1}(9)$ leads to $Q_b \sim 170\,kJmol^{-1}$, which compares well with grain boundary diffusion data for $\alpha Fe(9)$.

It also now becomes apparent why the specimen exhibiting the greatest degree of damage (Fig.3c) did not necessarily result in the lowest ductility. The implication is that this test was conducted at conditions corresponding to unconstrained cavity growth where overall specimen extension contributes to the fracture strain (eq 2).

PREDICTION OF VERY LONG-TERM DUCTILITY BEHAVIOUR

The above model has shown how the experimental observations can be semi quantitatively explained by creep cavitation theory provided that a stress dependent cavity spacing term is included in the analysis. The model is supported by microscopical observations. An important aspect however is the implication for design and life assessment and the need for more quantitative comparisons. Indeed estimation of long-term rupture ductility without the need for extensive creep testing would be highly beneficial. For this reason further work is being carried out with the aim of using the model to predict long-term ductility in creep-resistant ferritic steels in general. This includes examination of the use of short-term higher temperature tests in conjunction with the trends illustrated in Fig.2 as a basis for extrapolation to lower temperature longer-term conditions

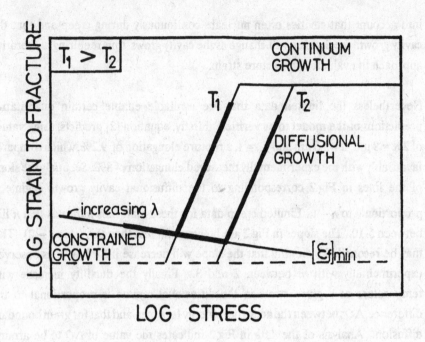

FIg.5: Ductility map for creep

CONCLUSIONS

The long-term rupture ductility behaviour of 12CrMoV steel in the range 525°C to 675°C has been investigated and the following conclusions drawn.

1) A minimum in the rupture ductility versus long-term rupture life relationship is exhibited with similar characteristics to that observed in other creep-resistant ferritic steels.

2) An important implication of the observed ductility minima is that as the rupture life increases beyond 100000 hours at 525°C rupture elongation will decrease to values below 10%.

3) The ductility minima can be modelled in terms of creep cavitation theory. This predicts a diffusional cavity growth controlled decrease in rupture ductility with decreasing stress at creep lives that are less than that at which the minimum occurs and a nucleation-controlled increase in rupture ductility at longer times.

4) Insufficient materials property data are currently available to allow a rigorous quantitative verification of the model though it has been possible to make certain predictions and these are in good agreement with the experimental data and metallographic observations.

5) Further work is underway to develop the model as a means of predicting the very long-term ductility behaviour of creep-resistant ferritic steels in general.

ACKNOWLEDGEMENTS

The paper is published with the permission of ERA Technology Ltd.

REFERENCES

1. Aplin P.F., Bullough C.K., Barlow D.J. Proc. Conf. Rupture Ductility of Creep Resistant Steels, York, 1990, Institute of Metals
2. Glen J., J. Iron and Steel Inst. 1955, 179, 320
3. Woodford D.A., Goldhoff R.M. Mat. Sci. Eng 1970, 5, 303
4. Cane B.J., and Williams J.A. Int. Met. Rev. 1987 32, 241
5. Hancock J.W. Metal Sci. 1976, 10, 319
6. Cocks A.C.F., Ashby M.F. Prog. Mat. Sci. 1982, 27, 189
7. Dyson B.F. Metal Sci. 1976, 10, 349
8. Anderson P.M., and Rice J.R. Acta Metall. 1985, 33, 409
9. Frost H.J., and Ashby M.F. Deformation-Mechanism Maps, 1982, Pergamon Press, Oxford

Failure Criteria on Creep Rupture of Mineral Salt Based on Micromechanical Mechanisms

E. Stein, U. Heemann [1]
Institute of Structural and Computational Mechanics,
University of Hannover, FRG

Abstract:
In the first part a theory for crack initiation in rock salt is reviewed basing on the model of Argon and Orowan of stress concentration by a blocked glide band, but furthermore making use of the concept of activation energy. Basing on these theoretical considerations on crack initiation as well as on experimental evidence the growth of cracks under compression can be assigned to a one-sided deformation of the crack, that means entering of edge dislocations. The stress field of the crack is shown to obstruct to further glide of dislocations in its environement. Failure is assumed to appear, when this back stress can not prevent further dislocation glide and resulting crack growth, which in polycrystalline structure leads to an interaction with cracks in neighbouring grains and build up of a damage band. The agreement with experimental results on failure in rock salt is very good.

Crack initiation model of Stroh and of Argon and Orowan

One of the oldest models of crack initiation has been developed by A.N. Stroh [1]. It assumes dislocations piling up at a barrier on a single glideplane under the action of shear stress (see fig. 1) leading to a high stress concentration which finally will split the crystal.

fig. 1: Crack initiation model of Stroh [1].

The developed theory requires a critical relation for the minimal number of dislocations which are able to split the crystal by their common strong tension field. Taking γ for the surface energy and b for the burgers vector under the action of a shear stress τ_{eff} the necessary number n of dislocations results as

$$n\tau_{eff} \geq 12\frac{\gamma}{b}. \tag{1}$$

Inserting the material constants of salt for the uniaxial compression test when failure is observed at pressure loads of about 20 MPa it results

$$n \geq 500. \tag{2}$$

Though this is a very conservative astimation and in reality should be much higher, such a high number of dislocations on a single glideplane has never been observed in salt and furthermore seems to be very improbable in other materials as well [2].

---

[1] Now at the GSF, Forschungszentrum für Umwelt und Gesundheit, Institut für Tieflagerung, Braunschweig, FRG

[2] E.g. Matucha [2] observed about 15 dislocations on a single glideplane in NaCl by means of an electron microscopical method.

Physically more realistical seems to be a Modell of Argon and Orowan [3] founded upon the action of a whole glide band. Because dislocations of orthogonal glide bands (or a grain boundary or inclusion) act as strong barriers, such an active band can be blocked there, so leading to incompatible deformations or an approximately equidistant row of dislocations respectively. In a dislocation model with continuised distribution the resulting stress field has logarithmical singularities at the ends of the rows.

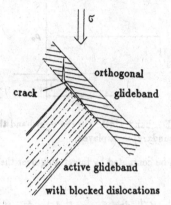

fig. 2: Simplified model of Argon and Orowan [3]: an active glide band is blocked by an orthogonal one and forms an - approximately equidistant - row of edge dislocations. Resulting from the high stress concentration a crack initiates.

Comparing the stress a distance of one burgers vector ahead of the row with the theoretical fracture stress and expressing the necessary number of dislocations by the local deformation, they got good agreement with experimental results.

Experimental investigations on single crystals

Artificially grown single crystals of NaCl of high purity and natural single crystals, received from the Asse salt dome and Gorleben test drilling cores, have been sawn or split to square specimens of about 1 to 3 cm lateral edge length. They have been polished mechanically and chemically. The crystals have been observed through a microscope during loading at constant creep rate ($\dot{\epsilon} \approx 1.7 * 10^{-4} \mathrm{sec}^{-1}$). The evolution of glide bands, consistent with the theory of Argon and Orowan, could be observed by means of polarised light. The evolution of cracks is shown in fig. 3.

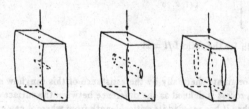

fig. 3: The crack inside a monocrystal in its first phase of evolution has a lancet like appearance, then widens in several small steps to the endfaces, and finally, after a pause, grows in hight again. In the near of the loading piston it comes to rest. The frontal endfaces deform inplane.

Etching of the surfaces after splitting did not reveal any great information because the density of the dislocations was so high that they could not be resolved by optical microscope. Etching of dislocations in specimens which where only deformed about 1%, did not show any sharp borders as it was assumed in the model of Argon and Orowan. As observation showed, the dislocations further have the ability to at least partly penetrate the orthogonal glideband.

Row of dislocations with smooth borderline

So consequently the theory has to be generalized for smooth density variations. In the sketch of fig. 4 the assumed density distribution inside a glideband is shown and the simplified model beside it.

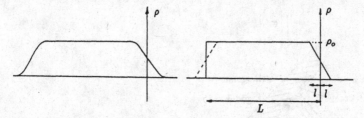

fig. 4: Sketch of the dislocation density ρ over glide band width. On the right hand the model of linear decrease of dislocation density (at one boundary) is displayed.

The stress on the line of the dislocation row can be computed by integrating over the stress field of the dislocations and the density.

$$\sigma(r) = \int \rho(x) \frac{D}{r-x} dx \approx \rho_o D \int_{-L}^{-l} \frac{dx}{r-x} + \rho_o D \int_{-l}^{l} \frac{1}{2} \frac{1-\frac{x}{l}}{r-x} dx$$
$$= \rho_o D \left[\ln \left| \frac{r+L}{r+l} \right| + \frac{1}{2} \ln \left| \frac{r+l}{r-l} \right| \left(1 - \frac{r}{l} \right) + 1 \right]$$

(3)

The nomenclature can be taken from the sketch. D is a material constant scaling the stress of a single dislocation ($\sigma \sim D/r$). The stress variation is plotted in fig. 5 for an arbitrary ratio of L/l.

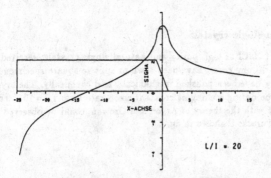

L/l = 20

fig. 5: Shape of stress over glide band width for the ratio $L/l = 20$.

In a first model the crack was assumed to originate thermally by the assistance of this (anyhow strong) stress field. The activation energy for this process is defined as the difference between the surface energy and the elastic energy release of the crack when it has reached its critical length from where it can grow by itself in an unstable manner under the action of the stress field. But as calculation by classical elasticity showed, only for unrealistic high ratios of $L/l \sim 10^7$ or very high dislocation densities respectively very great deformations which have not been observed in experiments, activation energies yielded of the order of $1eV$ which would give the stress field a reasonable chance to split the crystal.

So there must be a further stress concentration which is supposed to be given by dislocation dipoles or higher multipoles. Taking into account the high dipole stress fields, the achieved activation energies now show a very steep drop from very high values, preventing any crack opening, to even negative ones, standing for immidiate unstable crack initiation and growth. The resulting deformation for crack initiation, directly connected with the number of accumulated dislocations, is nearly independent of the assumed spacing of the dipole, but about a factor of 3 higher than the observed macroscopic deformation. But the calculation does not take account of higher multipoles leading to higher stresses and bases on a guessed ratio of $L/l = 20$ which never could be measured in a fundamental way. Furthermore the underlying shear is the shear inside the glideband which necessarily is higher than the macroscopic deformation and probably varying from case to case. So the result seems to be quite satisfying instead.

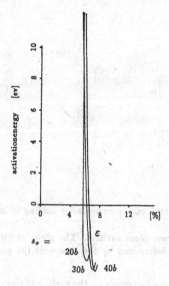

fig. 6: Activation energy as a function of local plastic shear $\varepsilon \sim \sigma_o$. The ratio $V = L/l$ amounts to ≈ 20.

Different to the model of Stroh here the crack has the possibility to grow in both directions being fed by the accumulated dislocations at its basis, till nearly the whole dislocation row is disappeared in the crack (see fig. 2). This might be ehanced even by glide band broadening.

Feature of damage

To get an insight into the appearence of creep fracture beyond the phenomenological stress - strain curves some cylindrical polycrystalline specimens of natural rock salt (achieved from the Federal Institute of Geoscience and Natural Resources (BGR) Hannover) which already had failed in a compression test, were cut by a saw parallel to the main pressure direction (fig. 7). The cone directly below the loading faces as a region of high hydrostatic pressure and low deviatoric stress naturally shows nearly no damage as can be seen in the sketch of fig. 7. It is confined by a narrow band of very dense cracks which by great parts already had coalesced to macrocracks. The cracks nearly without exception run transgranularly and are strongly aligned with the main loading direction (which coincides with the axis of symmetry). This is found to be right for the cracks in the remaining region as well, but here typically only one to three cracks are located within one grain. Looking through a microscope at the cracks, it could be verified that they follow the main {100} split planes of the individual crystals, but enabled the general alignement by changing the planes so getting a zig - zag form.

Further measurements in single crystals

As experiments on single crystals showed (see fig. 2), during loading there always were two opposing surfaces which remained plane while the two lateral surfaces bowed out. This is due to a strong blocking effect of active glide bands on orthogonal dislocations. This effect so leads to a deformation in one direction only (denoted as block gliding) such that one plane is uneffected from deformation. Block gliding is known to appear in single crystals as well as in the grains of polycrystalline material. Furthermore the crack

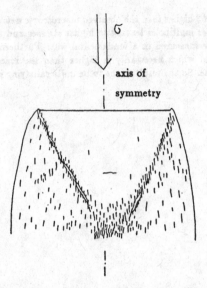

fig. 7: Sketch of the upper half of a sawn polycrystalline rock salt specimen after loading to failure.

developement was found to take place perpendicular to these plane surfaces. The plane of deformation (and of the connected perpendicular crack) could be predetermined by the choose of the geometrical dimensioning of the crystal.

These special circumstances enabled the measurements of displacements on the surface of single crystals with the aid of a laser - speckle -method. This method makes use of the surface bound speckles under laser light illumination, originating from interference effects on the rough surface. When the body to be measured is photographed before and after a certain deformation on the same film plate, the displacements and their inclination can be obtained in a successive optical evaluation. Reflection and interference of the laser light had to be rendered possible by coating the transparent surface with a graphit spray. In the numerical evaluation the local displacements of neighbouring surface points have been associated with simple shear. The degree and inclination of shear has been calculated by fitting it to the displacement data by means of a computer program. The results of this procedure are shown in the plots of fig. 8.

The hooklike symbols represent orientation and magnitude of shear in the lower parts of the single crystal. They show that shearing on the surface of the crystal (and obviously within it) on a {110} plane is very homogeneous, at least in a macroscopic point of view - the measured points are about 1 mm apart -, except for some small regions where deformation goes to zero, or for distortions in the orientation resulting from the fit. The region without deformation can be associated with blocking of the active glide bands. As can be seen from the second plot, exactly in that region a crack originates. It has to be realised that on one side of the crack the further deformation is zero or very small while on the other side it continues. So obviously more dislocations enter the crack and during following deformation the crack grows further.

The unilateral deformation can be explained by the great influence even small differences of hardening and stress have on deformation rate and the first plot doubtlessly shows hardening in this region.

Calculation of orientation and growth of cracks

As has been shown in fig. 7 the cracks in natural polycrystal rock salt are aligned with the direction of greatest pressure (or perpendicularly to lowest pressure respectively highest tension) but microscopically follow the individual split planes of the grain crystal resulting in a sig-zag crack. It can be shown by theory that the effect of the entering dislocations on the stress intensity factor (SIF) is the same irrespective of the relative position inside the crack surface and can be described by the theory of Stroh [1], where dislocations enter the crack only at its basis. So in fig. 9 for simplicity the dislocations, usually entering the crack in a broad glide band, are placed at the "basis" of the crack only.

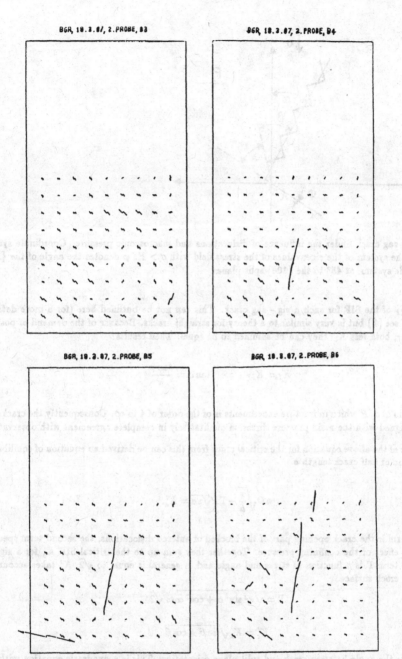

BGR, 10.3.87, 2.PROBE, B3

BGR 10.3.87, 2.PROBE, B4

BGR, 10.3.87, 2.PROBE, B5

BGR, 10.3.87, 2.PROBE, B6

fig. 8: Shearing here is displayed by hooklike arrows. In general block gliding apears at an angle of 45° (on {110} - planes). Furthermore cracks, as far as they were visible, are shown in the plots. Chronological sequence: B3, B4, B5, B6. Further explanations in text.

σ and P represent the main pressures in the eigen system of stress with $\sigma > P$, φ is the relative orientation of the glide system, α is the angle between crack and main pressure. In a cylindrical specimen σ corresponds with axial, P with confining pressure. During growing the components K_1 and K_2 of the SIF always switch between 0 and K_c. A theory now has been developed to describe the both components

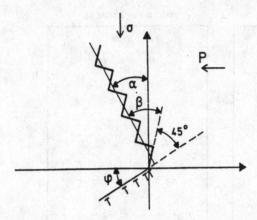

fig. 9: Zig-zag crack under the influence of dislocations and macroscopic pressure. Coordinate system is the system of the eigenvalues of the stress field with $\sigma > P$. φ denotes the angle of the {110} glide system, at 45° to the {100} split planes.

K_I and K_{II} of the SIF for such a zig - zag crack. This can not be outlined here (for a more detailed description see [4]) but is very similar to a theory for straight cracks. Because of the demand of positive K_I and K_{II}, both less K_c, they can be assumed to be equal. Then results

$$K_I = K_{II} \quad \Rightarrow \quad \tan\alpha = \frac{\tan\varphi}{\kappa}. \tag{4}$$

κ is the ratio of σ/P which in fracture experiments is of the order of 4 to ∞. Consequently the cracks are strongly aligned with the main pressure direction qualitatively in complete agreement with observation.

Making use of the above equation for the critical crack from this can be derived an equation of equilibrium as a function of half crack length a.

$$nD\sqrt{\frac{\pi}{a}} - \Gamma P\sqrt{\pi a} = K_c'. \tag{5}$$

The first term is the crack opening part of the blocked or entered dislocations, the second term specifies the closing effect of the confining pressure. Together they sum up to the critical SIF K_c' for a zig-zag crack. The term Γ is a function of stress and angle and in general is equal to $\sqrt{2}$, K_c' takes account of the greater crack surface.

$$\Gamma = \sqrt{\kappa^2 \sin^2\alpha + \cos^2\alpha} \approx \sqrt{2} \tag{6}$$

$$K_c' = K_c\sqrt{\sin\beta + \cos\beta} \tag{7}$$

β represents the angle between crack and split plane orientation. This is a quadratic equation with the solution

$$a = \frac{nD}{P\Gamma} + \frac{K_c'^2}{2\pi P^2\Gamma^2} \mp \frac{K_c'}{2\sqrt{\pi}P\Gamma}\sqrt{\frac{nD}{P\Gamma} + \frac{K_c'^2}{2\pi P^2\Gamma^2}} \tag{8}$$

for half crack length a. For pressure (positive P) only the negative sign is correct.

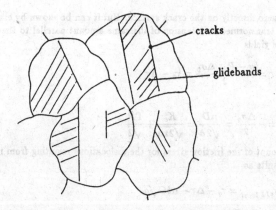

fig. 10: Representation of crack developement in polycrystalline structure in consequence of blocked glide
bands. Within one grain two cracks have the same alignement because of same stresses and dislocation glide geometry. So they cannot coalesce. But they can interact if they are in neighbouring
grains with different inclination thus becoming a greater crack with decreasing stability.

Failure of material
Theory for failure

Fig. 10 illustrates, that cracks can interact only if they have gained the length of grain diameter approximately. Two cracks within one grain can't coalesce because they have the same alignement.

In the following analysis the simplification has been made that the crack is parallel to main pressure
direction. A perpendicular pressure P tries to close the crack while dislocations enter it under 45° at a
global shear stress $\tau_g = (\sigma - P)/2$. Crack growth is possible as long as the stresses inside the glideband
allow further glide. Therefore the stresses of the crack have to be taken into account as well. Failure
should be determined by the capability of the cracks to grow to the grain boundaries or beyond them
where they can interact with the cracks of a (tilted) neighbour grain and coalesce.

From the simple relation of local shear with the number n of the passed (or blocked) dislocations with
burgers vector b, distributed over the glide band width L

$$\epsilon = \frac{nb}{L} \tag{9}$$

in conjunction with the equilibrium condition (5) the necessary local shear can be valuated.

$$\epsilon \approx \frac{nb}{L} \geq \frac{b\sqrt{2a}}{LD} \left(\frac{K_c}{\sqrt{2\pi a}} + \frac{\Gamma P}{\sqrt{2}} \right) \tag{10}$$

Crack length $2a$ can be equated with grain diameter $\Phi \approx 0.7$ cm. Setting $K_c \approx 5 MPa\sqrt{cm}$ [5]and
$L \approx 0.9$mm

$$\epsilon \geq \frac{9.76MPa + 2.86P}{367MPa} \geq 3\% \tag{11}$$

This is based on the assumption that not only the dislocations of a single glide band contribute to crack
growth, but a whole troop of them instead, being blocked at a common barrier.

The stresses of the crack opposing to a further running in of dislocations are smallest in the middle of
the crack so that they are a measure for further growth. Following Stroh [1] they can be calculated as

$$\Delta\sigma_y = -\sqrt{2}\frac{nD}{a}. \tag{12}$$

The lateral pressure P is reduced to sero directly on the crack surface. But it can be shown by elasticity theory that the lateral pressure P is transformed to a tension of the same amount parallel to the crack. Thus the maximal acting shear stress yields

$$\tau = \frac{\sigma - P + \Delta\sigma_y}{2} = \tau_g - \Delta\tau, \tag{13}$$

with

$$\Delta\tau = -\frac{\Delta\sigma_y}{2} = \frac{nD}{\sqrt{2}a} = \frac{K_c}{\sqrt{2\pi a}} + \frac{\Gamma P}{\sqrt{2}}. \tag{14}$$

The effective shear stress, taking account of the friction stress for the dislocations resulting from mutual elastic interaction (e.g. see [6]) so results as

$$\tau_{eff\ local} = \tau_g - \Delta\tau - \alpha b G\sqrt{\rho_t}, \tag{15}$$

ρ_t is the total dislocation density, $\alpha \approx 0.4$ is a constant and G shear modulus. If τ_{eff} locally (at the crack) gets sero, no further dislocations run into the crack and it can't grow. Because backstress gets smaller! with growing crack length (see eq. (12)) the crack should keep on growing when it reaches critical length and should be able to coalesce with neighbouring cracks. So eq. (15) for $\tau_{eff} = 0$ is another necessary condition for stability or failure respectively.

The limit of stability in loading tests among others is characterised by the transition of stationary creep to failure. The total dislocation density in this case just has reached its stationary limit described by

$$\alpha b G\sqrt{\rho_t} = z\tau_g \tag{16}$$

with $z \approx 0.65$ (see e.g. [7, 8, 6]). So the critical global shear stress can be calculated.

$$\tau_{eff\ local} = 0 \quad \Rightarrow \quad \tau_g = \frac{\Delta\tau}{1 - z} = \frac{3.4\text{MPa} + P}{0.35} = 9.76\text{MPa} + 2.86P \tag{17}$$

Comparison with experimental results

Transforming the stresses to the octahedral stress τ_o and the hydrostastic pressure σ_o for the compression, torsion and extension case (Lode parameter m = −1, 0, +1), a comparison with the experiments of Hunsche/Albrecht [9] on the limit of stability shows very good agreement for lower shear stresses (see fig. 11).

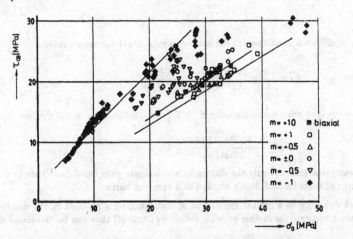

fig. 11: Extend of the theoretical limit of stability in rock salt for the three loading pathes m= −1, 0, +1, in comparison with the experimental triaxial results of Hunsche/Albrecht [9].

The main reason for the deviation at high shear stresses should essentially be found in the fact that the cross section in the region of the damage zone the cross section is mainly given by the end faces of the cylindrical specimen. So the true shear stresses should be higher there than the specimen mean values suggest which are the basis of fig. 11. So for great deformations as they are found for high hydrostatic pressures at failure the true values for shear stress are higher than shown in fig. 11 and should result in a closer agreement with the linear relation (17).

Experimental investigation on failure developement

Because coalescence of cracks is essentially determined by grain structure, an experiment for the observation of crack evolution in a polycrytalline specimen has been conducted. For this purpose half cylindrical specimens were put into a specially designed apparatus. Through a high security glass which simulated the existence of the other halfe, the specimen could be examined in its "inner parts".

The specimens have been tempered at 150 °C and 40 MPa hydrostatic pressure for a week to remove any preexisting damage. The plane of observation has been darkened with graphite and lightened from the back. In experiment the displacement controlled rate of deformation usually amounted to 10^{-4}sec^{-1}. The deformation periodically has been interrupted by quick deloading and loading again to measure elastic stiffness of the whole specimen (see fig. 12).

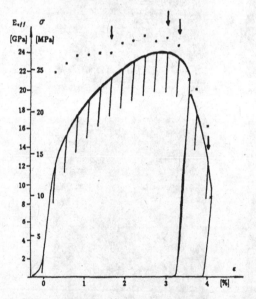

fig. 12: Stress as a function of deformation, interrupted by short phases of quick deloading, and from that calculated (fictitious) elastic modulus (small crosses).

In result:
Cracks always initiated in sudden bursts and then didn't show appreciable change during further raising of load. Only the number of cracks increased strongly and led to a general milkiness of the (uncovered) specimens. This is in accordance with equations (10,11) allowing initiation of cracks but which not necessarily fullfill condition (17). When they are too small, their backstress will prevent further deformation in their environment, so fixing the cracks except that the global shear stress rises sufficiently. The path of the cracks was always (with only very few exceptions) transcrystalline. This was true even for some specimens which have not been healed by tempering and showed great damage (but without directional preference), or have been soaked with oil along grain boundaries and cracks before as a result of inadequate treatment. These "bad" specimens even showed the same failure stress as the good ones.

From the inclination of the (elastic) deloading phases a fictitious E-modulus has been calculated which also is incorporated in fig. 12. The evolution of elastic stiffness even shows a small increase till failure (exceeding of maximum load), although the specimen nearly completely has been filled with cracks (see fig. 13a, the moments of photographing are indicated in fig. 12 by arrows). The increase of stiffness shurely is not real and should mostly account for the inelastic portion of deformation which can not be excluded, especially in the little hardened specimen.

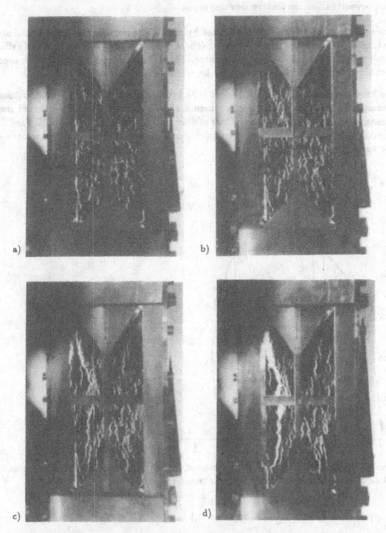

fig. 13: Run of crack evolution on the blackened surface of a half cylindrical specimen during loading (times of photographes are marked in fig. 12 by arrows).

Only after surpassing load maximum a dramatic loss of stiffness sets in. This decrease clearly is connected with the coalescence of cracks (fig. 13b,c) leading to a paring of the outer parts of the (half-)cylindrical specimen (fig. 13c,d) and so does not show more than the loss of bearing cross section. The reduction of stiffness to about 1/3 suggests a reduction of cross section by the same amount, in accordance with appearance (fig. 13d).

The shape of the fracture zone has a distinct influence on the maximum load. A symmetric fracture zone was connected with an about 15% higher failure stress than in case of an asymmetric (diagonal) one. This can be explained from the remaining cross section which beares the main portion of the load in the symmetric case, while in the asymmetric one there is only friction except for some remaining connections. Eventually this is a further important reason for the decrease of maximal stress for high hydrostatic pressure (relative to the theoretical limit of equation (24)), because the specimens are taking up a great deformation then and asymmetry evolves more easily. In the frame of these experiments only one asymmetric fracture appeared (among 8).

The cracks don't lead to an appreciable loss of elastic stiffness because they are mostly parallel with the main pressure axis and hence are only poorly influenced by it. Measuring the lateral stiffness by stress variation (which was not done here), the result shurely would have been very different. The specimen so developes a strong elastic anisotropy. This phenomenon might be a help when the state of damage shall be determined, e.g. below-ground.

Acknowledgement: The above investigation is part of a research project financed by the Minister of Research and Technology, FRG. Without the help and assistance of Dr. Alheid (Federal Institute Geoscience and Natural Rescources (BGR), Hannover), Dr. U. Hunsche (BGR), Mr. F. Schulte (BGR), Dr. F. Bäßmann of the Institute of Metrology, Univ. Hannover, Dr. K.P. Groß[1] and Dipl.-Ing. W. v. Cramon-Taubadel of the Institute of Concrete Constructions this work would not have been possible and is kindly acknowledged.

References

[1]   A.N. Stroh: The Formation of Cracks as a Result of Plastic Flow. Proceedings of the Royal Society, A223, 404-414, 1954

[2]   K.-H. Matucha: Elektronenmikroskopie von Gleitlinien auf NaCl. physica status solidi 26, 291-310, 1968

[3]   A.S. Argon, E. Orowan: Crack Nucleation in MgO Single Crystals. Philosophical Magazine Serial 8, 9 (102), 1023-1039, 1964

[4]   U. Heeman: Transientes Kriechen und Kriechbruch im Steinsalz. Doctoral thesis, Hannover, 1989

[5]   M. Breucker: Der Einfluß plastischer Verformungen auf die Rißausbreitung in Kochsalzeinkristallen. Doctoral thesis, Dortmund, 1983

[6]   U. Heemann, E. Stein: to be published in Philosophical Magazine.

[7]   W. Blum: private communication

[8]   N.Dieckmann, U. Hunsche, D. Meister: Über das geomechanische Verhalten von Steinsalz bei erhöhten Temperaturen. Zeitschrift der deutschen geologischen gesellschaft 137, 29-46, 1986

[9]   U. Hunsche, A. Albrecht: Results of True Triaxial Strength Tests on Rock Salt. International Conference on Fracture and Damage of Concrete and Rock, Vienna, Austria, July 4-6, 1988

---

[1] Now at the University of Bremen

VOLUME CHANGE AND ENERGY DISSIPATION IN ROCK SALT DURING TRIAXIAL FAILURE TESTS

U. HUNSCHE
Federal Institute for Geosciences and Natural Resources
W-3000 Hannover, Stilleweg 2, Germany

ABSTRACT

A constitutive equation for natural rock salt should take into consideration stress, strain, volume change until failure, as well as failure itself, under various conditions.

Therefore, equipment for precise true triaxial tests on cubic specimens and a special test procedure for quasistatic failure tests on rock salt have been developed. Deformation, volume change, and irreversible stress work were calculated from the empirical data as a function of time. Failure strength was determined as well.

The results of seven compression tests at various mean stresses are discussed. The volume change at failure is between 2 and 7 % in these tests. The calculated stress work at failure that is related to volume change (opening of micro-cracks) is between 0.5 and 1 MJ/m², whereas the stress work related to the change of shape (creep, initiation and propagation of microcracks) is dependant on the mean stress and is between 1.5 and 9 MJ/m².

The results will be used to derive an elasto-viscoplastic constitutive equation.

INTRODUCTION

Knowledge of the geomechanical behaviour of rock salt is important for the dimensioning and safety analysis of mines,

caverns, and repositories for toxic and radioactive wastes in
salt domes. Therefore, much experimental and theoretical work
has been carried out in this field during the last two decades. Test results and laws for the creep behaviour have been
published, e.g. by Heard (1977), Menzel & Schreiner (1977),
Albrecht & Hunsche (1980), Carter & Hansen (1983), Munson &
Dawson (1984), Langer (1984), Wawersik & Zeuch (1986), Wawersik (1988). Failure has been described, e.g. by Dreyer (1974),
Georgi, Menzel & Schreiner (1975), Wawersik & Hannum (1980),
Wallner (1983), Diekmann, Hunsche & Meister (1986), Desai &
Varadarajan (1987), Hunsche & Albrecht (1990), Hunsche
(1990a,b).

None of published equations for failure can be considered
as really satisfactory because the relevant deformation and
failure mechanisms are not yet fully understood. This is important, however, especially for repositories in salt domes.

PURPOSE OF THIS PAPER

Nearly all failure laws describe only the stresses at failure
but not the related deformations and volume changes. Both are
important for model calculations, however. Therefore, a rig
for true triaxial tests on cubic rock salt specimens, a soph-
isticated test procedure and an evaluation method have been
developed to provide reliable data for derivation of a constitutive equation for rock salt that will include stress,
strain, and volume change until failure, as well as failure
itself, as a function of time. A possible basis for further
analysis is the method of Cristescu (1989). It contains a procedure for obtaining an appropriate elasto-viscoplastic constitutive equation. Data on volume change, irreversible stress
work due to deformation (energy dissipation) during a test,
and failure strength, which will be discussed in this paper
together with a description of the tests, are required to
derive this equation.

The total stress work W^I_g (i.e energy used for deformation
minus the elastic energy) consists of two parts: The work W^I_v
related to volume change and the work W^I_d used for changing
shape.

$$W^I_v(t) = \int_0^t \sigma \cdot d\varepsilon v(t) - \sigma^2/(2 \cdot K) \qquad (1)$$

$$W^I_d(t) = \int_0^t \mathbf{s} \cdot \dot{\mathbf{e}} \cdot dt - \mathbf{s} \cdot \mathbf{s} /(4 \cdot G) \tag{2}$$

$$W^I_g = W^I_v + W^I_d \tag{3}$$

$\sigma = \sigma(t)$
$\mathbf{s} = \sigma - \sigma 1$: tensor of stress deviator
$\mathbf{e} = \varepsilon - \varepsilon 1$: tensor of strain deviator
$G = 11.8$ GPa (shear modulus)
$K = 21.7$ GPa (compression modulus)

W^I_v is caused by volume change and is consequently related to the closure or opening of microcracks, which finally grow together and cause failure. W^I_v can be used as a damage parameter, too. W^I_d is caused by the stress deviator and is related to creep, initiation and propagation of microcracks, as well as the further opening of microcracks. W^I_d supplies the energy consumed by microcrack opening, which is expressed by negative values for W^I_v (energy gain). Therefore, W^I_g only describes creep and microcrack initiation and propagation, which cannot be distinguished without model calculations.

The variables used to describe stress, deformation, and volume change are shown in Table 1. Compression is indicated by a positive value. The Lode parameter m for stress geometry has not been used because only compression tests are considered.

TABLE 1
Variables discussed in this paper

$\sigma = 1/3(\sigma_1 + \sigma_2 + \sigma_3)$	mean stress, octahedral normal stress
$s_i = \sigma_i - \sigma$	stress deviator
$\tau = [1/3(s_1{}^2 + s_2{}^2 + s_3{}^2)]^{\frac{1}{2}}$	octahedral shear stress
$m = \dfrac{3\,s_2}{s_1 - s_3}$; $s_1 > s_2 > s_3$	stress path, Lode parameter
$\varepsilon_i = (l_0 - l_i)/l_0$	relative deformation
$\varepsilon v = (l_1 \cdot l_2 \cdot l_3)/V_0$ + correction	relative volume change
$\varepsilon_{eff} = (\varepsilon_1{}^2 + \varepsilon_2{}^2 + \varepsilon_3{}^2)^{\frac{1}{2}}$	effective deformation

EQUIPMENT AND TEST PROCEDURE

To obtain a reliable constitutive equation, experiments have
to be performed with high precision. The test rig used for
these tests consists of a rigid frame in which three pairs of
double-acting pistons are arranged orthogonally opposite each
other about the center of the frame. A load is applied to the
cubic specimens via square steel plates (platens) which can be
heated. A detailed description of the apparatus has been published by Hunsche & Albrecht (1990) and Hunsche (1990a). The
most important specifications are as follows:

- Maximum force (per axis):   2000   kN
- Sample size (edge length), this study: 53   mm
                              maximum: ca. 200   mm
- Platen size in standard tests: 50   mm
- Temperature   maximum: 400   °C
- Friction of piston: < 1 % of load
- Independant load or deformation control along the 3 axes
- Loading rates in standard tests:
      hydrostatic phase: $\dot{\sigma}$ = 7.6 MPa/min
      deviatoric phase: $\dot{\tau}$ = 21.4 MPa/min
- Deformation rate at failure, standard tests: ≈ 0.007 s^{-1}
- Digital data acquisition, standard interval: 1 s
- Lubrication with paraffin wax or graphite.

All tests in this paper are load controlled, the stress
is first applied hydrostatically. When the desired σ level has
been reached, the three principal stresses, and thus τ, are
changed linearly with time so that σ is held constant (deviatoric phase). This is done until the octahedral shear stress τ
reaches the fracture strength of the sample and fracturing
occurs (Fig. 1). Fracture strength τ_B is defined as the local
maximum of τ in a test.

Deformation and forces in the three principal directions
are measured during the tests. The equipment used for these
measurements have been very precisely calibrated. Especially
the calibration for the determination of the volume changes εv
has been very difficult. It takes into account the deformation
of the platens and pistons, and the small amount of salt which
is squeezed out at the edges of the sample (Fig. 2). A
precision of better than 0.3 % has been achieved for εv.

Only the results of seven compression tests are given in
this paper. They were carried out at 30°C using different
hydrostatic stress levels σ (Fig. 1). This is only a small part
(series 9) of our test program. The samples were prepared from
rock salt from a salt dome in Lower Saxony (Germany). The rock

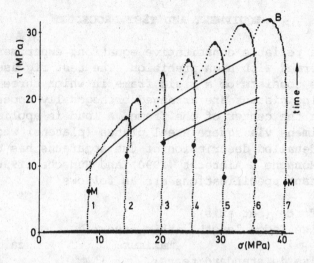

Figure 1: Seven true triaxial compression tests at 30 °C;
spacing of points: 1 s; M: minimum volume; B: failure

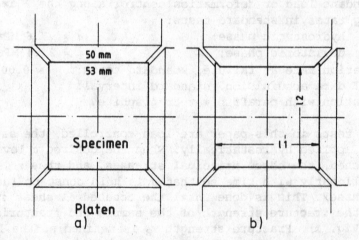

Figure 2: Cross section of the arrangement of specimen and
platens (to scale); a) at beginning; b) at failure (test 3)

salt consisted of very clean halite with less than 1 wt% impurities, mainly $CaSO_4$ (anhydrite). Grain size is about 10 mm.

EVALUATION AND RESULTS

The principal stresses σ_i and principal strains ε_i were calculated taking into account the changing dimensions of the sample. These values were used to calculate τ and σ (Fig. 1).

A conservative equation for failure strength has been derived on the basis of eight test series (215 tests) of the same kind (Hunsche & Albrecht 1990, Hunsche 1990a,b). Failure strength τ_B has been described as a function of σ, m, and temperature T. The resulting function for compression and extension is shown in Figure 1 for comparison. The failure strengths obtained in this test series lie within the upper part of the range obtained in the previous test series.

The three principal strains, ε_i, are the basis for the calculation of the volume change εv. Examples of the strains and volume change as a function of time or τ are shown in Figure 3. Failure stress τ_B and the point with minimum volume εv_M are marked. A considerable volume compaction of 0.5 to 1 % occurs during the hydrostatic loading phase. This is caused by loosening of the specimen during coring and machining. A small initial decrease of the volume can be observed during the deviatoric phase. The volume increases nonlinearly after the minimum point M due to microcrack opening and reaches several percent at failure (see Table 2).

In the next step, the irreversible stress work for volume change and for shape change have been calculated using Equations (1) and (2), where integration starts after compaction at the beginning of the deviatoric phase. The results for the three kinds of irreversible stress work as a function of time or of τ are presented in Figure 4. It has to be stated that the W^I are nonlinear functions of t or τ.

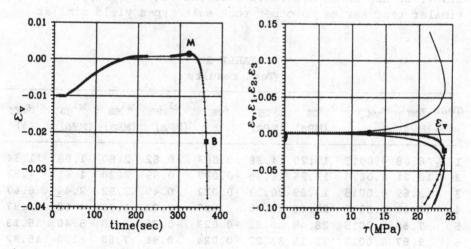

Figure 3: Deformations in test 3
a) Volume change εv vs. time
b) Volume change εv and deformations ε_1, ε_2, ε_3 vs. τ

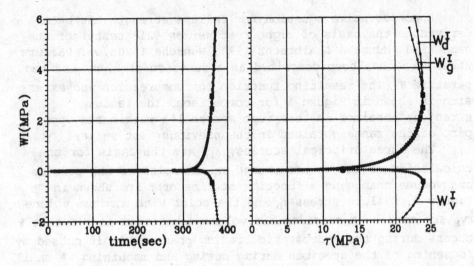

Figure 4: Irreversible stress work in test 3
  a) W^I vs. time; b) W^I vs. τ; ●: minimum volume; ■: failure

A summary of the results of all seven tests is given in Table 2. The initial volume decrease at minimum point M is εv_M = 0.1 to 0.2 % at an octahedral shear stress τ that is half of the failure strength τ_B or less. The volume increase at failure εv_B is of great interest; it is plotted in Figure 5. Obviously the volume increase is much larger at small mean stresses than above about 18 MPa, where it is between 2 and 2.5 %. It has to be mentioned that the ongoing analysis of similar test series on other rock salt types yield similar

TABLE 2
Test results

Test	τ_{OM} (MPa)	ε_{vM}	τ_{OB} (MPa)	σ_{OB} (MPa)	ε_{vB}	W^I_{vB} (MPa)	W^I_{dB} (MPa)	W^I_{gB} (MPa)	ε_{eff} (%)
1	6.08	0.0015	17.29	14.38	-0.063	-0.82	2.80	1.98	11.34
2	10.21	0.0010	19.95	16.08	-0.037	-0.59	2.20	1.61	7.52
3	12.69	0.0015	23.89	20.33	-0.022	-0.49	2.92	2.44	8.47
4	12.11	0.0015	25.88	24.40	-0.027	-0.72	4.90	4.18	12.97
5	7.68	0.0015	28.48	28.82	-0.023	-0.75	6.18	5.40	15.13
6	9.87	0.0017	31.15	33.22	-0.024	-0.91	7.80	6.90	15.82
7	7.06	0.0020	32.10	37.47	-0.021	-0.93	9.83	8.90	18.87

M: values at minimum volume; B: values at failure

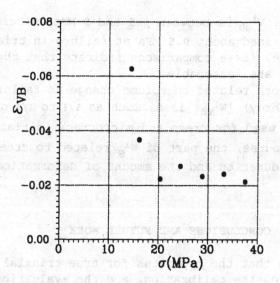

Figure 5: Volume increase at failure, $-\varepsilon v_B$, vs. mean stress σ

results. They are also in good agreement with the results obtained by Wawersik & Hannum (1980) (εv_B = 2.8 to 7 %). The volume increase at failure for different kinds of hard rock is between 1 and 3 % (e.g. Cristescu, 1989).

The different kinds of irreversible stress work (Equations 1, 2, & 3) at failure are shown in Figure 6. The energy gain W^I_{vB} lies between 0.5 and 1 MPa (1 MPa = 1 MJ/m³). This is comparable to the energies measured by Cristescu (1989, p.160)

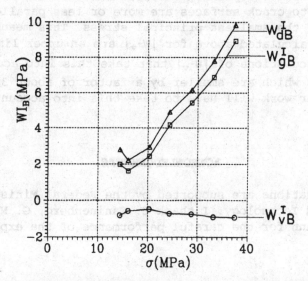

Figure 6: Irreversible stress work W^I at failure vs. σ

in uniaxial tests. W^I_{gB} is between 1.5 and 9 MPa; Alheid &
Rummel (1976) obtained about 0.5 MPa at failure in triaxial
tests on sandstone. These comparisons indicate that the results shown above are reasonable.

The stress work related to volume change at failure
(crack opening energy) $|W^I_{vB}|$ is as much as 1/3 to 1/9 of the
total energy W^I_{gB} used for creep and microcrack initiation and
propagation. Of course, the part of W^I_g related to creep is a
function of test duration and the amount of deformation.

CONCLUSIONS AND FUTURE WORK

It has been shown that the apparatus for true triaxial tests,
the test procedure, the calibration, and the evaluation method
yield reliable results. The volume changes and the irreversible stress work derived in the triaxial tests on rock salt
are reasonable. Therefore, the results will be a sound basis
for further work.

Further analysis will investigate the influence of load
path, temperature, load or deformation rate, and rock salt
type on the experimental results. The aim is to derive a constitutive equation for strength, deformation, and volume
change of rock salt. A preliminary formulation has already
been made by Cristescu (1989) and Cristescu & Hunsche (1991).

The calculation of W^I_v by Equation (1) presumes that crack
opening occurs isotropically. It is, however, well known that
the normals to crack surfaces are more or less parallel to the
direction of the smallest principal stress. This means that
the values calculated above for $|W^I_v|$ are an upper limit. Pre-
liminary calculations of $|W^I_v|$ that take this into account
yield values which are smaller by a factor of about 3. Therefore, further work will have to take this into account also.

ACKNOWLEDGEMENTS

The investigations are supported by the Federal Ministry of
Research and Technology. I thank C. Caninenberg, G. Notbohm,
and R. Baumann for the careful performance of the experiments.

REFERENCES

ALHEID & RUMMEL (1976): Energieregelung bei Druckversuchen an Gesteinen.- Sonderforschungsbereich 77 -Felsmechnik-, Universität Karlsruhe, Jahresbericht 1975, Karlsruhe.

ALBRECHT, H. & U. HUNSCHE (1980): Gebirgsmechanische Aspekte bei der Endlagerung radioaktiver Abfälle in Salzdiapiren unter besonderer Berücksichtigung des Fließverhaltens von Steinsalz.- Fortschr. Miner., v. 58 (2), 212 - 247; Stuttgart.

CRISTESCU N. (1989): Rock Rheology.- Kluwer Acad. Publ., Dordrecht.

CRISTESCU N. & U. HUNSCHE (1991): A constitutive equation for salt.- In: Proc. of 7. Int. Congr. on Rock Mech., Aug. 1991, Aachen, Germany. (In preparation)

CARTER, N. L. & F. D. HANSEN (1983): Creep of rock salt.- Tectonophys., v. 92, 275 - 333.

DESAI, C. S. & A. VARADARAJAN (1987): A constitutive model for quasi-static behavior of rock salt.- J Geophys. Res., v. 92, B 11, 11445 - 11456.

DIEKMANN, N., U. HUNSCHE & D. MEISTER (1986): Über das geomechanische Verhalten von Steinsalz bei erhöhten Temperaturen.- Z. dt. geol. Ges.; v. 137, 29 - 56; Hannover.

DREYER, W. (1974): Gebirgsmechanik im Salz.- 205 p.; F. Enke Verlag, Stuttgart.

GEORGI, F., F. MENZEL, W. SCHREINER (1975): Zum geomechanischen Verhalten von Steinsalz verschiedener Salzlagerstätten der DDR, Teil I: Das Festigkeitsverhalten.- Neue Bergbautechnik, Jg. 5, no 9, 669-676.

HEARD, C. H. (1972): Steady-state flow in polycrystalline halite at pressure of 2 kilobars.- In: Geophysical Monograph Series v. 16, Flow and Fracture of Rocks; Editors: H. C. Heard, I. Y. Borg, N. L. Carter & C. B. Raleigh; p. 191 - 209. AGU, Washington.

HUNSCHE. U. (1990a): On the fracture behavior of rock salt.- In: Constitutive Laws of Plastic Deformation and Fracture, Proc. 19th Canadian Fracture Conf., Ottawa (Canada) 1989; Editors: Krausz, A. S., J. I. Dickenson, J.-P. A. Immarigeon,

& W. Wallace; 155-163. Kluwer Acad. Publ., Dordrecht.

HUNSCHE, U. (1990a): A failure criterion for natural poly-cristalline rock salt.- In: Advances in Constitutive Laws for Engineering Materials, Proc. Int. Conf. on Constitutive Laws for Eng. Mat. (ICCLEM), Chongquing (China) 1989; Editor: Fan Jinghong vol. II, 1043 - 1046. Int. Academic Publ., Beijing.

HUNSCHE, U. & H. ALBRECHT (1990): Results of true triaxial strength tests on rock salt.- Engineering Fracture Mechanics, v. 35, No. 4/5, 867 - 877.

LANGER, M. (1984): The rheological behavior of rock salt.- In: The Mechanical Behavior of Salt, Proc. of the First Conf., University Park (USA) 1981; Editors: Hardy, H. R. Jr. & M. Langer; p. 201 - 240. Trans Tech Publ., Clausthal.

MENZEL; W. & W. SCHREINER (1977): Zum geomechanischen Verhalten verschiedener Salzlagerstätten der DDR, Teil II: Das Verformungsverhalten.- Neue Bergbautechnik, Jg. 7, no 8, 565-571.

MUNSON, D. E. & P. R. DAWSON (1984): Salt constitutive modeling using mechnism maps.-In: The Mech. Behavior of Salt, Proc. of the First Conf., Univ. Park (USA) 1981; Editors: Hardy, H. R. Jr. & M. Langer; p. 717 - 737. Trans Tech Publ., Clausthal.

WALLNER, M. (1983): Stability calculations concerning a room and pillar design in rock salt.- Proc. 5th Int. Congr. on Rock Mechanics, Melbourne 1983, v. II, p. D9 - D15, A. A. Balkema, Rotterdam.

WAWERSIK, W. R. (1988): Alternatives to a power-law creep model for rock salt at temperatures below 160 C.- In: The Mechanical Behavior of Salt II, Proc. of the Second Conf., Hannover (FRG) 1984; Editors: Hardy, H. R. Jr. & M. Langer; p. 103 - 128. Trans Tech Publ., Clausthal.

WAWERSIK W. R. & D. W. HANNUM (1980): Mechanical behavior of New Mexico rock salt in triaxial compression up to 200°C.- J. Geophys. Res., v. 85, B2, 891 - 900.

WAWERSIK,W. & D. H. ZEUCH (1986): Modelling and mechanistic interpretation of creep of rock salt below 200°C.

NON–STATIONARY CAVITY NUCLEATION – A LIMITING PROCESS FOR HIGH–TEMPERATURE STRENGTH AND SUPERPLASTICITY IN CERAMICS

HERBERT BALKE, HANS–ACHIM BAHR AND WOLFGANG POMPE
Central Institute of Solid State Physics and Material Research
Helmholtzstr. 20, Dresden 0–8027, Germany

ABSTRACT

The non–stationary nucleation theory is used to explain the formation of creep pores at grain boundaries as a result of transient stress concentrations in rapid sliding processes. The numerically calculated critical stresses for persistently growing pores are the base for the analytical determination of critical stresses at other values of temperature, surface energy, and grain size. By means of the numerical results and the analytical transformation rule for the critical stresses, limits for the deformation rates are given below which superplasticity is possible without pore development.

INTRODUCTION

Pores at grain boundaries are one of several potential sources of damage in poly-crystalline materials under high temperature creep conditions [1]. Pore formation proceeds by nucleation. It has been shown that a stationary nucleation theory would require exceedingly high levels of stress for pore growth [2] and therefore fails describing reality. A possible explanation for this may be derived from the fact that the stationary theory does not take into account transient local stress concentrations. Short–lived normal stress concentrations at grain boundaries may arise from sudden sliding at shear–stressed boundaries and subsequent boundary diffusion [2]. In [3], attention is drawn on pore formation as a non–stationary process under transient stress states. A solution of the problem by means of a quasi–stationary approximation and analytical estimates was attempted [4,5]. However, the above mentioned discrepancy could not be completely removed by this approach. This had been the reason for solving the non–stationary problem numerically [6,7], taking into account the transient stress concentrations after [2]. The result was satisfactory as it provided realistic values of critical stress for pore growth.

In this paper we discuss the implications of the numerical results of the non–stationary nucleation problem [6,7] for criteria of superplasticity. Superplasticity in the sense of the model in [2] means the fact that there is no pore growth due to transient stress concentrations below a critical applied load. Use is made here of a transformation law which finds the critical stress at other temperatures, surface energies, and grain sizes from the particular numerical results. The stresses found in this way serve as limit loads for Coble creep.

TRANSIENT STRESS CONCENTRATION MODEL

Let the polycrystalline material be loaded by an applied stress σ_∞. Among the grain boundary configurations there are such ones as drawn schematically in Figure 1 [2].

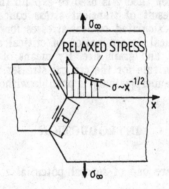

Figure 1. Plane model for transient stress concentration.

An assumedly sudden sliding at t = 0 along shear–stressed grain boundaries gives rise to transient stress concentrations at normally stressed boundaries of the type [2]

$$\sigma(x,t) = \kappa \, \sigma_\infty \, d^{1/2} \cdot f(x/(\alpha t)^{1/3})/(\alpha t)^{1/6} \quad , \quad x \geq 0$$

$$0.18 \leq \kappa \leq 0.41 \quad , \quad \alpha = G\Omega D_b\delta_b/(2kT(1-\nu)) \tag{1}$$

d – graine size, G – shear modulus, ν – Poisson's number, k – Boltzmann's constant, Ω – atomic volume, T – temperature, $D_b\delta_b$ – grain boundary diffusion coefficient. G and $D_b\delta_b$ are dependent on the temperature. The function $f(z)$ is given by

$$f(z) = \left\{ \begin{array}{ll} 0.74 & z = 0 \\ z^{-\frac{1}{3}} & z \rightarrow \infty \end{array} \right\} \quad \text{for} \qquad (2)$$

The initial stress singularity ($z = \infty$, solid line in Figure 1) relaxes by grain boundary diffusion (dashed line). Figure 2 shows the calculated stress distribution at the grain boundary for several times (solid lines) [7].

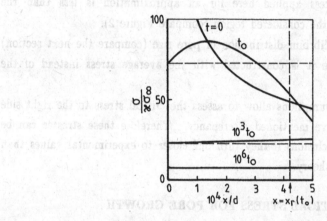

Figure 2. Decay of stress concentration at grain boundary.

In order to apply a transient nucleation theory without undue effort, nucleation is assumed to start at a certain time t_0 [7]. Also nucleation is assumed to be confined to a fraction of the grain boundary between the junction at $x = 0$ and a certain $x_r(t_0)$. That region is chosen such that it contains a sufficiently large number of atoms, and that the stress within the region does not much depend on position (horizontal dashed lines in Figure 2). A value x_r which meets these conditions is found by

$$x_r(t_0)/(\alpha t_0)^{1/3} = k_1 \quad , \quad x_r(t_0) = k_2 \, \Omega^{1/3}$$

$$k_1 \approx 4 \quad , \quad k_2 > 5 \qquad (3)$$

From (1),(2) and (3) we find

$$(\alpha t_o)^{1/3} = \Omega^{1/3} k_2/k_1 \quad , \quad \sigma(t) = 0.74 \kappa \sigma_\infty d^{1/2}/(\alpha t)^{1/6} \tag{4}$$

Figure 2 has been drawn for Al_2O_3 with $k_1 = 4$, $k_2 = 12$, $d = 10^{-5}m$ and $\Omega = 42.5 \cdot 10^{-30} m^3$ [6,7].

As a consequence of the restrictions (3), $k_2 = 12$, four contributions to nucleation are neglected so that the critical stress comes out too high:

— Nucleation up to t_o is neglected.

— Nucleation beyond x_r is neglected (however, nucleus sizes greater than x_r are admissible).

— The constant stress applied here for an approximation is less than the average stress in the considered region (compare Figure 2).

— The average equilibrium distribution of pore size (compare the next section) is too small since it is constructed with the average stress instead of the variable stress.

All these simplifying assumptions allow to assess the critical stress to the right side with respect to the above mentioned discrepancy. Therefore these stresses can be used for subsequent conclusions if they turn out closer to experimental values than those of the stationary theory.

CRITICAL STRESS FOR PORE GROWTH

Nucleation is governed by the equilibrium distribution of pores of size n (measured in atomic volumes)

$$Z_n^e = Z_0 \exp(-\Delta G_n/kT) \tag{5}$$

where the free enthalpy

$$\Delta G_n = -n\Omega\sigma(qt) + (18\pi\gamma^3\zeta)^{1/3}(n\Omega)^{2/3} \quad , \quad q = 2\pi D_b\delta_b/\Omega \tag{6}$$

is the decisive quantity. Z_0 is the number of available atoms

$$Z_0 = d \cdot x_r(t_0)/\Omega^{2/3} = k_2 d/\Omega^{1/3} \tag{7}$$

γ is the surface energy, and ζ is a pore shape parameter [2].

The strongly time–dependent potential (6) (compare Figure 3) affects the differential equation for the pore size distribution $Z_n(t)$ via equilibrium distribution (5) [4,5,6]

$$dZ_n(t)/dt = I_{n-1}(t) - I_n(t)$$
$$I_n(t) = q \, (Z_n(t) - Z_{n+1}(t) \cdot Z_n^e/Z_{n+1}^e) \tag{8}$$

with the intial condition

$$Z_n(t_0) = 0 \quad , \quad n=2,...,n_{max} \tag{9}$$

and also the boundary conditions

$$Z_1(t) = Z_0 \exp \left(-\Delta G_1(t)/kT\right) \quad , \quad Z_{n>n_{max}}(t) = 0) \tag{10}$$

where $n_{max}(t)$ is the maximum pore size at any given time t.

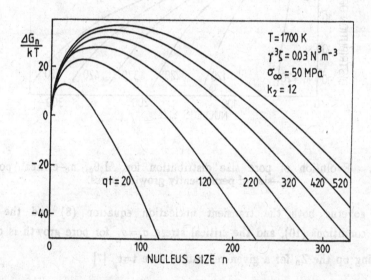

Figure 3. Free enthalpy versus nucleus size with time as parameter.

Because of the time dependence of ΔG_n, the numerical solutions $Z_n \, (qt)$ of (8), (9),(10) in Figure 4 are non–monotonic and thus non–stationary [7]. This feature of the solutions could not be found by analytical approximations [4,5].

From these solutions the critical pore growth stresses have been derived in [7]. In the following, relations are given by means of which numerical results for one set of parameters can be transformed into those for other sets.
The local stress $(4)_2$ can be rewritten with $(1)_3$ and $(6)_2$ as

$$\sigma(qt) = g(qt) \cdot \left[\frac{T}{G(T)}\right]^{1/6} \sigma_\infty d^{1/2} \tag{11}$$

(For the present purpose it is not necessary to know the function g.) Then the potential (6) reads

$$\frac{\Delta G_n}{kT} = -\frac{n\Omega}{k}\, g(qt) \left[\frac{T}{G(T)}\right]^{1/6} \frac{\sigma_\infty d^{1/2}}{T} + \frac{(18\pi)^{1/3}}{k}\,(n\Omega)^{2/3}\,\frac{(\gamma^3\zeta)^{1/3}}{T} \quad (12)$$

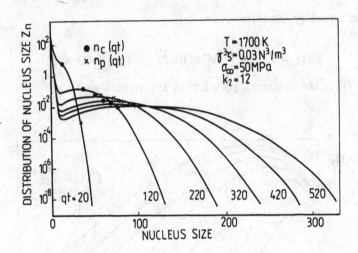

Figure 4.   Evolution of pore size distribution for Al_2O_3, n_c–critical pore size, n_p–size of permanently growing pores.

As (12) governs both the transient nucleation equation (8) and the related boundary conditions (10), and the critical stress $\sigma_c = \sigma_\infty$ for pore growth is obtained by summing up the Z_n for a given relaxation time $t = t_R$ [7]

$$\sum_{n > n_p(qt)}^{\infty} Z_n(qt, \sigma_\infty = \sigma_c) = 1 \quad (13)$$

other sets of values $\sigma_c d^{1/2}$, T, $\gamma^3\zeta$ can be derived from the numerical result denoted by $\{\}_1$ in the following way

$$\left\{\frac{(\gamma^3\zeta)^{1/3}}{T}\right\}_1 = \frac{(\gamma^3\zeta)^{1/3}}{T} \quad (14)$$

$$\left\{\left[\frac{T}{G(T)}\right]^{1/6} \cdot \frac{\sigma_c d^{1/2}}{T}\right\}_1 = \left[\frac{T}{G(T)}\right]^{1/6} \frac{\sigma_c d^{1/2}}{T} \quad (15)$$

The equations (14) and (15) make a rule of transformation for the critical stress σ_c. Critical stresses calculated in this way have been plotted versus temperature in Figure 5 for three values of $\gamma^3\zeta$. They are based on high temperature data of Al_2O_3 from [2,8]

$$G = 1.55 \cdot 10^5 \left(1 - 0.35 \frac{T - 300 \text{ K}}{2320 \text{ K}}\right) \text{ MPa} \quad , \nu = 0.25 \quad , \kappa = 0.41$$

$$D_b\delta_b = 10^{-9}\exp(-Q/RT)m^3/s \ , \ Q = 400 \text{ kJ/mol} \ , \ R = 8.314 \text{ J/mol·K}$$

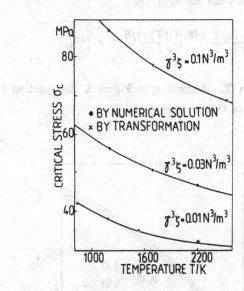

Figure 5. Numerically and analytically calculated critical stresses versus temperature.

By using three values of $\gamma^3\zeta$ and one value of d two parameter sets $(\sigma_c, T, \gamma^3\zeta)_{2,3}$ may be analytically obtained from one numerically calculated parameter set $(\sigma_c, T, \gamma^3\zeta)_1$ accordingly to (14), (15).

(15) includes an important relation between critical stress and grain size, which is known as the Hall—Petch relation:

$$\sigma_c \propto 1/\sqrt{d}$$

at given parameters T, $\gamma^3\zeta$.

CONCLUCSIONS

Micropore formation is supposed to be one of the mechanisms precluding superplasticity of ceramics. Consequently, avoiding micropore formation is considered a necessary condition for superplasticity. Micropores cannot form if the applied stress is kept subcritical.

Stationary creep of Al_2O_3 loaded as shown in Figure 5 is governed by grain boundary diffusion (Coble creep, [8], [9]).

$$\dot{\epsilon} = 14 \ \pi \ \frac{D_b \delta_b \Omega \sigma_{diff}}{kTd^3} \tag{16}$$

From (14), (15), (16) one obtains for $\sigma_{diff} \leq \sigma_c$

$$\dot{\epsilon} d^{7/2} \leq 14 \pi \Omega \sigma_{c1} d_1^{1/2} \frac{D_b \delta_{b}}{kT} [\frac{T}{T_1}]^{5/6} [\frac{G(T)}{G(T_1)}]^{1/6} = F \ (\gamma^3 \zeta, T) \tag{17}$$

Note that $D_b \delta_b$ depends on T. According to Figure 5 the function $F \ (\gamma^3 \zeta, T)$ was calculated for three values of the parameter $\gamma^3 \zeta$.

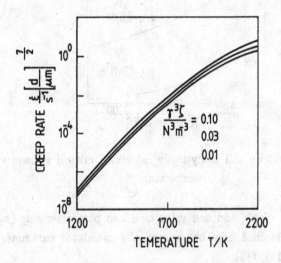

Figure 6. Region of stable hot working of Al_2O_3.

The region of stable hot working is below the curves in Figure 6. The curves show the essential result that the creep rate increases with temperature. This is not trivial in this case since it is brought about by a superposition of two counteracting effects: The critical stress for pore growth is lower with higher temperature (Figure 5). However, the creep rate for constant stress (16) rises with temperature so strongly that the net effect comes out as seen in Figure 6.

It must be mentioned that in considering creep rates one should take into account grain growth. This would mean setting up an extended model.

ACKNOWLEDGEMENT

The authors thank H.–J.Weiss for detailed discussions.

REFERENCES

1. Riedel, H., Fracture at high temperatures, Springer–Verlag, Berlin, Heidelberg, 1987.

2. Evens, A.G., Rice, J.R. and Hirth, J.P., Suppression of cavity formation in ceramics: prospects for superplasticity. J. Amer. Ceram. Soc., 1980, 63, 368–375.

3. Thouless, M.D. and Evans, A.G., Nucleation of cavities during creep of liquid–phase–sintered materials. J. Amer. Ceram. Soc., 1984, 67, 721–727.

4. Trinkaus, H. and Yoo, M.H., Nucleation under time–dependent supersaturation. Phil. Mag., 1987, A55, 269–289.

5. Trinkaus, H. and Yoo, M.H., Cavity nucleation under time–dependent stress concentration. Phil. Mag., 1988, A57, 543–564.

6. Bahr, H.–A., Balke, H., Pompe, W. and Werdin, S., Instationäre Porenbildung an Korngrenzen. Neue Hütte, 1989, 34, 201–207.

7. Bahr, H.–A., Balke, H., Pompe, W. and Werdin, S., Non–stationary nucleation of cavities at grain boundaries. 4–th IUTAM Symposium Creep in Structures, Sept. 1990 Cracow, to appear at Springer–Verlag.

8. Frost, H.J. and Ashby, M.F., Deformation–mechanism maps, Pergamon Press, Oxford, 1982.

9. Venkatachari, K.R. and Raj, R., Superplastic flow in fine–grained alumina. J. Amer. Ceram. Soc., 1986, 69, 135–138.

SUPERPLASTIC PHENOMENA IN SOME STRUCTURAL OXIDE CERAMICS

JAN LUYTEN, WILLY HENDRIX, JOS SLEURS AND WILLY VANDERMEULEN
Materials Research Unit
Vlaamse Instelling voor Technologisch Onderzoek (VITO)
Boeretang 200, B-2400 Mol, Belgium

ABSTRACT

Recently, superplasticity has been demonstrated in some structural ceramics. In this paper, the superplasticity of tetragonal zirconia partially stabilized by 3 mol.% Yttria (3Y-TZP) and MgO-doped Al_2O_3 is considered. The deformation characteristics of these superplastic structural ceramics are evaluated in 3-point bending creep tests. The creep rate is influenced by the stress, grain size, deformation temperature, the presence of a second phase and the amount of dopants.
Materials with a finer grain size or with a small amount of glassy phase offer possibilities to obtain more practical conditions (high deformation rate at the lowest temperature possible) for the application of super-plastic deformation as a forming technique.

INTRODUCTION

Superplasticity is a well known phenomenon in a special class of metallic alloys [1-3] but, in recent years, this deformation mechanism was also demonstrated in some ceramic systems [4-5].

For ceramic materials the superplastic deformation mechanism shows some similarities with diffusional creep ; Nabarro-Herring (lattice diffusion) , Coble (grain boundary diffusion) and interface-controlled diffusion creep had to be envisaged [6-7]. The definitive identification of specific mechanisms has not been successful, however.

The structural superplasticity is characterized by a generalized, unusually large sliding of grains with respect to each other. An accommodation mechanism makes this grain boundary sliding possible without cavity formation. Diffusion or dislocation motion can be considered as accommodation mechanism [8-9].

The conventional requirements for a superplastic deformation are :
- a fine grain size (< 5 μm)
- a high deformation temperature (> 0.5 Tm)
- a rather low deformation rate

To design a superplastic structural ceramic, i.e. a material allowing a high deformation rate at the lowest deformation temperature possible, we need a microstructure with a fine grain size in the first place. Secondly, the diffusion accommodation mechanism has to be improved by increasing lattice diffusion, grain boundary diffusion or by introducing a glassy phase constituting a fast diffusion path.

Another major concern is grain growth and cavity formation during superplastic deformation.

Some second phase and grain boundary segregating dopants can be effective to limit the grain boundary migration without lowering the grain boundary cohesive strength [10]. The aim of the study is not to determine very carefully the superplastic deformation parameters or mechanisms, but to develop a ceramic material that can deform at a practical temperature (1000°C) and moderate speed.

It is attempted to assess the superplastic behaviour of two oxide type ceramic materials : a Y_2O_3 stabilized ZrO_2 and a MgO-doped Al_2O_3. The deformation characteristics of these materials, known to have superplastic deformation capability, are evaluated in a simple 3-point creep test.

In a future approach, these materials will be modified to realize practical and useful values for the deformation rate and temperature.

EXPERIMENTAL PROCEDURE

A fine ZrO_2 powder was mixed with 5.15 wt. % Y_{2O3} (3Y-TZP), cold isostatically pressed in rectangular bars and sintered at 1500°C. The section of the bars was 3.5 mm by 4.5 mm after sintering.

To show the influence of the amount of dopant, two slightly different compositions of Al_2O_3 were synthesized : one with a commercially available Al_2O_3, containing already 400 ppm MgO, and another with the same powder but with an extra addition of 0.21 % MgO.

MgO is usually added to Al_2O_3 for grain size refinement and diffusion enhancement. The cold isostatically pressed bars are sintered under different conditions of time and temperature to obtain different grain sizes.

After synthesis, the ceramics were thoroughly characterized using SEM, TEM, XRD and ceramography.

Table 1 summarizes sintering conditions and characteristics of the ceramic materials used in this study.

TABLE 1

The ceramic materials

Composition	Sinter-conditions	Grain size after sintering μm	X.R.D. obtained cryst.phases
1. 3 Y T Z P ZrO_2 + 5.15 wt.% Y_2O_3 (3 mol Y_2O_3)	2h at 1500°C	0.6	Tetragonal + FCC, Moncl
2. Al_2O_3 + 0.04 % MgO	2h at 1650°C	4.5	α-Al_2O_3
3. Al_2O_3 + 0.04 % MgO	20' at 1550°C	1.1	α-Al_2O_3
4. Al_2O_3 + 0.25 % MgO	40' at 1650°C	3.8	α-Al_2O_3 + $MgAl_2O_4$
5. Al_2O_3 + 0.25 % MgO	20' at 1550°C	1.0	α-Al_2O_3 + $MgAl_2O_4$

Finally, the superplastic deformation capacity of those materials, was shown by 3-point bend creep tests at different temperatures. The tested length of the specimens was 30 mm. The deformation behaviour of the material in the 3-point bend tests was characterized by the deflection rate measured under the load. The load conditions were expressed as bending stress at the specimen centre. This stress has of course only a physical meaning at the start of the test. The deformation below the loading point was estimated from the radius of curvature. The average strain rates quoted in the text were obtained by dividing this strain by the test duration. For 3Y-TZP, creep bending tests at 1400, 1300 and 1200°C were executed. The creep distance of the different Al_2O_3 bars was evaluated at 1400, 1450 and 1500°C.

RESULTS

The bending tests on different Al_2O_3 bars illustrate the influence of grain size, temperature and amount of dopant (figures 1 and 2).

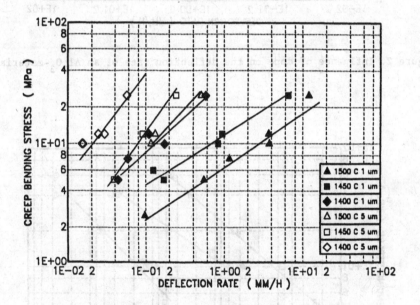

Figure 1. Effect of deformation temperature and grain size on the deflection rate of an Al_2O_3 (0.04 MgO)-material.

A fine grain sized Al_2O_3 with the extra amount of MgO gives the best superplastic deformation at 1500°C ($\dot{\varepsilon} = 10^{-4} s^{-1}$). Nevertheless, for this highly doped 1 µm material, there is a limit to the deformation rate ; with load above 12 Mpa all the bars are broken.

The Al_2O_3 with a grain size above 4 µm has a much higher creep resis-

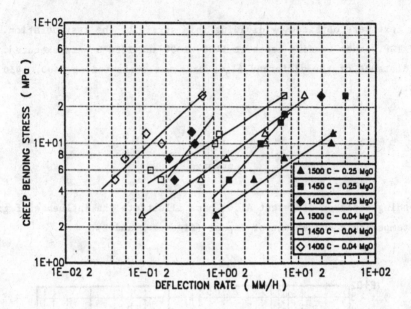

Figure 2. Influence of dope on the deflection rate of an Al_2O_3-material.

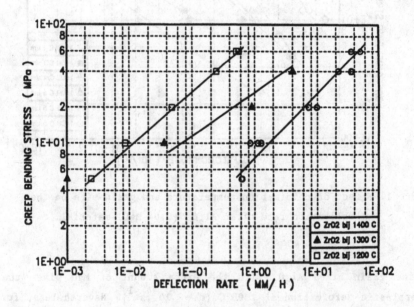

Figure 3. Effect of deformation temperature on the deflection rate of an
3Y-TZP-ceramic material.

tance and the bars of the Al_2O_3 with an extra addition of MgO and grains larger than 3 μm are broken with loads above 5 Mpa. It should be noted that all the tests were stopped when the deflection reached the limit of 4.5 mm.

As to the 3Y-TZP material, the results of the creep tests at 1400, 1300 and 1200°C are shown in figure 3. The specimens have a very low creep resistance at 1400°C and can be deformed rather fast (5 min) to maximum deflection ($\bar{\varepsilon} = 10^{-4}s^{-1}$). The deformability reduces when the temperature decreases and at 1200°C the deflection rate is low.

The grain sizes measured by the intercept line method on SEM-fractographies are summarized in table 1.

Grain size measurements, before and after deformation, reveal that the grain growth during superplastic deformation for the ceramic materials tested is negligible. The different TEM-micrographs of the Al_2O_3-0.25 MgO ceramics, sintered respectively at 1650°C and 1550°C and both deformed on 1400°C for ± 10 %, show only locally a significant difference in grain size. They demonstrate clearly that the grain boundary sliding deformation is not completely accommodated by other deformation mechanisms.

The TEM-observations revealing residual pores with pinned dislocations, triple point voids, some lattice and grain boundary dislocations are in agreement with previous observations (Figure 4) [11-12]. The $MgAl_2O_4$ spinel precipitates indentified by the TEM-diffraction patterns and XRD-analysis, are rather heterogeneously distributed and can be associated with cavitation formation (figure 5).

DISCUSSION

Although the fact that we use a conventional manufacturing route, which has still to be optimized, the average strain rates observed during the bending tests of our 3Y-TZP bars at 1400°C are comparable with the values of simular materials described in the literature [10-13-14-15-16]. A comparable deformation rate ($\bar{\varepsilon} = 10^{-4}s^{-1}$) can be obtained for a fine grained and highly doped Al_2O_3 at 1500°C only. Here, some cavity formation may be enhanced by the heterogeneous precipitation of the $MgAl_2O_4$-spinel particles. This is possibly also an explanation for the observed upper loading limit.

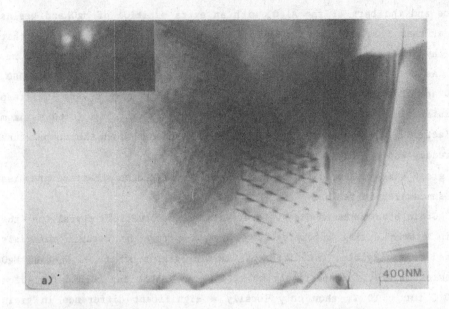

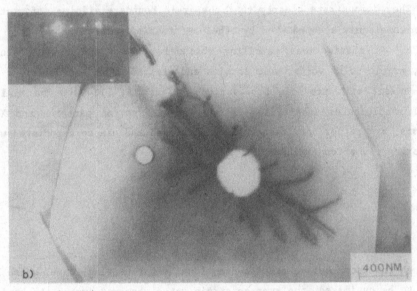

Figure 4. Dislocation configurations in MgO doped Al_2O_3
    a) some lattice dislocations
    b) residual pores with pinned dislocations

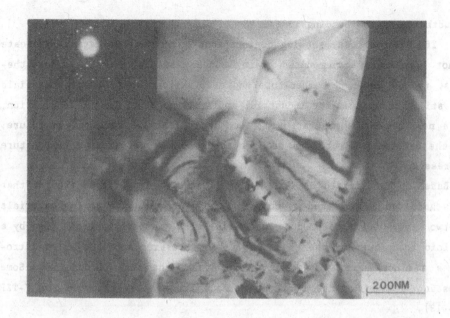

Figure 5. The MgAl$_2$O$_4$ spinel particles and cavitation formation.

TEM micrographs of this highly doped Al$_2$O$_3$ show cavity formation, grain boundary and lattice dislocations and indicate that the accommodation for grain boundary sliding is rather critical for Al$_2$O$_3$. This difference in superplastic behaviour between Y-TZP and Al$_2$O$_3$ (MgO) is explained by the possible existence of a small amount of glassy phase around the grains of the ZrO$_2$-material, and the higher cohesive strength of the ZrO$_2$ grain boundaries [10]. The glassy phase enhances the grain boundary diffusion and directly accommodates the grain boundary sliding. The higher cohesive strength of the ZrO$_2$ grain boundary opposes the cavitation formation and allows higher strain rates.

CONCLUSION

A superplastic behaviour could be demonstrated for a ZrO$_2$(Y$_2$O$_3$) - at 1400°C and for an Al$_2$O$_3$(MgO)-material at 1500°C. The influence of grain size, amount of dopant and temperature on the deformation rate is clearly illustrated by the bending tests.

TEM-observations, showing dislocations and cavity formation, indicate a not completely accommodated grain boundary sliding in Al_2O_3. Nevertheless, the deformation temperature and deformation rates of both materials are still not optimum for industrial application. Superplastic deformation, as a near-net-shape manufacturing technique [17] will have only a future, if the deformation rate can be increased and the deformation temperature decreased.

Based on our limited results and on literature data, it is thought that this gain can be achieved by adapting the microstructure of the materials in two ways. One is to produce a material with a superfine grain size by a colloid technique and the other one is a processing approach which introduces a minor amount of glassy phase around the grain boundaries. Some substantial improvements by these methods are already reported for Y-TZP [14-18].

Acknowledgements

The authors wish to thank the Characterization Group headed by P. Diels for the careful characterization of the materials. They are also indebted to J. Cooymans and G. Verreyt for technical assistance. Their acknowledgement also includes Mieke Dierckx for word processing.

REFERENCES

1. Pearson, C.E., The viscous properties of extruded eutectic alloys of lead-tin and bismuth-tin. J. Inst. Met., 1934, 54 [1], 111.

2. Backofen, W.A., Turner I.R., and Avery D.H., Trans A.S.M., 1964, 57, 980.

3. Edington, J.W., Melton K.N., and Cutler, C.P., Superplasticity. Prog. Mat. Sci., 1976, 21(2), 61-170.

4. Wakai, F., Sakaguchi, S., and Matsuno, Y., Superplasticity of Yttria-stabilized tetragonal ZrO_2 polycrystals. Adv. Ceram. Mater., 1986, 1(3), 259-63.

5. Carry, C., and Mocellin, A., Examples of superplastic forming fine-grained Al_2O_3 and ZrO_2 ceramics. High Tech. Ceramics ed. P. Vincenzini, Elsevier Science publishers B.V., Amsterdam, 1987, 1043-1052.

6. Lessing, P.A., and Gordon, R.S., Creep of polycrystalline alumina, pure and doped with transition metal impurities. J. Mat. Sci., 1977, 12, 2291.

7. Mohamed, F.A., and Laugdon, T.G., Deformation mechanism maps based on grain size. Met. Trans., 1974, 5, 2339.

8. Ashby, M.F., and Verrall, R.A., Diffusion-accommodated flow and super-plasticity. Acta Metall., 1973, 21, 149-63.

9. Gifkin, R.C., Grain Boundary sliding and its accommodation during creep and superplasticity. Metall. Trans. A, 1976, 7A(8), 1225-32.

10. Wei Chen, I., and Liang An Xue, Development on superplastic structural ceramics. J. Am. Ceram. Soc., 1990, 73(9), 2585-2609.

11. Heuer, A., Tighe, N., and Cannon, R.M., Plastic deformation of fine grained alumina (Al_2O_3) ; II-Basal slip and nonaccommodated grain boundary sliding. J. Am. Ceram. Soc., 1980, 63 (1-2), 53.

12. Fridez, J.D., Carry, C., and Mocellin, A., Effects of temperature and stress on grain boundary behaviour in fine grained alumina. Adv. Ceram., 1985, 10, 720.

13. Chin-Mau James Hwang and I-Wei Chen, Effects of a liquid phase on superplasticity of 2-mol.%-Y_2O_3-stabilized tetragonal Zirconia poly-crystals. J. Am. Ceram. Soc., 1990, 73(6), 1626-32.

14. Duclos, R., and Crampon, J., High temperature deformation of a fine-grained zirconia. J. Mat. Sci. letters, 1987, 905-908.

15. Wakai, F., and Nagono, T., The role of interface-controlled diffusion creep on superplasticity of yttria-stabilized tetragonal ZrO_2 poly-crystals. J. Mat. Sci. letters, 1988, 7, 607-609.

16. Wakai, F., Sakaguchi, S., and Matsuno, Y., Superplastizität von Yttrium-stabilisiertem TZP. Keram. Zeitschrift, 1987, 39, 7, 452-455.

17. Sheppard, L.M., Fabrication of Ceramics : the challenge continues. Am. Ceram. Bull., 1989, 68, 10, 1815-1820.

18. Yu-ichi Hoshizawa, and Taketo Sakuma, Role of grain boundary glass phase on the superplastic deformation of tetragonal zirconia polycrystal. J. Am. Ceram. Soc., 1990, 73(10), 3069-73.

"CREEP OF POWDER METALLURGICAL CHROMIUM"

ECK R., W. KÖCK, G. KNERINGER
Metallwerk Plansee GmbH
A-6600 Reutte, Austria

ABSTRACT

In contrary to molybdenum, tungsten or tantalum and
niobium, chromium has not yet achieved wide application
as a refractory metal. Melting at 1920° C chromium is
the lowest melting metal out of the VI A group of elements. Its cubic body centered crystalline structure
and its behaviour against interstitial elements makes
it a brittle material near room temperature.

This investigation deals with the analysis of powder
metallurgical 3N7 chromium examined under tensile conditions applying high temperature testing and performing
short time creep testing under constant load.
Phenomena of high temperature tensile testing according
to standardized conditions will be compared to the
creep results. Evaluation of tests performed supplies
mechanical data and creep data for chromium up to 1100° C
in comparison to data from literature. These data and
fracture modes relevant to high temperature testing will
be discussed and compared to existing deformation maps.

1. INTRODUCTION

Looking at the history of investigations performed on pure
chromium one finds, that this metal was of great interest to
research people since many years which led to two books
about the metal chromium (1, 2). The preparation of the
specimens for the determination of mechanical and physical
properties, impurity content and corrosion data was primarily
performed applying various methods of melting. An extensive
study published in 1959 describes the preparation and properties of high purity chromium with purities between 99,9 and
~99,998 % Cr melting electrolytic and "Iodide" chromium (3).
Main problem was brittleness of chromium near room temperature. Disregarding cost of production it has been veryfied
that remelted chromium starting from "highest" purity
"Iodide" chromium, can be ductile down to - 40° C tested by
bend testing, if a sufficient low content of impurities has
been achieved (4).

It is commonly accepted that the main embrittling elements
for chromium are N_2, O_2, C, H_2 and S, nitrogen being the
most effective and the most difficult element to control
during production and to avoid during application. Out of
the VI A group of elements Cr has the highest solubility
for N. Fig. 1 shows that about 1 at % N are soluble in Cr
at 1400° C (5). All other elements have a less pronounced
effect on ductile-brittle fracture of Cr (6).

2. POWDER METALLURGICAL PRODUCTION OF HIGH PURITY CHROMIUM

Various melting techniques have been chosen first to produce
the purest possible chromium. The powder metallurgical route
of preparation has been evaluated in comparison to melting,
but at that time a suitable pure powder production, adequate
processing and the fabrication of a porefree final product

were not possible (3). In recent years a new approach was started to produce semifinished and finished products following the powder metallurgical steps of comminution of electrolytical Cr, powder preparation, compaction, sintering and plastic forming processes to economically produce 99,97 % pure wrought chromium (7, 8). The main impurity elements of the remaining 300 µg/g are W, Mo, Fe and Si, each of them being below 100 µg/g.

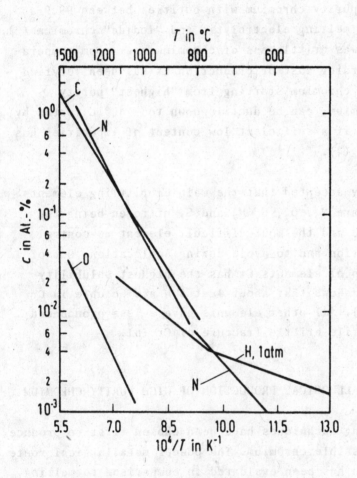

Figure 1. Solubility of C, N, O and H in Cr.

For this investigation metal sheets were produced in a range
of thickness between 2 and 8 mm. The main tests were performed on 3 mm thick sheet with the following interstitial
element content given in µg/g:

$$O_2 = 20 - 30 \qquad C = 10 - 15 \qquad N_2 < 5 \qquad H_2 = 2 - 2,5$$

3. TENSILE PROPERTIES OF CHROMIUM SHEET

As a basis for the determination of creep properties tensile
tests have been performed. Fig. 2 gives ultimate tensile
strength and elongation of 99,97 % Cr sheet for a thickness
range of 2 to 8 mm produced under standard conditions up to
1100° C at a testing speed of 2 mm/min in a ground surface
condition. Rate sensitivity of straining Cr is well known.
An acceleration of testing speed by a factor of 10 for instance increases strength and elongation by about 10 % at
750° C for the same sheet (9).

Dynamic strain aging peaks attributed to N have been measured over the temperature range now in discussion for Cr
containing more than 20 µg/g N (10). Temperature spacings
of 300° C as for these tensile tests are too large to discuss DSA effects and this topic was not primary interest of
this investigation. Furthermore we assume that a N-content
of the 3 mm sheet of below 5 µg/g, which is our detection
limit, has little effect on strengthening by strain aging.

Fig. 2 shows that the ductile-brittle fracture temperature
lies near room temperature for powder metallurgically produced Cr-sheet.
Strain or work hardening is not effective in this special
case due to high forming temperatures.

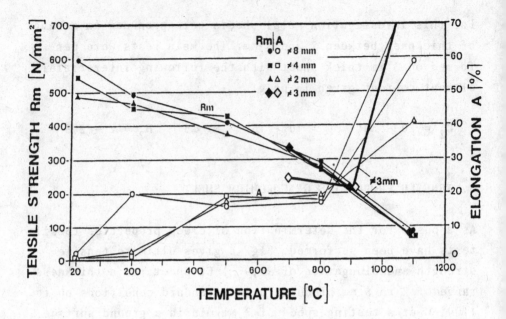

Figure 2. Tensile strength and elongation of Cr sheet up to 1100° C tested at 2 mm/min in vacuum.

4. CREEP TESTING OF CHROMIUM

Besides superior corrosion resistance, high temperature strength of the refractory metal chromium with a melting point of 1920° C is an important property as a material for constructions to be in high temperature use.

Distorsion of Cr under elevated temperature cyclic or static stresses depends on creep properties. Creep data and mechanisms are important for life estimation or life prediction of constructions at high temperatures.

Creep tests performed under vacuum are listed in table 1 in comparison to tensile tests at 700°, 900° and 1100° C.

TABLE 1.

Results of tests performed on Cr 99,97 sheet 3 mm thick under vacuum in ground surface condition.

Testtemperature (°C)	Tensile tests			
	HV 10	$R_m(N/m^2)$	$R_{p\ 0,2}(N/m^2)$	A (%)
700	190	330	320	26
900	175	235	220	23
1100	140	68	47	90

Testtemperature (°C)	Creep tests		
	sec creep rate (h^{-1}) at stresses of		
	100 (N/m^2)	200 (N/m^2)	250 (N/m^2)
700		$3,9.10^{-4}$ Fig. 5	$8,7.10^{-3}$ Fig. 6
900	$5,6.10^{-3}$ Fig. 7		
1100			

Geometry of creep specimen used is shown in Fig. 3.

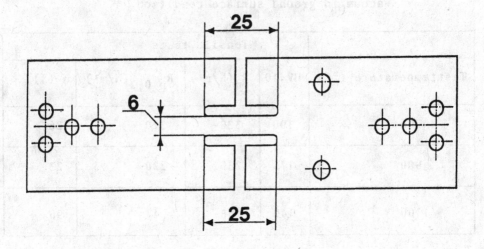

Figure 3. Dimensions of creep test specimen.

Creep rupture data of Table 1 are also plotted in Fig. 4
and compared to creep tests performed on 2 mm Cr sheet
99,7 % pure in air and to data from literature evaluated
from tests in Argon (10). The main impurity of 99,7 % Cr
is Fe, which has no pronounced effect on rupture strength.
Also 1100° C rupture strength of 3 mm sheet tested in this
serie is in good agreement with creep rupture data evaluated from tests performed on arc cast specimens which
were taken from literature (11).

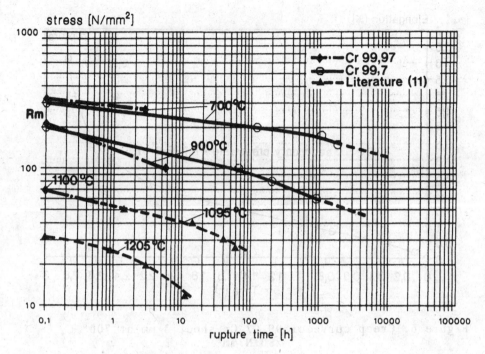

Figure 4. Creep rupture time vs. creep stress of chromium
for temperatures up to 1200° C.

Actual short time creep curves registered at 700° and 900° C
at stresses up to 250 N/mm are shown in Figs. 5, 6 and 7.

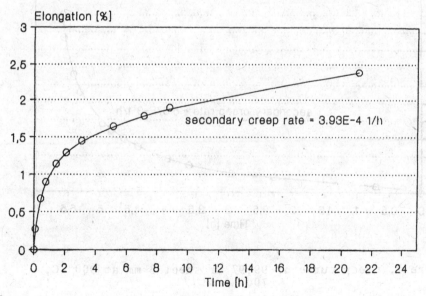

Figure 5. Creep curve of 99,97 Cr sheet 3 mm at 700° C,
200 N/mm²

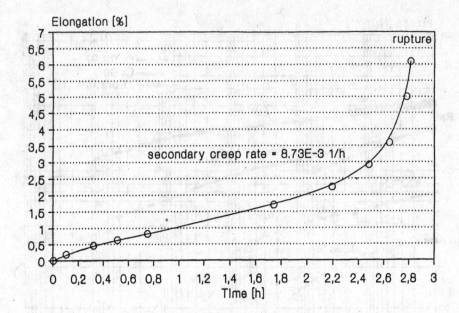

Figure 6. Creep curve of 99,97 Cr sheet 3 mm at 700° C,
250 N/mm²

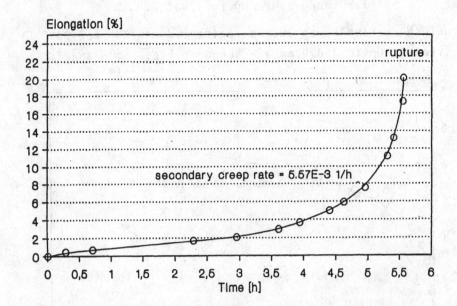

Figure 7. Creep curve of 99,97 Cr sheet 3 mm at 900° C,
100 N/mm²

Scaling of the diagrams is different for all three curves.
They show the typical three stage behaviour as expected, with
the first stage pronounced for the 700° C 200 N/mm² specimen.
The second stage creep rate is very pronounced and could be
calculated in all cases. Short time creep tests are terminated
after 20 hours, so rupture time of the 700° C 200 N/mm² sample
was not determined.

A metallographic section of the deformed und ruptured area of
the creep specimen tested at 900° C and 100 N/mm² at low magnification is shown in Fig. 8.

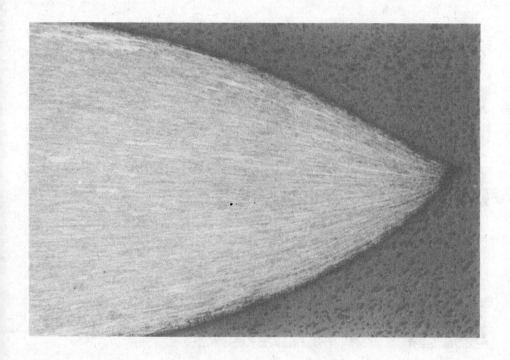

Figure 8. Metallographic section of short time creep specimen
from sheet 3 mm ≠ tested at 900° C 200 N/mm² in
vacuum. x 32

Grain structure and Vickers hardness of 170 show that re-
cristallization just started after 5,5 hours at 900° C.
The finally ruptured areas of tensile and creep test specimens
at 700° C and 900° C were examined by SEM and are shown in
Figs. 9 to 12 at a magnification of 240. The appearance of the
fractured surface is similar independend of rate of tensile
or creep testing. After a reduction in area near to 100 %
final rupture forms deep dimpls and grooves, typical for ductile fracture. We can not identify any pronounced differences
between the 700° C and 900° C fractures.

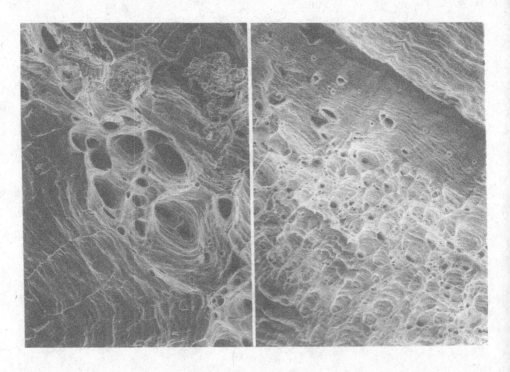

Figure 9. Fractured surface of   Figure 10. Fracture surface of
          tensile test specimen   short timme creep
          tested at 700° C in   specimen tested at
          vacuum. x 240   700° C 250 N/mm²
                                                 in vacuum. x 240

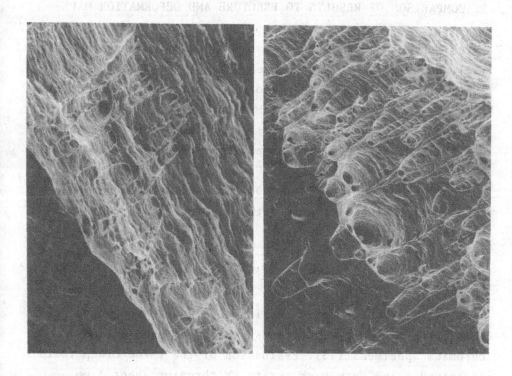

Figure 11. Fractured surface of
tensile test specimen
tested at 900° C in
vacuum. x 240

Figure 12. Fracture surface
of short time
creep specimen
tested at 900° C
200 N/mm² in
vacuum. x 240

A peculiarity of dimple fracture of PM chromium are spherical
particles visible at the bottom of the craters or ends of the
grooves, visible in Fig. 12. An analytical proof that these
inclusions are Cr-oxides was not possible, but looking at the
analysis of the 3 mm sheet 20 - 30 µg/g oxygen is sufficient
to form Cr-oxides. We assume, that these oxides are the starting point of the gliding processes in connection with grain
structure to form the fracture surface geometry shown on the
fractographs.

5. COMPARISON OF RESULTS TO FRACTURE AND DEFORMATION MAPS

Established in 1978 fracture maps for chromium are available
(12). We assume that data chosen for this map have been evaluated from tensile tests performed on arc cast chromium or
chromium that has been melted by other methods in most cases.
We were interested in a comparison of these data to results
of this investigation, evaluated from powder metallurgical
chromium of very high purity. Fig. 13 shows normalized tensile stress of our tests at 700, 900 and 1100° C over homologous temperature drawn into the fracture map for Cr.
As fracture analysis has shown ductile fracture is typical
for these conditions.

Deformation maps for chromium, summarizing slow processes
are also available and as mentioned we assume that also for
this creep map most experiments evaluated, used melted
chromium specimens (13). Evaluation of the three creep tests
performed using very high purity PM chromium sheet 3 mm
thick at 700° and 900° C is shown in Fig. 14 for normalized
shear stress. Comparing rate data of Table 1 for secondary
or steady state creep parameters to creep rates given
in this diagram one finds differences. Smaller size and
shape of PM chromium grains, variations in final chemical
composition and finally uncertenties in shear moduli data
at these temperatures we assume are obvious explanations for
these deviations.

6. CONCLUSION

Tensile tests and short time creep tests of high purity
99,97 % PM-chromium containing 20 - 30 µg/g O_2, 10 - 15 µg/g
C, 5 µg/g N and 2 - 2,5 µg/g H_2 show ductile fracture at
temperatures of 700° and 900° C.

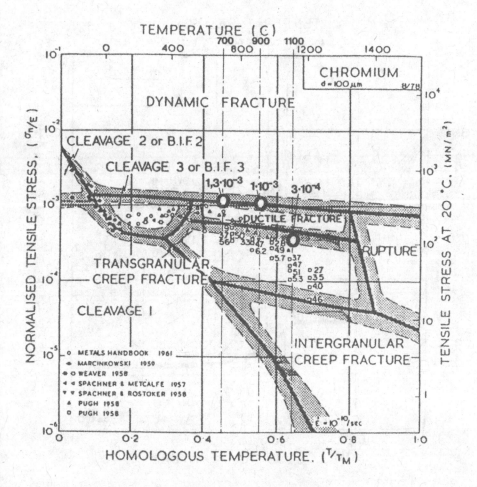

Figure 13. Deformation map for tensile fracture for chromium (12).

PM chromium of a purity of 99,7 containing max. 0,3 % Fe shows no significant difference in tensile and short time creep data. 1100° and 1200° C creep data for melted chromium from literature agree with data for powder metallurgically fabricated chromium.

A comparison of tensile data from high purity PM chromium to fracture maps established on melted Cr, show agreement

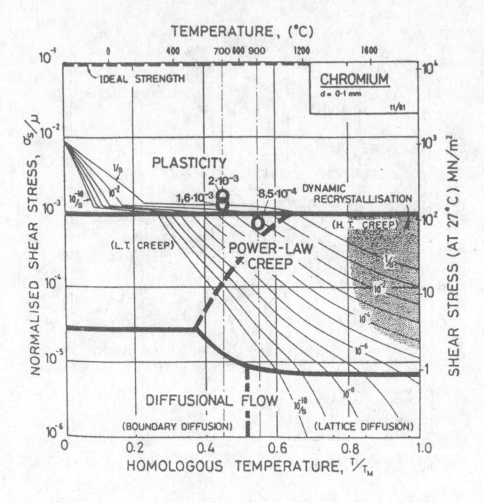

Figure 14. Deformation map for creep fracture for chromium.
(12)

for temperatures between 700° and 1100° C. Deformation maps
for creep data of melted Cr and high purity PM chromium
show differences which can be explained by differences in
grain structure and chemical composition of the tested material primarily.

LITERATURE

(1) SULLY A.H.: "CHROMIUM" Butterworths, London 1954

(2) "Ductile Chromium and its Alloys" Conference Proceedings
     by Office of Ordnance Research 1955 ASM Cleveland
     Ohio 1957

(3) EDWARDS A.R., J.I. NISH and H.L. WAIN: "The Preparation
     and Properties of High Purity Chromium" Metallurgical
     Reviews, vol. 4, 1959

(4) Datasheet W.C. Heraeus Hanau Germany

(5) FROMM E., E. GEBHARDT: "Gase und Kohlenstoff in
     Metallen", Springerverlag 1976, p. 528

(6) CAIRNS R.E. and N.J. GRANT: Trans AIME vol. 230 (1964),
     p. 1150

(7) ECK R., J. EITER, W. GLÄTZLE: "Powder Metallurgy
     Chromium" Intern. Powder Metallurgy Conference PM '86,
     Düsseldorf 1986

(8) ECK R., J. EITER, G. KNERINGER: "Wrought Chromium"
     World Conference on Powder Metallurgy PM '90,
     London 1990

(9) Metallwerk Plansee: Internal Report 1989

(10) WILKINSON W.P.: "Properties of Refractory Metals"
     Gordon and Breach 1969, p. 247

(11) PUGH O.W.: ASM Trans. 50 (1958), p. 1072

(12) GANDHI C., M.F. ASHBY: Acta metall., 27, (1979)
     p. 1565 - 1602

(13) FROST H.J., M. ASHBY: Deformation Mechanism Maps,
     Pergamon Press (1982), Oxford

EFFECTIVE PROPERTIES IN POWER-LAW CREEP

PEDRO PONTE CASTAÑEDA
Mechanical Engineering and Applied Mechanics
University of Pennsylvania
Philadelphia, PA 19104, U.S.A.

ABSTRACT

Recently developed variational principles are used to predict the effective properties of power-law creeping composites with two distinct phases in prescribed volume fractions. The new variational principles serve to estimate the effective properties of nonlinear composites in terms of the corresponding properties of linear composites with the same microstructural distributions of phases. The nonlinear composite is further assumed to possess overall isotropic symmetry. General results are given for composites with arbitrary creep exponents and reference stresses in a form that is almost explicit involving the solution of a nonlinear algebraic equation. Explicit results are given for the specific cases where the second phase is either rigid or void, corresponding to rigidly reinforced and porous materials, respectively. For the case of rigid particles the effect of particle shape is briefly explored by means of the well known self-consistent estimates for linear materials with isotropic distributions of spheres, disks and rods.

INTRODUCTION

New variational principles allowing the estimation of the effective properties of composite materials with phases exhibiting nonlinear constitutive behavior have been proposed recently by Ponte Castañeda [11, 13]. These variational principles suggest a straightforward and versatile procedure expressing the effective properties of the nonlinear composite in terms of the effective properties of a family of linear composites with distribution of phases identical to that of the nonlinear composite. Thus, bounds and estimates for the effective properties of linear composites can be translated directly into corresponding bounds and estimates for the nonlinear composite. The new procedure was applied in [11] to composite materials containing a nonlinear isotropic matrix either weakened or reinforced by isotropic distributions of voids or rigid particles, respectively. This procedure was also applied to a porous Ramberg-Osgood material in [14]. The case of a ductile matrix reinforced by incompressible elastic particles was studied in [12]. The Hashin-Shtrikman bounds obtained in [11] via the new method directly from the Hashin-Shtrikman [6] bounds for the linear comparison material were found to be an improvement over the corresponding bounds obtained in [15] for the same class of nonlinear materials using the nonlinear extension of the Hashin-Shtrikman variational principle proposed by Talbot and Willis [16]. Recently, however, Willis [17] has shown that the bounds obtained via the new method can also be generated by the Talbot-Willis method with a better choice of their comparison material. More generally, the new procedure can make use of other bounds and estimates for the linear comparison material to yield corresponding bounds and estimates for a broad class of nonlinear heterogeneous materials. In fact, the new procedure can be shown [12, 13] to yield exact results for nonlinear composites with special microstructures.

In this paper we apply the procedure to a general composite containing two phases with different reference stresses and power-law creeping exponents in given volume fractions in such a way that the composite is isotropic in an overall sense. We assume that the phases are perfectly bonded to each other, incompressible and isotropic. Additionally, the size of the typical heterogeneity is assumed to be small compared to the size of the specimen and the scale of variation of the applied loads. Explicit results in the forms of bounds and estimates are given for the specific cases where the second phase is either rigid or void, corresponding to rigidly reinforced and porous materials, respectively. For the cases of rigid particles, the effect of particle shape is briefly explored by means of the well known self-consistent estimates for linear materials with isotropic distributions of spheres, disks and needles.

One of the goals of this work is to compare the predictions of the new method, particularly for the bounds, with the results of other investigators using either *ad hoc* methods, or computations based on periodic microgeometries. One popular method [4, 5, 9] for estimating the effective properties of composites is to model the composite by a finite shell of the matrix material surrounding a solid sphere of the inclusion phase subject to uniform loading conditions. If the matrix is infinite in extent, the results obtained by this method are rigorous in the dilute limit (of inclusions) corresponding to negligible interactions of the inclusions in the actual composite. This approach was used for example by Lee and Mear [9] for the rigidly reinforced material, and by Gurson [5] for the porous material. If, on the other hand, the matrix is finite in extent corresponding to a finite shell, as for example in the model of Cocks [4] for porous materials, then a sound physical interpretation of this model is not available as it is not known whether such an approximation would model adequately the interaction effects between the inclusions in actual composites. We note, however, that for the case of pure hydrostatic loading (but not more generally!) such an interpretation exists in terms of the well-known composite spheres assemblage. Thus, we observe that bounds can also serve a useful role in testing the physical validity of these and other micromechanical models. Another procedure avoiding the above potential difficulty involves complete specification of the microgeometry, as in periodic composites [1, 3], to yield rigorous results for that (and only that) specific geometry. However, in practice this approach also has its limitations, because approximations are usually needed to insure overall isotropy in the application of this method to materials with nonlinear constitutive behaviors, and also because these methods provide uncertain information about alternative microgeometries occurring in practice that may not be periodic. The work described in this paper does not suffer from these limitations, because it is concerned with the effective properties of a family of composites which is only required to be isotropic in an overall sense. On the other hand, by relaxing the prescribed information about the microgeometry, a price is paid in terms of the accuracy of the results for specific composites, particularly, if the contrast between the properties of the two phases is extreme. It is anticipated that a complete understanding of this very complex problem of the effective constitutive behavior of nonlinear composites will arise eventually from the synthesis of the results of many different approaches. The approach favored in this paper is only one such possibility.

EFFECTIVE PROPERTIES

Consider a two-phase heterogeneous material occupying a region in space of unit volume Ω. We characterize the local stress potential $\phi(\sigma, \mathbf{x})$ of the material in terms of the homogeneous phase potentials $\phi^{(r)}(\sigma)$ via the relation

$$\phi(\sigma, \mathbf{x}) = \sum_{r=1}^{2} \chi^{(r)}(\mathbf{x}) \phi^{(r)}(\sigma), \tag{1}$$

where the $\chi^{(r)}$ $(r = 1, 2)$ are the characteristic functions of the two phases. We will assume that the two phases are nonlinear, incompressible and isotropic, so that the potentials $\phi^{(r)}(\sigma)$ depend only on the effective stress $\sigma_e = \sqrt{\frac{3}{2}\mathbf{S}\cdot\mathbf{S}}$, where $\mathbf{S}$ is the deviator of σ. Thus, we write

$$\phi^{(r)}(\sigma) = f^{(r)}(\sigma_e), \tag{2}$$

where the scalar-valued functions $f^{(r)}$ are chosen to have the power-law form

$$f(\sigma_e) = \frac{1}{n+1}\dot{\varepsilon}_o\sigma_o\left(\frac{\sigma_e}{\sigma_o}\right)^{n+1} \tag{3}$$

with n, σ_o and $\dot{\varepsilon}_o$ denoting material constants with different values in each phase.

The local constitutive relation for the creeping material is given by

$$\dot{\varepsilon} = \frac{\partial\phi}{\partial\sigma}(\sigma, \mathbf{x}), \tag{4}$$

where $\dot{\varepsilon}$ is the rate-of-deformation (strain-rate) tensor.

To define the effective properties of the heterogeneous material we introduce, following Hill [7], a uniform constraint boundary condition

$$\sigma\mathbf{n} = \Sigma\mathbf{n}, \qquad \mathbf{x} \in \partial\Omega, \tag{5}$$

where $\partial\Omega$ denotes the boundary of the composite, $\mathbf{n}$ is its unit outward normal, and Σ is a given constant symmetric tensor. It follows that the average, or overall, stress is precisely Σ, and we *define* the average strain-rate $\dot{\mathbf{E}}$ in an analogous manner by the relation

$$\dot{\mathbf{E}} = \int_\Omega \dot{\varepsilon}(\mathbf{x})\, dx. \tag{6}$$

The effective, or overall, behavior of the heterogeneous creeping material follows from the principle of minimum complementary energy, which can be stated in the form

$$\Phi(\Sigma) = \min_{\sigma \in S(\Sigma)} \int_\Omega \phi(\sigma, \mathbf{x})\, dx, \tag{7}$$

where $S(\Sigma) = \{\sigma | \sigma_{ij,j} = 0 \text{ in } \Omega, \text{ and } \sigma_{ij}n_j = \Sigma_{ij}n_j \text{ on } \partial\Omega\}$ is the set of statically admissible stresses, and where we have assumed convexity of the nonlinear potential $\phi(\sigma, \mathbf{x})$. Thus, we have that

$$\mathbf{E} = \frac{\partial\Phi}{\partial\Sigma}, \tag{8}$$

where Φ is known to be convex.

In this work, we are interested in predicting the effective behavior of two-phase heterogeneous materials, as described above. We further restrict our consideration to composites with phases in given volume fractions

$$c^{(r)} = \int_\Omega \chi^{(r)}(\mathbf{x})\, dx, \tag{9}$$

that are additionally isotropic in an overall sense. We note that because the microstructure is not completely specified, the effective properties of such systems are not uniquely determined, but rather must lie within a certain range of values. This is in contrast with unit-cell computations for composites with periodic microstructures [1, 3] for which the microstructure is completely determined, and with computations based on the finite-shell model which do not correspond to *actual* microstructures. Thus, in the context of our problem, it makes sense to seek bounds and estimates for the effective potential Φ. The results will not only be useful on their on right, but should also serve to guide and test the results of micromechanically based models.

NEW VARIATIONAL PRINCIPLES

New variational principles for determining bounds and estimates for the effective properties of nonlinear composites in terms of the effective properties of linear composites were proposed recently by Ponte Castañeda [11, 13]. In this section, we summarize the statement of the complementary energy formulation of the variational principle for the case where both phases are incompressible.

The new variational principles are based on a representation of the potential for the nonlinear isotropic material in terms of the potentials of a family of linear *comparison* materials. This representation is obtained with the help of the Legendre transformation. Thus, for a homogeneous nonlinear material with "superquadratic" growth in its potential ϕ, and certain additional convexity hypothesis, we have that

$$\phi(\sigma, \mathbf{x}) = \max_{\mu > 0}\{\phi_o(\sigma, \mathbf{x}) - V(\mu, \mathbf{x})\}, \tag{10}$$

where

$$V(\mu, \mathbf{x}) = \max_{\sigma}\{\phi_o(\sigma, \mathbf{x}) - \phi(\sigma, \mathbf{x})\} \tag{11}$$

and where ϕ_o is the potential of a linear comparison material with shear modulus μ. Note that in the above optimizations, the position vector $\mathbf{x}$ is fixed.

The new variational principle is obtained by means of relation (10) and the complementary energy principle (7) (see [11, 13] for details), and can be expressed by

$$\Phi(\Sigma) = \max_{\mu(\mathbf{x}) \geq 0}\left\{\Phi_o(\Sigma) - \int_\Omega V(\mu, \mathbf{x}) \, dx\right\}, \tag{12}$$

where

$$\Phi_o(\Sigma) = \min_{\sigma \in S(\Sigma)} \int_\Omega \phi_o(\sigma, \mathbf{x}) \, dx \tag{13}$$

is the effective potential of a linear heterogeneous comparison material with local potential ϕ_o and arbitrarily variable shear moduli $\mu(\mathbf{x})$.

The variational principle described by (12) and (13) can be given the interpretation of first solving a linear problem for an arbitrarily heterogeneous *linear* material, and then optimizing with respect to the variations in moduli to account for the nonlinearity in the actual material. Thus, relations (12) and (13) are the global (effective) counterparts of the local (at a fixed point $\mathbf{x}$) relations (10) and (11). This suggests that if the fields happen to be constant over the nonlinear phases in the actual composite, then the variable moduli $\mu(\mathbf{x})$ can be replaced by constant moduli over each phase, *i.e.* by $\mu^{(1)}$ and $\mu^{(2)}$ in phase 1 and phase 2, respectively. More generally, however, we have the following lower bound for Φ, namely,

$$\Phi_-(\Sigma) = \max_{\mu^{(1)},\mu^{(2)} \geq 0} \{\Phi_o(\Sigma) - c^{(1)}V^{(1)}(\mu^{(1)}) - c^{(2)}V^{(2)}(\mu^{(2)})\}, \tag{14}$$

where $V^{(1)}$ and $V^{(2)}$ correspond to relation (11) evaluated for each of the homogeneous phase potentials $\phi^{(1)}$ and $\phi^{(2)}$, and Φ_o is the effective energy of an isotropic two-phase linear composite with phase moduli $\mu^{(1)}$ and $\mu^{(2)}$ in volume fractions $c^{(1)}$ and $c^{(2)}$, respectively. This latter expression has the advantage over the previous more general relation that it involves only a finite-dimensional optimization problem, and can be further used in conjunction with bounds and estimates for the linear composite to induce corresponding results for the nonlinear composite. We explore these possibilities in the next section for general two-phase incompressible composites, as well as rigidly reinforced and porous materials.

BOUNDS AND ESTIMATES

The general two-phase incompressible composite

The estimate (14) of the previous section for the effective energy of the nonlinear incompressible composite can be used in conjunction with the well-known Hashin-Shtrikman [6] bounds for the effective properties of the linear comparison composite to yield estimates for the effective potential of the nonlinear composite. It is shown in reference [13] that the result of this calculation takes on the simple form

$$\Phi_{HS}(\Sigma) = \min_{\omega \geq 0} \{c^{(1)}f^{(1)}(S^{(1)}) + c^{(2)}f^{(2)}(S^{(2)})\}, \tag{15}$$

where $f^{(1)}$ and $f^{(2)}$ are given by relation (3) applied to each phase with distinct material constants, and $S^{(1)}$ and $S^{(2)}$ are simple expressions of ω involving the phase volume fractions $c^{(1)}$ and $c^{(2)}$, and the overall effective stress applied to the composite Σ_e. For instance, if $\mu^{(1)} \geq \mu^{(2)}$ and the Hashin-Shtrikman lower bound is used for Φ_o, the corresponding expressions for $S^{(1)}$ and $S^{(2)}$ become

$$S^{(1)} = \sqrt{\left(1 - c^{(2)}\omega\right)^2 + \frac{2}{3}c^{(2)}\omega^2}\, \Sigma_e,$$

and

$$\hspace{8cm} (16)$$

$$S^{(2)} = \left(1 + c^{(1)}\omega\right)\Sigma_e.$$

Depending on the relative strengths of the two nonlinear potentials $f^{(1)}$ and $f^{(2)}$, the result of putting (16) into (15) would either lead to a rigorous lower bound for the nonlinear potential Φ, or alternatively, to an upper "estimate" (see [12]) for Φ. The surprising feature of the result is that it only involves one optimization in place of the four optimizations implicit in expression (14). Further, the result holds not only for the pure-power potentials described by relation (3), but more generally for any nonlinear isotropic potentials $f^{(1)}$ and $f^{(2)}$. We will not discuss the results for the general incompressible two-phase composite in any more detail. We will simply refer the reader to reference [11] for a discussion of explicit results for the case where both phases have the same creep exponent but different reference stresses, and to reference [12] for the case where one phase has power-law behavior and the other is linear. In the next subsection, we consider the special case of a rigidly reinforced composite.

The rigidly reinforced composite

This special case of the general incompressible composite is obtained by letting the potential of one of the phases, say phase 2, approach zero. Thus, expression (15) specializes to

$$\Phi_{HS}(\Sigma) = (1-c)f(S), \tag{17}$$

where we have dropped the superscript for phase 1 (except that c stands for the volume fraction of rigid particles), and S reduces to

$$S = \frac{1}{\sqrt{1+\frac{3}{2}c}} \Sigma_e. \tag{18}$$

We note that for this choice of phase potentials, the Hashin-Shtrikman expression (17) corresponds to an "upper estimate" for Φ. The rigorous lower bound for Φ in this case vanishes because one of the phases is rigid. Other estimates for the nonlinear potential can be obtained by making use of alternative estimates for the linear comparison composite. These estimates all have the same form (17), but the expressions for S are different, although linear in Σ_e. Rather than give those results here in the form of (17) and (18), we will present our results in terms of a more standard form involving a "reference stress" for the rigidly reinforced material Σ_n such that

$$\Phi(\Sigma) = \frac{1}{n+1} \dot{\varepsilon}_o \Sigma_n \left(\frac{\Sigma_e}{\Sigma_n}\right)^{n+1}. \tag{19}$$

Thus, the results of making use of the linear Hashin-Shtrikman (HS) upper estimate [6], the self-consistent estimates for spheres (SCS) [2, 8], the self-consistent estimate for needles (SCN) [18], and the differential self-consistent [10] can be expressed respectively in the forms

$$\left(\frac{\Sigma_n}{\sigma_o}\right)_{HS} = \frac{\left(1+\frac{3}{2}c\right)^{\frac{n+1}{2n}}}{(1-c)^{\frac{1}{n}}}$$

$$\left(\frac{\Sigma_n}{\sigma_o}\right)_{SCS} = \frac{(1-c)^{\frac{n-1}{2n}}}{\left(1-\frac{5}{2}c\right)^{\frac{n+1}{2n}}} \qquad \left(c \le \frac{2}{5}\right)$$

$$\left(\frac{\Sigma_n}{\sigma_o}\right)_{SCN} = \frac{(1-c)^{\frac{n-1}{2n}}(4+c)^{\frac{n+1}{2n}}}{(4-9c)^{\frac{n+1}{2n}}} \qquad \left(c \le \frac{4}{9}\right)$$

$$\left(\frac{\Sigma_n}{\sigma_o}\right)_{DSC} = (1-c)^{-\frac{3n+7}{4n}} \tag{20}$$

These results, which of course agree with the well-known linear results for $n = 1$, are plotted in Figure 1 for two values of n, namely 3 and ∞, corresponding to a weakly nonlinear matrix and a perfectly viscous (or plastic) matrix, respectively. By reference to these Figures, we can make the following observations. The HS estimate is a lower bound (an upper bound in the Figure!) for all of the above results. The HS and DSC estimates approach the perfectly rigid limit only as the volume fraction of rigid particles approaches unity ($c \to 1$). In the perfectly viscous limit, interestingly, the HS estimate does not approach a rigid limit. More realistically, from a physical point of view, the self-consistent estimates for spheres (SCS) and needles (SCN) approach an early percolation limit (i.e. the rigid limit) at the same values as the corresponding linear composites (e.g. 2/5 and 4/9, respectively). Also, we note that, as in the linear case, the self consistent (SC) estimate for randomly oriented rigid disks agrees with the

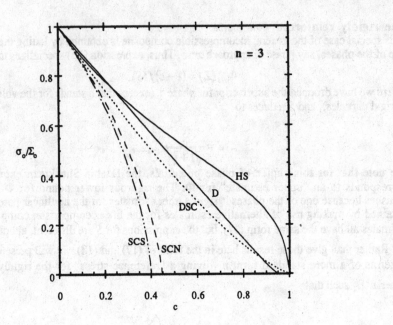

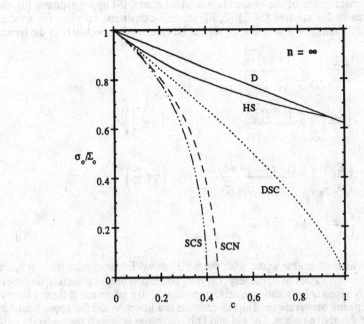

Figure 1 a, b. Plots of the reference stress Σ_n for the rigidly reinforced material

HS upper bound (infinite in this case), and thus is not presented. In general, we observe, as expected, that the effect of softening the matrix materials (by increasing n) leads to softer behavior for the composite; however, this effect is more pronounced for the HS and DSC estimates than for the SCS and SCN estimates.

Additionally, we present in Figure 1 dilute (D) results based on computations for a sphere in an infinite matrix of the power-law material from references [1] and [9]. They are given by

$$\left(\frac{\Sigma_n}{\sigma_o}\right)_D = 1 + \left(\frac{g(n)}{n}\right)c \quad \text{as } c \to 0,$$ (21)

where $g(n)$ is such that $g(1) = 5/2$, $g(3) \approx 3.21$ and $g(10) \approx 6.09$ [9], and $g(n) \to 0.38n$ as $n \to \infty$ [1]. These results do not compare very favorably with the corresponding dilute limits from (20): 5/2, 4.00, 9.25 and $0.75n$, respectively, and this is depicted in the Figures by noting that the dilute estimates fall outside the range of the HS estimate for small c. This is in agreement with our analysis that does not yield the HS estimate as a rigorous bound, but simply an estimate in this case. As we will see in the next section, the situation is different for the case of porous materials. None the less, the results of (19) with (20) may provide reasonable estimates for larger values of the volume fraction of the rigid inclusions. Finally, we note that, interestingly, for the perfectly viscous case, the dilute (D) and HS estimates are not very different for the whole range of volume fractions of rigid particles.

The porous material
This case involving a vacuous phase is not covered by relation (15) for the incompressible composite (since the porous material is compressible), but an analogous result is available for a general compressible composite [13]. When this result is specialized to the porous material, it yields a relation of the form (17), except that S is now given by

$$S = \frac{1}{1-c}\sqrt{\left(1 + \tfrac{2}{3}c\right)\Sigma_e^2 + \left(\tfrac{9}{4}c\right)\Sigma_m^2} ,$$ (22)

where c is a measure of the porosity (see also [11, 14]). Note that, unlike the previous result, the effective behavior of the porous material is compressible as indicated by the dependence of S on the overall mean stress Σ_m. This is in agreement with our physical intuition. Also, in contrast to the previous case dealing with the rigidly reinforced composite, expression (17) together with (22) yield a rigorous lower bound for the potential of the porous creeping material Φ.

Once again, for the sake of easier comparison with the results of other investigators, we opt to present our results in terms of a "yield function" $F(\Sigma_e, \Sigma_m; c, n)$ such that

$$\Phi(\Sigma) = \frac{1}{n+1}\dot{\varepsilon}_o\sigma_o\left(\frac{F}{\sigma_o}\right)^{n+1} .$$ (23)

The reason for the "yield-function" terminology is motivated by the perfectly viscous limit ($n \to \infty$), which is well-known to correspond to an ideally plastic material. In fact, one of the models against which we will compare our results is the Gurson model for porous ductile (ideally plastic) metals.

Thus, the above result for the HS bound, a result based on the SC approximation for spherical voids [2, 8], a spherical shell model due to Cocks (C) [4], Gurson's model (G) for perfectly plastic materials ($n = \infty$, only) [5], and the spherical shell model under pure hydrostatic loading (X) are given by the following expressions

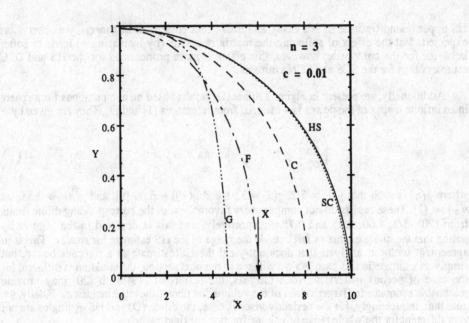

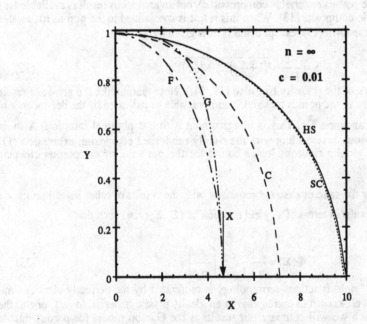

Figure 2 a, b. Plots of the yield surfaces $F = 1$ for the porous material (low porosity c)

$$F_{HS} = \left(\frac{1}{1-c}\right)^{\frac{n}{n+1}} \sqrt{(1+\tfrac{2}{3}c)Y^2 + cX^2}$$

$$F_{SC} = \sqrt{\left(\frac{1}{1-c}\right)\left(\frac{1-c/3}{1-2c}\right)\left(Y^2 + \frac{c}{1-c}X^2\right)}$$

$$F_C = \left(\frac{1}{1-c}\right)^{\frac{n}{n+1}} \sqrt{Y^2 + \frac{2nc}{(n+1)(1+c)}X^2}$$

$$F_G = \sqrt{1 + 2c\frac{\cosh(X)}{Y^2} - \left(\frac{c}{Y}\right)^2}\, Y$$

$$F_X = \frac{X}{\left[n\left(c^{-\frac{1}{n}}-1\right)\right]^{\frac{n}{n+1}}} \tag{24}$$

where $Y = \frac{\Sigma_e}{\sigma_o}$ stands for the normalized overall effective stress, and $X = \frac{3}{2}\frac{\Sigma_m}{\sigma_o}$ denotes the normalized overall hydrostatic stress.

These "yield functions" are depicted in Figures 2 in the form of "yield surfaces" $F(X,Y) = 1$ for different values of the porosity c and the creep exponent n. In these Figures, we give results for representatives values of c (1% and 25%) and n (3 and ∞). The following observations and comparison can be made. In the form of a yield surface, the HS bound is a rigorous upper bound for all the other yield surfaces. Indeed the SC estimate lies below the HS bound for all values of n and c, and significantly below for the larger value of c.

In the low-triaxiality domain (X small), we observe that for sufficiently large porosity and small enough creep exponent, the estimate of Cocks [4, eq. 34] based on the finite shell model violates the bound. Correspondingly, the Gurson model violates the bound for sufficiently large porosity, and sufficiently large creep exponent. Of the two effects, the porosity is more important. As pointed out by Cocks [4], this effect is due to the use of too crude a trial field in the shell model for small triaxiality. Actually, Cocks [4] proposed an *ad hoc* modification to his model to alleviate this deficiency. His proposal actually yields the same coefficient for the Y term of his yield function as the HS bound function given in [11]. However, it can be shown that this modified result still violates the HS bound, if only slightly.

In the high-triaxiality domain (Y small), we find that the prediction of both the Gurson and Cocks models lie well within the HS bound. This suggests that the bound may be weak in that limit. In these plots, we have also included the pure hydrostatic loading results based on a finite shell model (the result is rigorous in this case). These are depicted by an arrow pointing to the appropriate value of X (hydrostatic load). Thus, we see that the smaller the porosity and the larger the creep exponent, the larger the overshooting by the HS bound in the high-triaxiality limit. We observe that the Gurson model agrees with the pertinent high-triaxiality limits (the arrows) for $n = \infty$ (because it was designed to do so), but it underestimates the high-triaxiality limits for the smaller value of n. In plasticity, it is commonplace to use Gurson's model for porous hardening materials by introducing an appropriate hardening mechanism, but otherwise leaving the yield surface unchanged. In view of the results presented here, this may introduce significant errors for large enough hardening, and it is not recommended. The model of Cocks performs better than the HS bound for the smaller value of the porosity, but not significantly better for the larger value. However, Cocks also gives special formulae for the high-triaxiality limits that are expected to perform better that the result plotted in our figures in that limit.

228

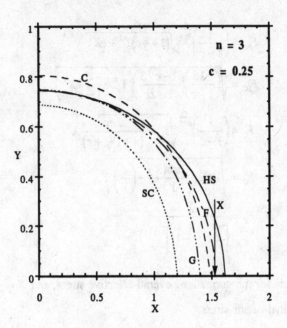

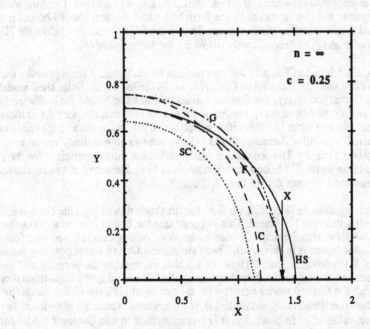

Figure 2 c, d. Plots of the yield surfaces $F = 1$ for the porous material (high porosity c)

Additionally, in Figures 2 we give results for an interpolation formula obtained by modifying the X coefficient in the HS bound formula to agree with the high-triaxiality limits. The result of this approximation (F) is

$$F_F = \left(\frac{1}{1-c}\right)^{\frac{n}{n+1}} \sqrt{(1+\tfrac{2}{3}c)Y^2 + \left[\frac{1-c}{n\left(c^{-\frac{1}{n}}-1\right)}\right]^{\frac{2n}{n+1}} X^2}, \tag{25}$$

and we can see that it performs significantly better in the high-triaxiality limit than the original HS prescription. However, the improvement in the high-triaxiality range seems to come at a cost in the low- to intermediate- triaxiality range for the larger creep exponents. However, this approximate prescription appears to be the best overall performer. This is not to say that other interpolations may not do better; the point is that until rigorous improved results are not available, we cannot really tell whether interpolation (25), or some other candidate is best.

Finally, we emphasize the *indirect* utility of the HS bound. Results based on micromechanical models are only as good as the accuracy of their predictions, and generally there is no *a priori* way to determine the accuracy of the predictions. For example, there is no reason to believe that the results of a calculation based on a finite shell subject to shear should correspond to the effective behavior of an arbitrary isotropic porous material for large values of the porosity. The results of this paper show that the results are in fact reasonably accurate. Thus, the rigorous bounds presented here, and other improved results that will no doubt be developed in the future, play a central role in the understanding of the constitutive behavior of composite materials, and should be considered equal partners to the physically based models in the search of better constitutive models of composite behavior.

ACKNOWLEDGEMENTS

This work was supported by the National Science Foundation under grant DDM-90-12230, and by the NSF/MRL program of the University of Pennsylvania under grant DMR-88-19885.

REFERENCES

[1] BAO, G., HUTCHINSON, J.W. and MCMEEKING, R.M. (1991) To appear.

[2] BUDIANSKY, B. (1965) *J. Mech. Phys. Solids* **13**, 223.

[3] CHRISTMAN, T., NEEDLEMAN, A. and SURESH, S. (1989) *Acta Metall.* **37**, 3026.

[4] COCKS, A.C.F. (1989) *J. Mech. Phys. Solids* **37**, 693.

[5] GURSON, A.L. (1977) *J. Engrg. Mater. Technol.* **99**, 2.

[6] HASHIN, Z. and SHTRIKMAN, S. (1962) *J. Mech. Phys. Solids* **10**, 335.

[7] HILL, R. (1963) *J. Mech. Phys. Solids* **11**, 357.

[8] HILL, R. (1965) *J. Mech. Phys. Solids* **13**, 213.

[9] LEE, B.J. and MEAR, M.E. (1991) To appear.

[10] McLAUGHLIN, R. (1977) *Int. J. Engng. Sci.* **15**, 237.

[11] PONTE CASTAÑEDA, P. (1991) *J. Mech. Phys. Solids* **39**, 45.

[12] PONTE CASTAÑEDA, P. (1991) In *Inelastic Deformation of Composite Materials* (ed. G.J. DVORAK), Springer-Verlag, New York, 216.

[13] PONTE CASTAÑEDA, P. (1991) To appear.

[14] PONTE CASTAÑEDA, P. and DE BOTTON, G. (1990) In *Developments in Theoretical and Applied Mechanics, Volume XV* (ed. S.V HANAGUD *et al.*), G.I.T., Atlanta, 653-660.

[15] PONTE CASTAÑEDA, P. and WILLIS, J. R. (1988) *Proc. R. Soc. Lond. A* **416**, 217.

[16] TALBOT, D. R. S. and WILLIS, J. R. (1985) *IMA J. Appl. Math.* **35**, 39.

[17] WILLIS, J. R. (1991) *J. Mech. Phys. Solids* **39**, 73.

[18] WU, T.T. (1966) *Int. J. Solids Struct.* **3**, 1.

DEFORMATION AND CREEP OF SILICON NITRIDE-MATRIX COMPOSITES

YURY GOGOTSI
Institut für Keramik im Maschinenbau, Universität Karlsruhe,
W-7500 Karlsruhe 1, FRG

DMITRY OSTROVOJ, VLADIMIR TRASKOVSKY
The Institute for Problems of Strength, Academy of Sciences,
Ukrainian SSR, 252014 Kiev, USSR

ABSTRACT

The stress-strain behaviour and creep resistance of sintered
ceramics containing free silicon and having structure similar
to self-bonded SiC, reaction-bonded ceramics with SiC, hot-
pressed TiN-containing ceramics and hot-isostatically pressed
Si_3N_4 were studied under varying stresses at temperatures up
to 1400°C in air, also with sodium sulphate present.
Interesting features concerning the high-temperature
properties of these materials are discussed.

INTRODUCTION

The application of ceramics as high-temperature materials is
proposed for many purposes, e.g. in engines and energy
systems. In order to replace metals for application under
these severe conditions outstanding properties are needed for
ceramics to be considered as alternative materials to
superalloys. In the high-temperature regime, the highest
strength values for structural ceramics are offered by the
various Si_3N_4 materials. Therefore the creep of traditional
reaction-sintered and hot-pressed Si_3N_4-based materials has
been investigated quite thoroughly [1-3]. Various mechanisms
have been suggested to explain the creep of glass-containing
ceramics such as sintered or hot-pressed silicon nitride.

These include dissolution-reprecipitation creep, dissolution-enhanced plasticity, and viscous flow [4].

But Si_3N_4-matrix composites have not been studied so well. For silicon nitride ceramics the reinforcement with SiC fibers or transformation toughening with ZrO_2 [5] are usually used. Recently increased attention has been devoted to particulate additions for the production of structural Si_3N_4-based ceramics [6].

In the earlier work on silicon nitrides and sialons, it has been also demonstrated that oxidation may have a profound influence on the microstructure of these materials by modification of the inter-granular phase [1, 2]. This phase plays a crucial role in the creep behaviour of these materials: in oder to reach the highest creep resistance, amorphous or glassy phases have to be minimized at grain boundaries and triple grain junctions. Hence creep tests in oxidizing, or more generally corrosive environments, may give rise to rather complex effects, since the microstructure of the material is not constant.

The main objective of this paper is to report some new results emanating from studies carried out on four Si_3N_4 materials from different routes in high temperature oxidizing environments. Creep as well as stress-strain measurements were performed with aims: to determine the high-temperature potential including critical temperature regimes, to understand the kinetics and mechanisms of the relevant microstructural processes at high temperatures and to evaluate the effect of particulate additions and environment on high-temperature properties of silicon nitride-based ceramics.

The study forms a part of a broader programme, focussed on the combined influence of mechanical loads and chemical environment on behaviour of non-oxide particulate composites, which are candidate materials for the most demanding high temperature applications.

MATERIALS

Four Si_3N_4 materials of laboratory origin were investigated, representing different technologies and group of additives. The materials tested in this study included samples of reaction-sintered silicon nitride (RBSN-SiC), post-sintered reaction-bonded silicon nitride (SRBSN-Si), hot-pressed silicon nitride (HPSN-TiN) and hot isostatically pressed silicon nitride (HIPSN). The composition and properties of ceramics under consideration are listed in Table 1.

TABLE 1
Characterization of materials

Material	Additive, vol.%	Sintering aid, wt.%	Phase composition	Density, g/cm^3	E, GPa	Ref.
SRBSN-Si	12 Si	Y2O3+MgO	β-Si$_3$N$_4$, Si, Y$_2$Si$_2$O$_7$	3.15	252	[7]
RBSN-SiC	30 SiC	MgO	β-Si$_3$N$_4$, α-SiC	2.51	190	[8]
HPSN-TiN	5 TiN	Al$_2$O$_3$	β-Si$_3$N$_4$, TiN	3.32	280	[9]
HIPSN		Y2O3+Al$_2$O$_3$		3.30		

Examples of the microstructure of the reaction-sintered materials are shown in Figure 1. All the ceramics contained sintering aids responsible for the glassy phase formed at the grain boundaries. The microstructure of SRBSN-Si is characterized by elongated β-Si$_3$N$_4$ grains growing through Si inclusions. The studies on the microstructure of RBSN-SiC and HPSN-TiN demonstrated that coarse SiC (light grains in Fig. 1a) and TiN grains were relatively uniformly distributed over the silicon nitride matrix (grey area in Fig.1a). The shape of TiN and SiC grains was fragmental, inequiaxial.

All specimens were cut from large plates, then ground with a diamond wheel to the desired size.

EXPERIMENTAL PROCEDURE

The creep and stress-strain measurements were conducted in four point bending on a precise jig that eliminated most of the extraneous twist upon loading [10]. The inner and outer span were 20 and 40 mm respectively. The specimen size was

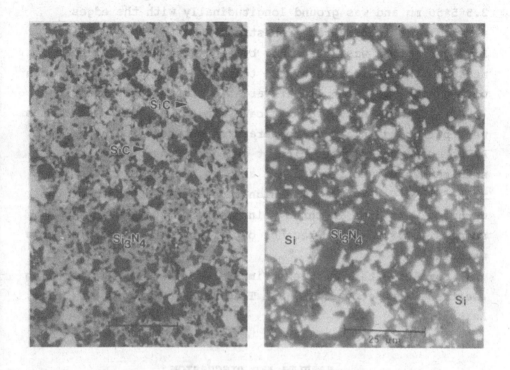

Figure 1. Microstructure of RBSN-SiC (a) and SRBSN-Si (b)

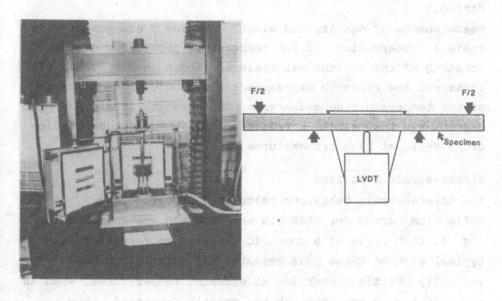

Figure 2. Test machine and schematic diagram of the jig used
to monitor load-deflection during flexure of
specimens in four point bending

3.5*5*50 mm and was ground longitudinally with the edges rounded. The loading rate for strength tests was 0.5 mm/min. Load-deflection was determined by a purpose built jig that measured the deflection of the tensile section of the specimen with a high precision linear variable displacement transducer (LVDT), this is shown schematically in Fig. 2. The creep tests were performed in air at temperatures between 1200 and 1300°C. All components of the equipment being in contact with the specimen were made from Si_3N_4 and the heating elements were made from SiC (Fig. 2). The high temperature strength of the ceramics were done in three point flexure (20 mm span) on samples 2.75*3.5*25 mm because of limitations of the available material.

Samples after creep experiments were examined by scanning electron microscopy (JEOL, JSM T-20) and metallography (Zeiss, Axiomat).

RESULTS AND DISCUSSION

Strength

Measurements of density and elastic modulus E are compared in Table 1. Observations of the temperature dependence of the strength of the various materials are shown in Fig. 3. In all instances the strength decreases above 1000°C. The comparison of the temperature dependences of the strength in air and argon shows the drastic effect of oxidation on the properties of RBSN-SiC at high temperatures (Fig. 3b).

Stress-strain behaviour

The stress-strain behaviour calculated from the load-deflection curves for HPSN-TiN and RBSN-SiC are compared in Fig. 4. Each curve at a specific temperature represents a typical plot of three measurements. All materials behave initially brittle manner and at elevated temperatures, when as shown in Fig. 3 the strength has greatly decreased, they become ductile (Fig. 5). At 1000°C all the materials were still brittle (stress-strain diagrams were linear).

The studies of SRBSN-Si strain and fracture at different

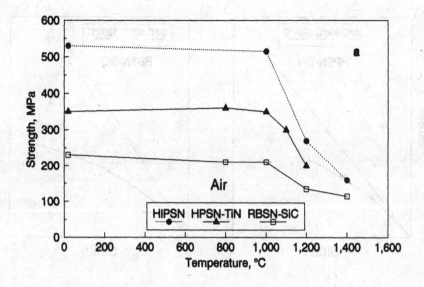

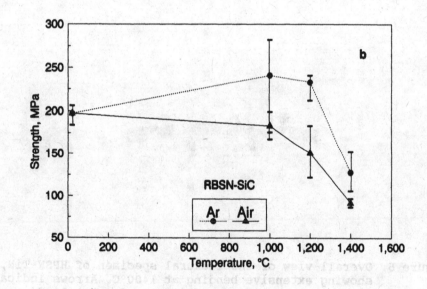

Figure 3. Bending strength as function of temperature

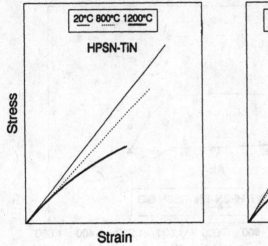
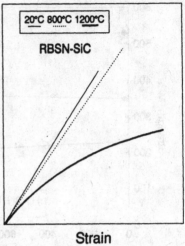

Figure 4. Stress-strain curves for HPSN-TiN and RBSN-SiC at different temperatures

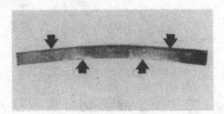

Figure 5. Overall view of the flexural specimen of HPSN-TiN, showing extensive bending at 1400°C. Arrows indicate approximate positions of inner and outer loading edges

loading rates (0.5 and 0.005 mm/min) have shown (Table 2) that the fracture of samples up to 1300°C occurs at a small ultimate strain which points to subcritical crack growth and weak creep tendency. At the same time the presence of sodium sulphate on the sample surface at a small strain rate results in an abrupt growth of deformability and a weak increase in strength, with horizontal sections on stress-strain diagrams. A flow plateau on the diagrams is indicative of creep activation under given experimental conditions.

TABLE 2
Mechanical properties of SRBSN-Si

Tempera-ture, C	Environ-ment	Loading rate, mm/min	Strength, MPa	Ultimate strain, %	E, GPa
20	Air	0.5	506	0.201	252
20	Air	0.005	510	0.202	252
1300	Air	0.5	390	0.457	140
1300	Air	0.005	191	0.671	111
1300	Na_2SO_4	0.5	410	0.427	189
1300	Na_2SO_4	0.005	222	1.572	117

Creep

As can be seen in Fig. 6a, at 1200°C a rather high creep rate of RBSN-SiC was observed at a load equal to 40% of its ultimate bending strength. At 60% it growth double-fold and at 80% it increases 7 times. For HPSN-TiN (Fig. 6b) the creep rate at a 40% load exceeds 2 times that of RBSN-SiC due to a higher content of additives in the former. At 60% the area of steady creep was absent at all, and already in several minutes after the load application the fracture of a sample occured. The specimen may break due to the phenomenon of slow crack growth (Fig. 7). While at a low stress level, rupture is controlled by a critical creep strain, creep rupture under high stresses is controlled by the subcritical growth rate of the dominating crack. Creep of these polyphase materials is always accompanied by damaging processes, e.g. cavitation and microcracking. The comparison of the curves otained with similar ones for ceramics without a filler (Fig. 6c and [2])

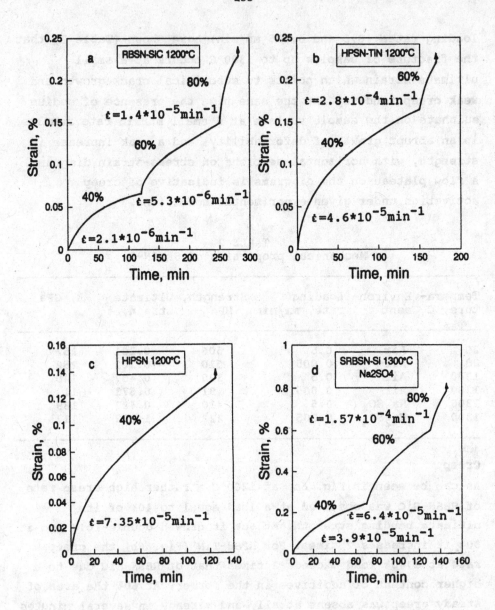

Figure 6. Creep curves at loads equal to 40, 60 and 80% of
ultimate bending strength at corresponding
temperatures

has demonstrated that particulate reinforcement does not practically influence creep resistance. It is associated with the fact that the mechanical behaviour of such ceramics in the temperature range investigated is determined, first of all, by the state of a grain-boundary phase. One can suppose that viscows flow is the mechanism responsible for creep in these materials [4]. Further evidence comes from microscopy studies of the samples before and after creep. These studies prove the contributions made by complex creep processes, including cavitation and microcrack formation.

Creep behaviour can be described by the relation

$$t = a\sigma^n exp(-Q/RT),$$

with a factor a and the stress exponent n. If one calculates the stress exponent n from the creep curves with changing stresses at 1200°C (Fig. 6), it results in values near 4. Such values are most commonly observed for ceramic materials [3, 11] and may be a consequence of accelerated creep damage processes.

Creep experiments (Fig. 6d) confirmed the effect revealed at short loading times for SRBSN-Si samples covered with Na_2SO_4. And at longer loading times stress corrosion cracking and strength decrease characteristic of RBSN-SiC and HPSN-TiN under similar conditions [8,9] were not observed. This phenomenon may be explained by the Ioffe effect (uniform dissolution of a defective surface layer in the soda-silica glass (Fig. 8b) being formed) or the diffusion of sodium ions along the grain boundaries into inner material layers resulting in higher plasticity of the grain-boundary phase and grain boundary sliding. In the latter case lower strength of the samples could be expected. The absence of this effect may be associated with free silicon present in the structure of the material. Viscous flow can cause crack retardation as it was observed for self-bonded SiC [12].

Consequently, there is still a deficit in the comprehensive treatment and the complete understanding of these important long-term-high-temperature phenomena in silicon nitride ceramics. A better understanding may help to improve the materials quality in future.

Figure 7. Overall view of fracture surfaces in creep specimens of HIPSN (a) and HPSN-TiN (b), showing slow crack growth regions

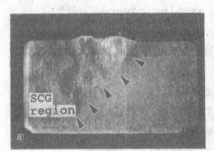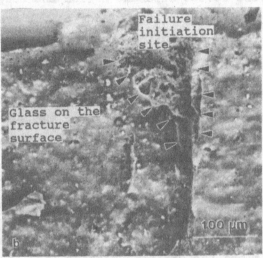

Figure 8. Typical fracture surfaces of SRBSN-Si specimen, showing slow crack growth region (a) and glass formation promoting grain boundary sliding and creep deformation (b)

ACKNOWLEDGEMENTS

Dr. Yury Gogotsi gratefully acknowledges the receipt of a Research Fellowship from Alexander von Humboldt Foundation.

REFERENCES

1. Thümmler, F. and Grathwohl, G., High temperature oxidation and creep of Si_3N_4- and SiC-based ceramics and their mutual interaction. In <u>MRS Int. Meeting on Advanced Materials</u>, Materials Research Society, 1989, **4**, pp. 237-53.

2. Van Der Biest, O., Valkiers, S., Garguet, L., Tambuyser, P. and Baele, I., Creep, oxidation and microstructure in silicon nitrides. <u>Brit. Ceram. Proc.</u>, 1987, **39**, 33-44.

3. Gürtler, M. and Grathwohl, G., Tensile creep testing of sintered silicon nitride, In <u>Proc. 4th Int. Conf. on Creep and Fracture of Engineering Materials and Structures</u>, The Institute of Metals, London, 1989, pp. 399-408.

4. Wilkinson, D.S. and Chadwick, M.M., Creep mechanisms in silicon nitride at elevated temperatures. In <u>Proc. 11th Risø Int. Symp. on Metallurgy and Materials Science</u>, ed. J.J. Bentzen et al., Risø National Laboratory, Roskilde, 1990, pp. 517-22.

5. Falk, L.K.L., Microstructure and oxidation behaviour of Si_3N_4/ZrO_2 ceramics. In <u>Proc. 11th Risø Int. Symp. on Metallurgy and Materials Science</u>, ed. J.J. Bentzen et al., Risø National Laboratory, Roskilde, 1990, pp. 277-82.

6. Pompe, R., Prospects for non-oxide particulate composites at high temperatures. In <u>Proc. 11th Risø Int. Symp. on Metallurgy and Materials Science</u>, ed. J.J. Bentzen et al., Risø National Laboratory, Roskilde, 1990, pp. 97-110.

7. Vikulin, V.V., Dorozhkin, A.I., Ogneva, I.V., Nikonov, Yu.P. and Ovchinnikov, V.I., Post-sintered reaction bonded silicon nitride and self-reinforced Si_3N_4-Si composite materials. <u>Ogneupory</u>, 1989, No. **9**, 29-33 (in Russian).

8. Gogotsi, G.A., Zavada, V.P. and Gogotsi, Yu.G., Strength degradation of Si_3N_4-SiC-based ceramics in salt environments. <u>Ceram. Int.</u>, 1986, **12**, 203-8.

9. Gogotsi, Yu.G., Zavada, V.P., Zudin, N.N., Ivzhenko, V.V. and Traskovsky, V.V., Strength of silicon nitride ceramics in various environments. <u>Sverkhtverdye Materialy</u>, 1990, No. **3**, 25-9 (in Russian).

10. Gogotsi, G.A., Zavada, V.P., Kutnyak, V.V. and Ostrovoj, D.Yu., A machine for determination of the mechanical properties of ceramics at high temperatures. <u>Strength of Mater.</u>, 1988, **20**, 558-62.

11. Kingery, W.D., Bowen, H.K. and Uhlmann, D.R., <u>Introduction to Ceramics</u>, Wiley, New York, 1975.

12. Gogotsi, G.A., Gogotsi, Yu.G. and Ostrovoj, D.Yu., Deformation and fracture of self-bonded silicon carbide at various loading rates. <u>Ogneupory</u>, 1989, No. **10**, 27-30 (in Russian).

LOCALIZATION AND DEVELOPMENT OF DAMAGE UNDER HIGH TEMPERATURE LOADING CONDITIONS

VÁCLAV SKLENIČKA, IVAN SAXL[+], JOSEF ČADEK
Czechoslovak Academy of Sciences, Institute of Physical
Metallurgy, Žižkova 22, 616 62 Brno, Czechoslovakia
[+]Mathematical Institute, Mánesova 94,12000 Prague,Czechoslovakia

ABSTRACT

The standard metallographic methods of the estimation of the degree of intergranular damage (cavitation) based on the quantitative microscopy are rather unsatisfactory since the typical feature of cavitation is its pronounced inhomogeneity. This work attempts to highlight the problem areas just in this respect. The attention is focused on the semiquantitative and quantitative investigation of inhomogeneities in the spatial distribution of cavities and the proposed approaches are illustrated by recent experimental results. The total of acquired information gives the detailed quantitative picture of cavitation in different stages of loading and can be used in constitutive models of creep fracture process. The critical amount of creep damage accumulation is defined and demonstrated experimentally. Finally, the application of the percolation model of cavitation to explain theoretically the critical amount of cavitation has been used.

INTRODUCTION

The static or slowly changing in time loading of materials of components in service at elevated temperatures leads to creep deformation and creep damage that limit the component's life. The creep damage results from: (a) microstructural changes of a loaded material producing to a continuous reduction in the creep strength during service, (b) intergranular damage (cavitation) |1|. The general consequence of the creep intergranular damage is a severe loss of ductility. This is why the phenomenon of creep cavitation is of considerable importance for high temperature technology.

The creep cavitation depends strongly on material and loading conditions; in some creep resistant materials cavities appear already in early stages in loading, whereas others cavitate in an observable amount only towards the end of their creep life t_f so that the life fraction t/t_f at which cavitation can be detected is variable. Nevertheless, some features of cavity accumulation are common to all situations and form a basis of the lower bound life prediction. ·Consequently, the estimation of the degree of cavitation based on the quantitative metallography of cavitation should be the starting point of any remanent life assessment route. Unfortunately, the standard methods |2| are rather unsatisfactory as the typical feature of cavitation is its inhomogeneity originating from different conditions of their nucleation and growth at various structural elements.

This work attempts to highlight the problem areas just in this respect and thus tries to establish basis for the solution of the problems outlined. The attention is focused on the semiquantitative and quantitative investigation of inhomogeneities of the spatial distribution of cavities and the proposed approaches are illustrated by recent experimental results obtained in the Institute of Physical Metallurgy of the Czechoslovak Academy of Sciences.

INTERGRANULAR CREEP DAMAGE DEVELOPMENT

Intergranular creep fracture involves the nucleation of intergranular cavities, their growth and linking to form cracks, and the eventual propagation of these cracks leading to the final fracture (Fig. 1).

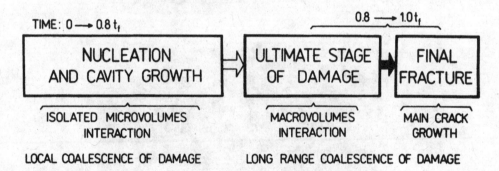

Figure 1. Intergranular creep fracture development.

For a long loading time, the development of intergranular creep damage takes place in mutually independent microvolumes (Fig. 2a). Independently of the operating mechanisms of the heterogeneous cavity nucleation, sooner or later the coalescence of cavities starts to take place (Fig. 2b). In its course, which already represents an advanced state of damage, the conditions are no more steady; depending on their stress sensitivities the operating processes speed up. Then the inhomogeneity of spatial distributions of cavities is made more

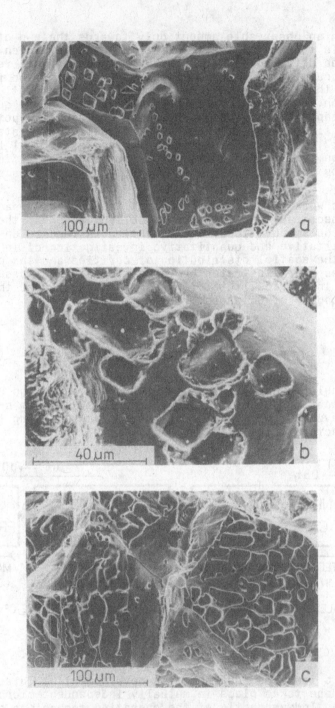

Figure 2. SEM micrographs showing (a) isolated creep cavities
(Ni 99.99%, 15 MPa, 1073 K),(b) flat coalesced cavities
(α-Fe, 20 MPa, 873K) and (c) heavily cavitated facets
in the ultimate state of creep damage (Ni 99.99%,
15 MPa, 1073 K).

profound and clearly observable, namely islands of coalesced cavities (Fig. 2c) and later also grain boundary microcracks appear first in the close vicinity of the free surface. Even when the formation of superficial cracks may proceed continually, the positions of those nucleated earlier are important as the predetermined places for the start of the last stage of the fracture process. This stage is characterized by the long range coalescence of damage inside the creeping body and by the junction of internal cracks with surface cracks. The final fracture takes place by a relatively quick propagation of one of long surface cracks - the "main" crack.

MICROSTRUCTURAL FEATURES FOR CREEP DAMAGE QUANTIFICATION

Due to the difficulties involved, precise experimental investigation of creep void localization and development are quite scarce in the literature. However, a detailed analysis of the creep damage can be performed using slightly modified approaches of quantitative metallography based on simple stochastic models of structure and structural processes |3-5|.

The important features of the multiconnected grain boundary surface are facets (two-grain junctions - faces of approximately polyhedral grains), edges (three-grain junctions - edges of grains) and corners (four-grain junctions - vertices of grains). All these structural constituents are possible nucleation sites for cavities but the actual nucleation and/or growth conditions may differ considerably depending on the external conditions.

The structural elements will be denoted as follows: α, β are the cavities and grain boundaries, resp., $\beta_i \subset \beta$ is the one-dimensional framework of the edges. Further, B^i is the two-dimensional sampling quadrat (window) moving in the chosen section plane F. Finally, the usual notation A and L is used for surface area and length, respectively. Thus $\beta \cap B$ are the grain boundary traces in the examined window, $\alpha \cap \beta$ are the cavity sections in grain boundaries, $\alpha \cap \beta \cap F$ are the cavity chords cut by the grain boundary traces in the section plane, $\beta_i \cap B$ are the point traces of edges etc.

INHOMOGENEITY OF CREEP DAMAGE

The loading conditions (stress, temperature) determine the processes participating in creep damage accumulation such as intragranular and intergranular deformation, volume, grain boundary and surface diffusion and, quite naturally, influence the number, shape and especially the spatial distribution of cavities.

The semiquantitative and quantitative investigation of the spatial distribution of cavities has been carried out on creep specimens of various metallic materials crept in a broad range of the loading conditions. It was found that there are four scale levels at which the inhomogeneity of cavitation manifests itself, namely (i) along and across the specimen body, (ii) at various grain boundary facets, (iii) at three-grain junctions and (iv) within the facet area. Moreover, also the cavity size, varies considerably and is another characteristic of the local accumulation of creep damage.

INHOMOGENEITY ALONG AND ACROSS THE SPECIMEN BODY

This macroscopic inhomogeneity can be most simply estimated by the value of the measure ratio (linear fraction of α on the traces $\beta \cap B$)

$$\xi = \frac{L(\alpha \cap \beta \cap B)}{L(\beta \cap B)} \qquad (1)$$

corresponding to different positions of the moving quadrat B of a suitably chosen (fixed) orientation: either perpendicular to ($B_{x\perp}$) or parallel with ($B_{y\parallel}$) the direction of the applied stress σ-see Fig. 3. Note, that for the isotropic stationary feature set $\alpha \cap \beta$ represents ξ the unbiased estimator of the areal fraction A ($\alpha \cap \beta$)/A(β).

The above mentioned measurements performed on creep fractured specimens of copper and some of its alloys and several creep resistant steels led to the following conclusions: (i) longitudinal inhomogeneity: In copper and its primary solid solutions with Al and Zn, the variation of $\xi_\perp$ along the gauge length is negligible in a broad range of stresses and temperature. The distribution of creep damage in creep resistant steels is more complicated. At the lowest stresses at which purely intergranular creep occurs, the homogeneous (Fig. 4 - σ = 70 MPa) spatial distributions of cavities were found. On the other hand, a pronounced stress and temperature dependent localization of creep damage in the vicinity of fracture surface occurs

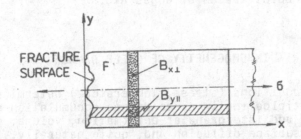

Figure 3. Schematic illustration of the selection of areas of the intergranular damage quantitative examination.

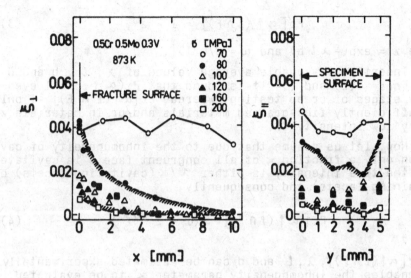

Figure 4. The inhomogeneties of cavitation along the gauge
length and at the cross-section of creep fractured
specimens tested at different applied stresses
(low alloy CrMoV steel).

especially in the region of a mixed mode (transgranular and
intergranular) of creep fracture |6|.
(ii) cross-sectional inhomogeneity: the extent of cavitation is
highest at the specimen surface and $\xi_{\parallel}$ gradually decreases (at
all stresses and temperatures) when moving $B_{y \parallel}$ into the
specimen interior |6|. In this case, no significant difference
between examined materials has been found.

INHOMOGENEITY AT VARIOUS GRAIN BOUNDARY FACETS

The extent of cavitation at various grain boundary facets can
be followed either by means of the serial sectioning technique
|7| or by inspection of their traces in the metallographic
section plane. For the latter case, a simple stochastic model
and method of estimation have been developed |3|.

Let us assume that all the facets are congruent an their
area is ℓb, where ℓ, b are the mean chord length and mean
width of the facet, respectively. Further, let the cavities be
spheres of fixed diameter d and the point pattern of their
centres on every facet is a random piece of Poisson point
process with the intensity λ. Then the probability of a randomly chosen facet f containing at least one cavity section is

$$Pr(\alpha \uparrow f) = 1 - z \tag{2}$$

and the probability that a facet trace $f \cap F$ in the section
plane hits a cavity is

$$Pr[\alpha \uparrow (f \cap F)] = 1 - z^\omega, \tag{3}$$

where $z = \exp(-\lambda \ell b)$ and $\omega = d/b$ |3|.

Inserting reasonable average values of λ, ℓ, b and d into Eqns (2) and (3) it is found that $Pr(\alpha \uparrow f) = 1$ even in early stages of creep testing, whereas $Pr[\alpha \uparrow (f \cap F)] < 1$ only in sufficiently fine-grained materials and/or in materials with a very low intensity λ.

Now, let us assume that due to the inhomogeneity of cavitation only a fraction $\varkappa$ of all congruent facets is cavitated. Then the true intensity is either $\lambda/\varkappa$ (cavitating facets) or 0 (remaining facets) and consequently

$$Pr[\alpha \uparrow (f \cap F)] \doteq \varkappa (1 - z^{\omega/\varkappa}). \tag{4}$$

As $Pr[\alpha \uparrow (f \cap F)]$, λ, ℓ and d can be estimated experimentally, Eq. (4) enables the inhomogeneity parameter $\varkappa$ to be evaluated. Then we can estimate also the areal fraction of cavity sections in cavitating facets, namely $\tilde{\xi} = \xi/\varkappa$. Even when this model oversimplifies the real situation it can be used at least for a semiquantitative estimation of $\varkappa$.

The mean values of damage parameters $\varkappa$ and ξ_f of the investigated materials are sumarized in Fig. 5.

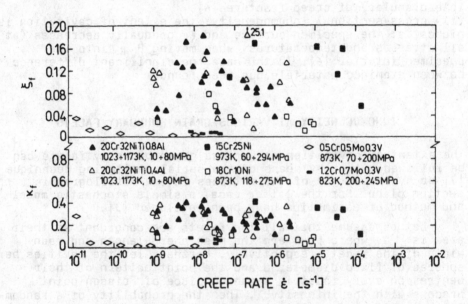

Figure 5. Fracture values of the areal fraction of cavities, ξ_f, and the fraction of cavitating grain boundaries, $\varkappa_f$ for various creep-resistant steels.

INHOMOGENEITY AT THREE-GRAIN JUNCTIONS

The grain edges and spherical cavities centred at them generate in the section plane two interpenetrating point processes, namely the points $\beta_1 \cap F$ and suitably chosen points of the sections $\alpha \cap F$ - the intensity of cavitation at the edge traces (points $\alpha \cap F \cap \beta_1$) is μ_t. If there is no preferential cavitation at the edges then (as just three facets traces meet in any $\beta_1 \cap F$) |4|

$$\mu_t = (3/2)\,\xi \,. \tag{5}$$

Thus the coefficient $C_e = 2\mu_t/3$ describes the inhomogeneity of cavitation at the grain edges; if $C_e > 1$ then pronounced cavitation at the edges takes place (Fig. 6). Close connection of the inhomogeneity of this type with s.c. wedge cracks has been firmly established |9|.

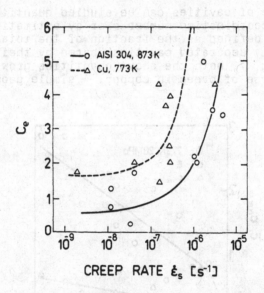

Figure 6. The rate dependence of the coefficient C_e disclosing the pronounced inhomogeneity of cavitation at three-
-grain junctions at higher creep rates.

INHOMOGENEITY WITHIN FACET AREAS AND CAVITY SIZE

A simple method of estimating the number of contacts between neighbouring cavities and the coordination number $\bar{c}$ (the mean number of contacts per cavity) from the planar sections has been proposed in a previous paper |5|. It was then found using the theory of random clumping (continuous percolations) that the experimentally observed pronounced increase of $\bar{c}$ with the areal fraction f can be explained assuming that cavities in the facet interior form linear chains. This assumption has been also confirmed experimentally using the serial sectioning technique |7|.

Closely spaced serial sections can also be interpreted as a three- dimensional probes (disector, selector) producing unbiased estimates of cavity size distribution and enabling to reconstruct the cavity shapes. The widths (projection heights) of cavities are of the order of $1 \div 10$ μm and their shape is frequently polyhedral (at three-grain junctions sometimes non- -convex) |10|.

COALESCENCE OF CAVITIES AND THE ULTIMATE STATE OF DAMAGE

Coalescence of cavities

The coalescence of cavities can be studied quantitatively by measuring the coordination number $\bar{c}$ or by estimating the planar contiguity $c_{\alpha\alpha}$ defined as the fraction of the total boundary area of formally separated cavities shared by their contacts. As shown in Fig. 7, both the above quantities grow monotonously during the course of creep in copper. A simple geometric

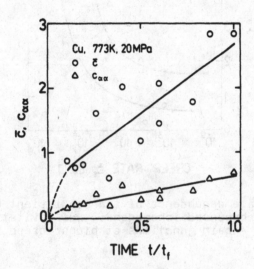

Figure 7. Time dependence of the coordination number $\bar{c}$ and the planar contiguity $c_{\alpha\alpha}$.

approach |5| enables not only to explain their mutual relation, namely $c_{\alpha\alpha} \sim \bar{c}$, but also to describe the kinetics of cavity coalescence and chaining. For time $t < 1/3 t_f$ isolated cavities and bicavities prevail ($\bar{c} \leq 1$), later on the chains of cavities

($\bar{c} \leq 2$) are formed (the linking of cavities occurs preferably along certain directions and finally, at $t > 2/3\ t_f$, the development of areal clusters (in the intersections of singular and multiple chains) and microcracks proceed ($\bar{c} > 2$).

The ultimate stage of damage

The final stage of fracture, namely the formation and propagation of the main crack, proceeds spontaneously in a very short time and escapes the observation. Consequently, we are confined to examine only the **"ultimate prefracture state"** in the vicinity of fracture surface or in the body of homogeneously cavitated specimen at the time immediately preceding the time of fracture. Numerous studies on several materials carried out in a broad range of experimental conditions have shown that this critical state of damage can be characterized by the value of the critical cavitation of cavitating facets $\tilde{\xi}_f$ = = 0.2 - 0.4 - see Fig. 8. Hence it can be seen that this value depends neither on creep conditions (temperature, stress) nor on the material investigated. Thus, $\tilde{\xi}_f$ can be considered to be a universal parameter characterizing the ultimate state of the creep damage by intergranular cavitation. Therefore also, the ratio $0 \leq \tilde{\xi} / \tilde{\xi}_f \leq 1$ is a universal damage parameter, which can be expected to be independent of material and testing conditions. To explain theoretically the critical value of $\tilde{\xi}_f$ the bond percolation theory can be used |4|.

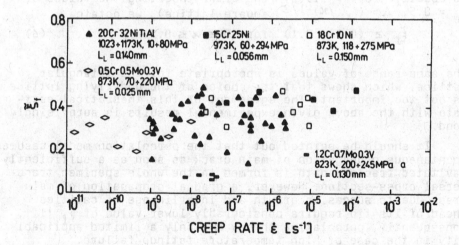

Figure 8. Fracture values of the areal fraction of cavities related to cavitating grain boundaries, $\tilde{\xi}_f$ for various creep-resistant steels. The values of $\tilde{\xi}_f$ were calculated from the data given in Fig. 5. (L_L is grain size).

The percolation theory |8| solves the problem of the existence of an infinite cluster or of an infinite path in an unbounded random graph, the vertices and edges (bonds) of which are open for connection with given probabilities $0 \leq p_v, p_e \leq 1$. It can be shown that there is a whole domain of values $p_v < 1$, $p_e < 1$, for which the probability of an infinite cluster $Pr_\infty(p_v, p_e) > 0$. The highest value of p_e, for which $Pr_\infty(1, p_v) = 0$ is called the bond percolation limit $\hat{p}_e$; for a random graph on a square lattice is $\hat{p}_e = 0.5$, on a triangular lattice $\hat{p}_e = 0.35$.

The application of this theory to the stability problem of cavitated structure is based on the assumption that the long--range coalescence of cavities necessary for the propagation of the main crack is analogous to the formation of an infinite cluster or path and that p_v can be related to the intensity of cavitation λ, whereas p_e is connected with the surface-to--surface distance between neighbouring cavities. To simplify the problem, we have assumed that $p_v = 1$ (all potential cavity sites are occupied) and set $p_e = \bar{d}/[\lambda(1-\gamma)]$. Here $\bar{d}$ is the mean cavity diameter, Λ is the centre-to-centre distance of cavities and $0 \leq \gamma \leq 1$ is the parameter describing the stability of "bridges" between neighbouring cavities; namely, at the length $\gamma\Lambda$ the bridge disrupts and cavities coalesce spontaneously. Now, the critical state of cavitation is achieved when the value of $\bar{d}_f = \hat{p}_e [\Lambda(1-\gamma)]$, to which corresponds the areal fraction of cavity sections in cavitating grain boundaries is $\tilde{\xi}_f = \pi\lambda(d_f)^2/4\varkappa$. In the experimentally observed populations is usually $\overline{(d)^2} = 1.5(\bar{d})^2$ and after inserting the values of $\hat{p}_e = 0.5$, $\Lambda = (\lambda/\varkappa)^{-1/2}$ (square lattice), we obtain

$$\tilde{\xi}_f \geq (0.3 \div 0.1) \quad \text{for} \quad 0 \leq \alpha \leq 0.5. \tag{6}$$

The same range of values is appropriate for the triangular lattice, which shows that the choice of the underlying lattice is not too important. The agreement of this theoretical estimate with the above given experimental results is surprisingly good.

It should be pointed out that the percolation model assumes spontaneous propagation of main crack as soon as a sufficiently cavitated fracture path is formed in the whole specimen transversal cross-section. However, a gradual propagation of main crack due to succesive growth and interlinkage of cavities ahead of its tip require considerably lower value of $\tilde{\xi}_f$ |11|. Consequently, percolation theory has only a limited applicability in the case of high temperature fatigue failure.

CONCLUDING REMARK

Creep damage mechanics has been developed as a continuum approach to the analysis of creep fracture. The damage is measured by the scalar parameter ω which varies from zero (no damage) to one (fracture). One interpretation of the Kachanov parameter ω is the decrease in load-area. In a metallographic assessment, this would correspond to the area of transverse grain boundaries occupied by cavities. Hence, the ratio $\tilde{\xi}/\tilde{\xi}_f$ is the microstructural equivalent of the damage parameter ω in the classical Kachanov-Rabotnov treatment |1|.

REFERENCES

1. Riedel, H., _Fracture_ _at_ _High_ _Temperature_, Springer-Verlag, Berlin, 1987, pp. 51 - 259.

2. Underwood, E.E. _Quantitative_ _Stereology_. Addison-Wesley, Reading (Mass.), 1970, pp. 80 - 108.

3. Saxl, I. and Stoyan,D., Application of stochastic geometry to the analysis of high temperature creep cavitation. _Kovové_ _mat_. (Metallic materials), 1985, **23**, pp. 298 - 308.

4. Saxl, I., Sklenička, V. and Čadek, J., Inhomogeneities of intercrystalline cavitation in metals. In _GEOBILD´_ _85_, Wissenschaft. Beiträge der F. - Schiller-Universität, Jena, 1985, pp. 159 - 65.

5. Saxl, I., Sklenička, V. and Čadek, J., Quantitative microscopy of intercrystalline cavitation. _Z_. _Metallkde_, 1981, **72**, 499 - 03.

6. Sklenička, V. and Čadek, J., Intergranular cavitation and prediction of creep life. In _High_ _Temperature_ _Alloys_ _for_ _Gas_ _Turbines_ _and_ _Other_ _Applications_ _1986_, ed. W. Betz, D. Riedel Publ. Co., Dodrech, 1986, Part II, pp. 1585 - 97.

7. Procházka, K., Saxl, I., Sklenička, V. and Čadek, J., On the formation of magistral crack in high temperature creep of copper. _Scr_. _Metall_. 1983, **17**, pp. 779 - 84.

8. Hammersley, J.M. and Welsh, D.J., Percolation theory and its ramification. _Contemp_. _Phys_. 1980, **21**, pp. 593 - 05.

9. Čadek, J., _Creep_ _in_ _Metallic_ _Materials_, Elsevier Sci. Publ., Amsterdam, 1988, pp. 290 - 92.

10. Saxl, I., Use of 3D probes in metallography. In _Proc_. _8th_ _Int_. _Conf_. _on_ _Stereology_, Irvine 1991 (to be published).

11. Sklenička, V., Lukáš, P. and Kunz, L., Intergranular Fracture in Copper under High Temperature Creep Fatigue and Creep-Fatigue Conditions. _Scr_. _Metall_. _Mater_., 1990, **24**, pp. 1795 - 00.

NON-LINEAR MODELS OF CREEP DAMAGE ACCUMULATION

VLADISLAV P. GOLUB

Mechanics of Creep Department, Research Institute of Mechanics,
Ukrainian Academy of Sciences,
Nesterova str., 3, Kiev-57, USSR 252057

ABSTRACT

The article deals with the problem of constructing adequate damage evolutionary equations under creep conditions. The main approaches to damage function identification are specified. A critical analysis of existing hypotheses of damage accumulation is presented. A damage accumulation process is shown to be essentially nonlinear both depending on stresses and time. The nonlinearity conditions and principles of nonlinear model construction are formulated. Only evolutionary equations which observe the "separability" principle are shown to lead to the nonlinear models. The balance of specific energies of temporal and momentary damage components is used to specify the "separability" principle. The loading history is accounted for within the framework of constructed nonlinear models. The residual lifetime of rods under stepwise loading are calculated.

INTRODUCTION

Two different approaches to damage term indentification have formed historically in mechanics of materials. The first one based on "partial measures" for damage evaluation was proposed by A. Palmgren [1], M.A. Miner [2] and E.L. Robinson [3]. In the simplest case of one-demensional creep the damage measure is written in the form

$$\Delta\omega = \frac{\Delta t_i}{t_{Ri}} \longrightarrow d\omega = \frac{dt}{t_R(\sigma)} . \qquad (1)$$

and the fracture condition for the discrete and continual laws of load variation is prescribed by the relation

$$\sum_{i=1}^{n} \frac{\Delta t_i}{t_{Ri}} = 1 \longrightarrow \int_0^{t_R} \frac{dt}{t_R(\sigma)} = 1 \qquad (2)$$

which is known in the literature as the linear cumulative damage rule. Here Δt_i and dt are the period of the i - th load activity ($i = 1, 2, \ldots$, n); $t_R(\sigma)$ represent the time to fracture at the activity of the same constant load.

The physical essence of "linearity" implied by law (2) is that equal intervals of partial time periods $\Delta T = \Delta t/t_R$ lead to equal damage $\Delta\omega$ increments. The damage kinetics is reflected by the straight line both in normalized time scale (Figure 1(a)), curve 1) and in the physical time scale (Figure 1(b), dash-dot lines). The total damage is determined from a simple summation, and fracture occurs when the total damage reaches 1. In the particular case of a two step loading the graphical interpretation of the linear cumulative damage rule is prescribed by the straight line on the axes of Figure 1.

The linear cumulative damage rule is obviously insensitive to loading history. The total normalized lifetime, in accordance with (2), will always be equal to 1 independently of the sequence of applying the loads. Actually, however, experimental data demonstrate a dependence of the residual time on the loading history. In particular, an incremental sequence of stress changes (additional loading) leads to the lifetime being less than 1 (Figure 1c, open symbols), while a decremental sequence of stress changes (unloading) leads to the lifetime being higher than 1 (Figure 1c, dark symbols).

Another approach to damage evaluation was proposed by L.M. Kachanov [4]. It is based on an interpretation of damage in terms of a function which is never decreasing in time and which characterizes irreversible changes of material properties. This approach was named Continuum Damage Mechanics [5].

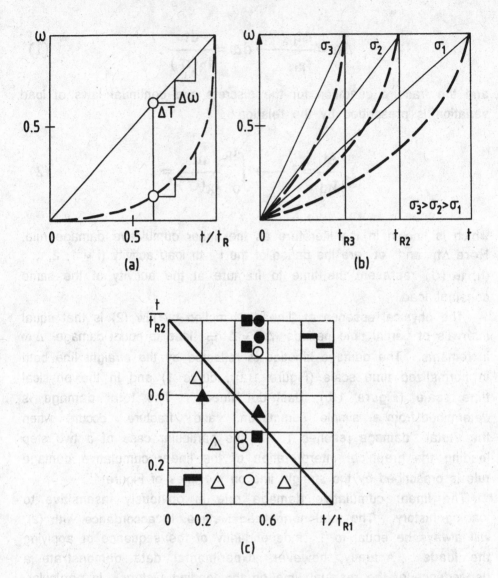

Figure 1. Kinetics of damage accumulation in normalized (a) and physical (b) time and under two stepwise loading conditions (c)

A scalar function ω is the most frequently chosen as the damage measure. The damage kinetics is prescribed by the differential equation

$$\frac{d\omega}{dt} = F(\sigma, \omega, \dots) \qquad (3)$$

with the initial (t = 0) condition and failure (t = t$_R$) condition expressed in the following form

$$\omega(t)\big|_{t=0} = 0 \quad \text{and} \quad \omega(t)\big|_{t=t_R} = 1 \tag{4}$$

The increment of the function ω over a certain short period of time dt depends evidently on the value of ω at the start of the period and on the applied load appropriate to this period.

To reflect the cumulative character of the damage process the F function should be non-decremental. All known specifications of F in equations like (3) may be generalized by the relation

$$\frac{d\omega}{dt} = C\left(\frac{\sigma}{1-\omega^z}\right)^m \cdot \frac{\omega^\beta}{(1-\omega)^q}, \tag{5}$$

where C, m, z, q, β are experimentally determined coefficients (C > 0; m $\geq$ 1; z $\geq$ 1; q $\geq$ 0; β $\geq$ 0). In particular, if z = 1 and β = q = 0, equation (5) is transformed to the equation given by L.M. Kachanov [4]:

$$\frac{d\omega}{dt} = C\left(\frac{\sigma}{1-\omega}\right)^m, \tag{6}$$

if z = 1 and q = 0 it transforms to the equation given by Yu.N. Rabotnov [6]

$$\frac{d\omega}{dt} = C\left(\frac{\sigma}{1-\omega}\right)^m \omega^\beta, \tag{7}$$

and if z = 1 and β = 0 it transforms to the equation proposed by J. Lemaitre - A. Plumtree [7]

$$\frac{d\omega}{dt} = C\left(\frac{\sigma}{1-\omega}\right)^m \left(\frac{1}{1-\omega}\right)^q. \tag{8}$$

The evolutionary equations (6)-(8) prescribe the damage kinetics in a non-linear form both in physical (Figure 1b, dashed lines) and normalized (Figure 1a, curve 2) time. The diagrams corresponding to equations (6)-(8) differ only in their curvature. However their adequacy in describing the real damage process is not ensured. They lead to the linear cumulative damage rule in the case of step-wise loading and give results coinciding with the straight line in Figure 1c (8,9).

NON-LINEAR MODELS OF DAMAGE ACCUMULATION

As mentioned above, conditions (I) and (2) are associated with linearity of damage accumulation. However equations (6)-(8) also lead to equation (2), despite the damage kinetics being prescribed as non-linear dependence with time. So let us define as "nonlinear" the cumulative damage laws which are different from those that are represented by (1) and (3) and do not lead to relations like (2). The nonlinearity conditions may be analytically represented in the following manner

$$d\omega = \psi\left(\frac{dt}{t_R}\right) \quad \text{and} \quad \sum_{i=1}^{K} \int_{o}^{t} \omega(t)dt \neq 1, \tag{9}$$

where ψ is a function which differs from linear.

For evaluating damage equations satisfying the nonlinearity conditions (9), the separability principle will be used.

The separability principle

The common feature of relations (1) and (3) resulting in a linear cumulative damage rule (2) is the following: These relations prescribe the unified damage curves as being independent of the stress level in the normalized time scale (Figure 1a). The condition, representing the damage kinetics which depends upon stresses in the normalized time we will define as the separability principle. The separability principle may be represented analytically by the relation

$$\frac{d\omega}{dt} = F_T(\sigma, \dots) , \tag{10}$$

which accounting for the nonlinearity condition (9) is reduced to the form

$$\left.\frac{d\omega}{dT}\right|_{\sigma=\text{const}} \neq \text{const} \quad \text{and} \quad \left.\frac{d\omega}{d\sigma}\right|_{T=\text{const}} \neq \text{const} \tag{11}$$

Here $T = t/t_R$ is normalized time; F_T - function depending on the stresses which is associated with the F function of equation (3).

Now, on the basis of condition (10) we will analyse the structure of the evolutionary equations (6), (7) and (8). Upon substituting the dependent variable T for t, we obtain, respectively

$$\frac{d\omega}{dT} = \frac{1}{(1+m)} \cdot \frac{1}{(1-\omega)} \tag{12}$$

$$\frac{d\omega}{dT} = \frac{\omega^\beta}{(1-\omega)^m} \int_0^1 \frac{(1-\omega)^m}{\omega^\beta} \, d\omega \tag{13}$$

$$\frac{d\omega}{dT} = \frac{\omega^\beta}{(1+m+q)} \cdot \frac{1}{(1-\omega)^{q+m}} \tag{14}$$

The structure of equations (12)-(14) leads to the conclusion that the normalized damage growth rate is not a function of stress. As a result the prediction of the initial evolutionary equations are equivalent to those obtained using the linear cumulative damage rule (2).

For the general case of evolutionary damage equations with the form of equation (3), which may be represented in the form

$$\frac{d\omega}{dt} = f_1(\sigma) \cdot f_2(1-\omega), \tag{15}$$

substituting T for t in accordance with (10) we obtain

$$\frac{d\omega}{dT} = t_R \cdot f_1(\sigma) \cdot f_2(1 - \omega) \tag{16}$$

Substituting the expression for lifetime $t_R [f_1(\sigma)]^{-1}$ into (16) we may keep the following relation

$$\frac{d\omega}{dt} = f_2(1 - \omega) \cdot \int_0^1 \frac{d\omega}{f_2(1 - \omega)}, \tag{17}$$

in which the damage rate is also independent of stress.

The Balance of Momentary and Temporary Damage

The nonlinearity condition (10) can be specified using the evolutionary equation (3) as the initial form of the differential inseparable equations, so that

$$\frac{d\omega}{dT} = F(\sigma(t), \ \omega(\sigma), \dots) \tag{18}$$

The practical realization of such an approach is however connected with insurmountable difficulties. As a result only empirical solutions are available at the present time.

An approach where the initial evolutionary equation contains stress dependent coefficients (8) seems to be more promising. The possibility of such an approach was pointed out in (10). It should only be remarked that coefficients of this type may be the power indices to the variable ω, as, for example β in (7) and q in (8). The coefficient in equation (6) is determined from the longterm strength curve and is independent of stress if a salient point is absent.

In order to construct the nonlinear models on the basis of equation (10) we identify the damage with the energy which is dissipating due to the formation of new microsurfaces. We also assume that momentary damage occurs at every instant the load is changed. In this connection, two scalar functions W_t and W_σ representing integral measures of temporary and momentary damage respectively are introduced. We also assume that energy spent on fracture due to total damage accumulation is a constant value for a given material and temperature

$$W_\sigma + W_t = W_R \longrightarrow \frac{W_\sigma}{W_R} + \frac{W_t}{W_R} = 1 \qquad (19)$$

where W_R is the specific energy of total fracture.

We identify the momentary damage W_σ with the specific work spent by the applied stresses σ during momentary fracture deformation, ε_R. The value of this work is determined from the momentary deformation diagram (Figure 2a, shaded area), so that

$$W_\sigma = \int_0^{\varepsilon_\sigma} \psi_0(\varepsilon)d\varepsilon + (\varepsilon_B - \varepsilon^\sigma)\sigma, \qquad (20)$$

where $\psi_0(\varepsilon)$ is the equation of momentary deformation on the diagram; ε^σ is the deformation corresponding to the momentary stress σ; ε_B is the deformation corresponding to the strength limit.

The temporary damage W_t is identified with the specific work of longterm fracture according to the damage diagram of Figure 2b (shaded area). In this case

$$\frac{W_t}{W_R} = \omega_m \bigg|_0^{t_R} = \frac{\int_0^{t_R} \omega(t)dt}{t_R} = \int_0^1 \omega(T)dT \qquad (21)$$

where ω_m is the mean value of the function ω in the interval $0 \le t \le t_R$; $T = t/t_R$ is the normalized time.

The value of W_R is also determined from the momentary deformation diagram, so that

$$W_R = \int\limits_0^{\varepsilon_R} \psi_0(\varepsilon)\,d\varepsilon, \qquad (22)$$

where ε_R is the deformation value at failure.

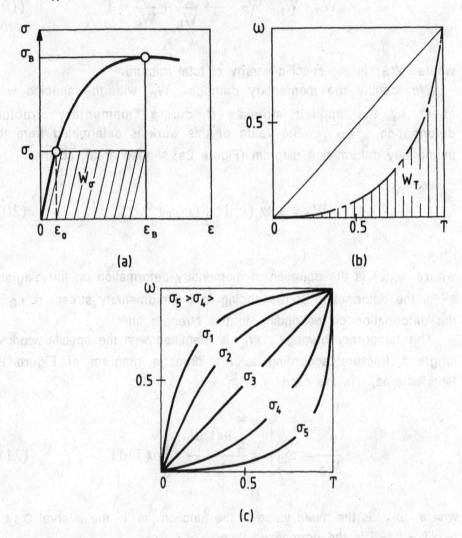

Figure 2. Integral measures of the momentary (a) and temporary (b) damage and damage kinetics (c) during stress variation.

The balance condition between the momentary and temporary damage components (19), taking relations (20)-(22) into account, and after some transformations can be written in the form

$$\left[1+\int_0^1 \omega\left(\frac{t}{t_R}\right)d\left(\frac{t}{t_R}\right)\right] \cdot \int_0^{\varepsilon_B} \psi_0(\varepsilon)d\varepsilon - \int_{\varepsilon^\sigma}^{\varepsilon_B}\left[\psi_0(\varepsilon)-\sigma\right]d\varepsilon = 1 \quad (23)$$

This expression may be considered as an energy criterion of longterm fracture under creep conditions.

The Damage Kinetics

The energy criterion (23) allows us to construct the nonlinear model of damage accumulation by determining the stress dependence of one of the coefficients in (5). As mentioned above, evolutionary equations containing two or more power indices in ω, may be used as the initial equations for $\omega(T)$. Let us consider a model based on equations (7) and (8).

In particular, for $\omega(T)$ from (7) we have

$$\frac{t}{t_R}(\omega) = \frac{\Gamma(2+m-\beta(\sigma))}{\Gamma(1+m).\Gamma(1-\beta(\sigma))} \cdot \int_0^\omega \frac{(1-\omega)^m}{\omega^{\beta(\sigma)}} a\omega . \qquad (24)$$

where

$$t_R = \frac{\Gamma(1+m).\Gamma(1-\beta(\sigma))}{\Gamma(2+m-\beta(\sigma))} \cdot \frac{1}{C\sigma^m}, \qquad (25)$$

is assumed. Here $\Gamma(.)$ is a gamma-function.

On substituting (24) into (23) and varying (σ) we may obtain the following expression for $\beta(\sigma)$

$$\beta(\sigma) = \frac{-(1+m)\int_0^{\varepsilon_B} \psi_0(\varepsilon)d\varepsilon}{\int_0^{\varepsilon_B} \psi_0(\varepsilon)d\varepsilon - \int_{\varepsilon^\sigma}^{\varepsilon_B}\left[\psi_0(\varepsilon)-\sigma\right]d\sigma} + (2+m) \qquad (26)$$

The graphical interpretation of the damage kinetics represented by equations (24) and (26) is shown in Figure 2c (dashed line).

From equation we have

$$\omega\left(\frac{t}{t_R}\right) = 1 - \left(1 - \frac{t}{t_R}\right)^{\frac{1}{1+m+q(\sigma)}} \qquad (27)$$

for $\omega(T)$, where it is assumed that

$$t_R = \frac{1}{(1+m+q(\sigma))C(\sigma)\sigma^m} \qquad (28)$$

On substituting (27) into (28) and varying σ we may obtain the following expression for $q(\sigma)$

$$q(\sigma) = \frac{\int\limits_{0}^{\varepsilon_B} \psi_0(\varepsilon)d\varepsilon}{\int\limits_{\varepsilon^\sigma}^{\varepsilon_B} \left[\psi_0(\varepsilon) - \sigma\right]d\varepsilon} - (2+m) \qquad (29)$$

The graphical interpretation of the damage kinetics represented by equations (27) and (29) is shown in Figure 2c (solid line).

DISCUSSION

The essential difference of the proposed models of damage accumulation from those previously developed is that the damage kinetic diagrams represented in coordinates "ω - T" is separated according to the stress parameter (see Figure 2c). With a rising stress they are displaced in the direction of decreasing temporary damage . The momentary damage value rises correspondingly. Traditional approaches give a single kinetic diagram for all stresses (Figure 1a).

The effect of separation of the damage kinetic diagrams in terms of stress allows us to take into account the loading history in lifetime computations under non-stationary conditions (Figure 3). As an example we consider the classical problems of additional loading $(\sigma_1 > \sigma_2)$ and partial unloading $(\sigma_1 > \sigma_2)$. The indices characterize the sequence of stress changes. To obtain solutions the simplest graphical representations (see Figure 3a) will be sufficient.

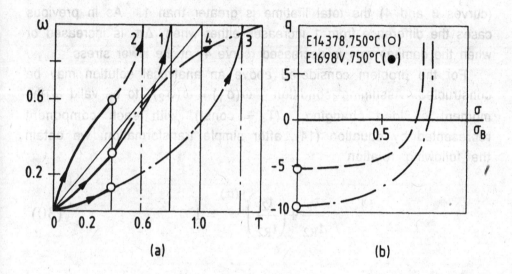

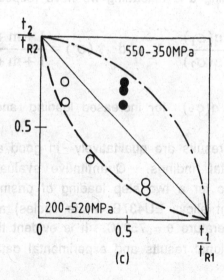

Figure 3. Kinetics of damage accumulation (a), stress function (b) and residual life (c) during two stepwise loadings.

The total normalized lifetime is less than 1 if the stress is changed from a low to a high level (curves 1 and 2 of Fig. 3a). When the stress gradient $\Delta\sigma$ is increased, particularly when damage accumulates with "decreasing rate" in the "ω - T" plane under lower stress, the normalized lifetime decreases (curve 2).

In the case of the stress changing from a high to a low level (curves 3 and 4) the total lifetime is greater than 1. As in previous cases the difference from 1 increases either when $\Delta\sigma$ is increased or when the damage rate is decreased (curve 4) at the lower stress.

For the problem considered above an analytical solution may be constructed. Assuming condition $\omega(\sigma_1) = \omega(\sigma_2)$ to be valid at the moment of load changing (T = const), with each component represented by equation (14), after simple transformation, we obtain the following relation

$$\frac{t_1}{t_{R2}} + \left(\frac{t_2}{t_{R2}}\right)^{\gamma(\sigma)} \doteq 1 \qquad (30)$$

For additional loading and unloading we have respectively

$$\gamma(\sigma) = \frac{1+m+q(\sigma_1)}{1+m+q(\sigma_2)} < 1 \quad \text{and} \quad \gamma(\sigma) = \frac{1+m+q(\sigma_1)}{1+m+q(\sigma_2)} > 1 \quad (31)$$

because $q(\sigma_1) < q(\sigma_2)$ for increased loading, and $q(\sigma_1) > q(\sigma_2)$ for unloading.

The obtained results are qualitatively in good agreement with well known experimental findings. Quantitative evaluations are given in Figures 3b and 3c for a two step loading of prismatic rods fabricated from heat resistant alloys EU437B (white circles) and EU698BD (dark circles) at a temperature $\theta = 750^\circ C$. It is evident that good agreement between computational results and experimental data is obtained.

REFERENCES

[1]   Palmgren, A., Die Lebensdauer von Kugellagern, Z. Vereines Deutscher Ing., 1924, **68**, N 14, pp. 339-341.

[2]   Miner, M.A., Cumulative damage in fatigue, J. Appl. Mech., 1945, **12**, N 1, A159-A164.

[3]   Robinson, E.L., Effect of temperature variation on the longtime rupture strength of steels, Trans. ASME, 1952, **74**, N 5, 777-780.

[4]   Kachanov, L.M., Time to failure under creep conditions (in Russian), Izv. Akad. Nauk SSSR, 1958, N 8, pp. 26-31.

[5]   Janson, S. and Hult, J., Fracture mechanics and damage mechanics a combined approach, J. Appl. Mech., 1977, **1**, N 1, 69-84.

[6]   Rabotnov, Yu. N., Creep in structure components (in Russian), Nauka, Moskow, 1966, p. 752.

[7]   Lemaitre, J. and Plumtree, A., Use of damage notion to rupture calculation in conditions of simultaneous fatigue and creep effect   (in Russian), ASTM, 1979, N 3, 124-134.

[8]   Golub, V.P. and Romanov, A.V., On damage kinetics of isotropic material under creep conditions (in Russian), Prikl. Mech., 1989, N 12, 107-115.

[9]   Golub, V.P. and Romanov, A.V., On problem of construction of nonlinear models of damage accumulation under creep condition (in Russian), Probl. Proch., 1990, N 6, 9-14.

[10] Bolotin, V.V., Service life prediction of machines and structures (in Russian), Mashinostroenie, Moskow, 1984, p. 312.

CALCULATION OF FAILURE PROBABILITY FOR BRITTLE MATERIALS

J. Smart and S.L. Fok
Department of Engineering
University of Manchester
Manchester, M13 9PL, UK.

ABSTRACT

The determination of the failure probability of brittle materials using Weibull statistics and finite element method is examined. Three cases are examined, a beam in pure bending, a tapered rod and a notched bar in bending.

The results show that if the postprocessing analysis of the Weibull probability data is made using 4 point Gaussian quadrature within each quadratic finite element the postprocessing errors are similar but generally better than the errors arising from the finite element analysis. Gaussian quadrature gives better results than the rectangular rule.

INTRODUCTION

For brittle materials, such as engineering ceramics, instead of a single failure load there is a probability of failure at a given load and the spread of failure loads for nominally identical specimens can be up to 100%. At the previous conference in this series, Sneddon and Sinclair [1] examined various ways in which ceramic probability data could be analysed. They concluded that, although the beta distribution is the most appropriate, the generalised pareto and Weibull distributions could be supported. However, the Weibull distribution is commonly used in the analysis of brittle materials and a review of the work in this area is given by Kittl and Diaz [2].

To predict the failure of engineering components the basic material properties must first be obtained. These are commonly obtained from either three or four point bending tests. The data from these tests is then generally analysed using simple bending theory [3] to obtain the material properties. The two material properties obtained define a "mean strength per unit volume" and a value, the

Weibull parameter, which defines the spread. Thus, two materials which had the same average failure load but different spreads would have one parameter the same and another which was different.

The basic material data can then be used to predict the failure probability of a component in which the stress varies throughout the body. The stress distribution can be determined using the finite element method and an estimate of the failure probability at a given load obtained using a post processor. There are commercial programs available for this analysis e.g. CARES which is written by NASA [4]. Essentially CARES is a postprocessor for using with finite element output from other programs such as NASTRAN or ANSYS. There are also other programs available such as that of the authors, BRITPOST [5].

When determining the failure probability an integral has to be determined which has the form $\int \sigma^m \, dV$ where m is the Weibull parameter and typically varies between about 5 and 20. To evaluate the integral, CARES "takes into account element stress gradients by dividing each brick element into 27 subelements and each quadrilateral into 9 subelements". Whilst not specifically stated it is assumed that these subelements are of equal size. BRITPOST, however, uses four point Gaussian quadrature, i.e. the element is subdivided into 16 non-equal sized subelements.

The purpose of this paper is to examine closely the different integration procedures to draw conclusions as to the most efficient and accurate method.

GAUSSIAN QUADRATURE

Gaussian quadrature is used for the numerical integration of stiffness matrices in finite elements because the weighting factors and sampling points have both been optimised. For a 1-D integration the Gaussian quadrature formula is

$$\int_{-1}^{+1} f(\xi) \, d\xi = \sum_{i=1}^{N} W_i \, f(\xi_i)$$

where W_i are the weighting factors, ξ_i are the sampling points and N is the quadrature order. The points have been chosen so that an order of N exactly integrates a polynomial of order (2N-1). In 2-D the integration is given by

$$\int_{-1}^{+1} \int_{-1}^{+1} f(\xi, \eta) \, d\xi \, d\eta = \sum_{i=1}^{N} \sum_{j=1}^{M} W_i \, W_j \, f(\xi_i, \eta_j).$$

Generally the sampling points will be the same in each direction so that N = M. However, this is not essential. Further details of the method and the sampling points and weighting factors are given in most finite element text books, e.g. Zienkiewicz and Taylor [6].

WEIBULL PROBABILITY THEORY

When making predictions of the failure probability of brittle materials there are various laws. For example, Thiemeier et al [7] examined seven different failure criteria when comparing four-point bending tests with ring-on-ring tests of aluminium nitride. However they all involve an integral of stress to the power m where m is the Weibull modulus. The integrals may also be either surface integrals or volume integrals. However, a typical integral to be evaluated is $\int \sigma^m \, dV$.

As the purpose of this paper is not to compare the various theories but to examine the integration procedures and the implications of the choices made, the various theories will not be discussed in further detail.

A BEAM IN PURE BENDING

As basic material data is often obtained from four-point bending tests, e.g. Thiemeier et al [7], the case of pure bending will be examined.

This results in a uniform linear stress distribution across the beam between the central supports. Also, if 8-noded isoparametric finite elements are used in the finite element analysis, the stress distribution within the element is approximately bi-linear and, as the element size becomes smaller, the stress distribution will tend to become bi-linear. Thus, a linear stress distribution will represent many analyses.

For the four-point bending test configuration shown in fig. 1, the moment within the central span is Pd and so the stress within the central span is given as $\sigma = Pdy/I$. If, as is commonly assumed, the compressive failure stress is much larger than the tensile failure stress so that compressive stresses can be ignored in the failure calculations, then in the central span the stress volume integral [8], $\sum$, is given by

$$\sum = \int \sigma^m dv = s \int_0^h \left(\frac{Pdy}{I} \right)^m dy$$

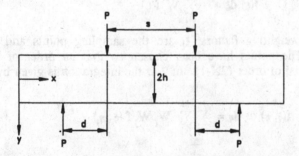

Figure 1. Beam in four point bending

However, to obtain reasonable numerical values within the integral, the stresses are divided by a nominal stress and in this case the most sensible value to choose is the outer fibre stress, σ_{nom} = Pdh/I. Thus the integral is now

$$\sum = s \int_o^h (y/h)^m \, dy = \frac{h.s}{m+1}$$

and, as mentioned earlier, typical values for m lie between 5 and 20.

The analytical value for the integral can be compared with values obtained using either Gaussian quadrature or with those obtained by, for example, sampling at the 3 x 3 subelement points. The beam has been divided into one, two and four elements through the depth and the results of the numerical integrations determined using various orders of Gaussian quadrature and sampling at regular intervals using the rectangular rule. This is illustrated in figs 2 and 3 for one and two elements across the beam and for 3 point Gaussian quadrature and 3 point subdivision using a 2-D rectangular rule.

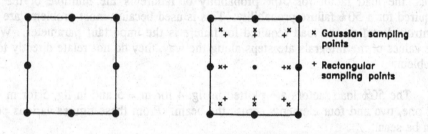

Figure 2. Beam with one element across depth and 3 point Gaussian and rectangular rule sampling points

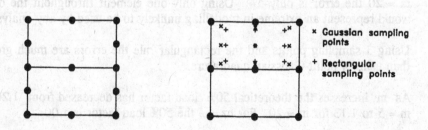

Figure 3. Beam with two elements across depth and 3 point Gaussian and rectangular rule sampling points

However, the integral is only one step in the evaluation of the probability of failure at a given load. For this loading case, as the only stress is σ_x, the probability of failure will be given by [8]

$$P_f = 1 - e^{\{-(\frac{1}{m}!)^m (\frac{\sigma_{nom}}{\bar{\sigma}_{fv}})^m \frac{1}{v} \sum\}}$$

where $(\frac{1}{m}!)$ is the Gamma function and $\bar{\sigma}_{fv}$ is the mean tensile strength for a unit volume of material v. Now, for sh = 2, v = 1 and $\bar{\sigma}_{fv} = \sigma_{nom}$ both $\sum$ and P_f have been evaluated.

However, rather than quote a probability of failure the load factor required to give a 50% probability of failure is quoted. For the case considered, the load is such that at the outer fibres of the beam, the stress is equal to the tensile strength per unit volume, i.e. $\sigma_{max} = \bar{\sigma}_{fv}$. For m = ∞, this will mean failure at this load but for lower values of m there will be contributions to the integral from areas within the beam which are not stressed so highly and the failure probability will be lower. Thus, the load factor for 50% probability of failure is the multiple of the load required for a 50% failure probability. This is used because most problems are load controlled and so the load required for failure is the important parameter. Whilst the values of the integrals are steps along the way, they do not relate directly to the problem.

The 50% load factors are plotted in fig. 4 for m = 5 and in fig. 5 for m = 20 for one, two and four elements across the beam. From these figures various points can be seen.

1) Gaussian quadrature is superior to the rectangular rule converging more quickly to the theoretical value. The exception is one sampling point when the two rules become identical.

2) As shown in [5] the convergence is quicker for a lower value of m.

3) Even with only one element across the beam, using 4 point Gaussian quadrature, the error is only about 0.1% in the 50% load factor for m = 5. For m = 20 the error is only 5%. Using only one element throughout the depth would represent an extreme in modelling unlikely to be used by any analyst.

4) Using 3 sampling points and the rectangular rule the errors are much greater than using 4 point Gaussian quadrature.

5) As m increases the theoretical 50% load factor has decreased from 1.26 for m = 5 to 1.13 for m = 20. For m = ∞ the 50% load factor is 1.00

6) Even though for m = 20 the integral to be valuated is y^{20}, once this has been incorporated into the equation to determine failure probability using 4 point quadrature, which would evaluate y^7 exactly, the errors are 5% or less depending on the number of elements used across the depth. If 5 point quadrature is used the error is less than 2% even for a single element.

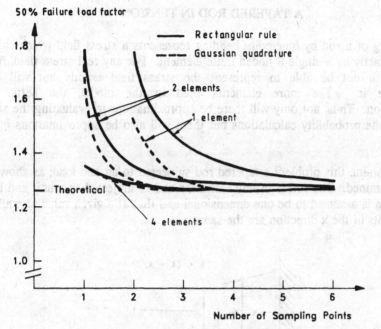

Figure 4. Variation of 50% failure load factor for m = 5 for one, two, and four elements across beam.

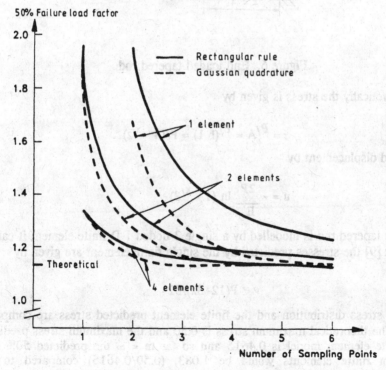

Figure 5. Variation of 50% failure load factor for m = 20 for one, two, and four elements across beam.

A TAPERED ROD IN TENSION

Pure bending of a rod by four point bending represents a stress field which can be modelled exactly by a single 8-noded finite element. For any real stress field, finite elements will not be able to represent the stress field exactly but will only approximate it. The more elements used in the model, the better the approximation. Thus, not only will there be approximations in evaluating the stress integral in the probability calculations but there will also be approximations in the stress field.

To examine this problem a tapered rod subjected to an end load, as shown in fig 6, is examined. The rod is of unit thickness and is subjected to a unit end load. The problem is assumed to be one dimensional and that at a given value of x all the displacements in the x direction are the same.

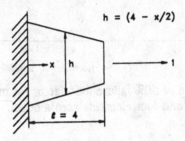

Figure 6. End loaded tapered rod.

Theoretically the stress is given by

$$\sigma = {}^P\!/_A = {}^P\!/(h.1) = P/(4 - {}^x\!/_2)$$

and the end displacement by

$$u = -\frac{2P}{E} \ln (1 - {}^x\!/_8)$$

If the tapered rod is modelled by a single 3-noded 1-D finite element it can be shown that [9] the stresses predicted by the single finite element are given by

$$\sigma = P(12+3x)/52$$

The actual stress distribution and the finite element predicted stress are compared in fig 7. The theoretical maximum stress is 0.50 and the maximum stress predicted by the finite element model is 0.4615 and so far m = ∞ the predicted 50% load factor from finite elements would be 1.083, (0.50/0.4615) compared to the theoretical 50% load factor of 1.00.

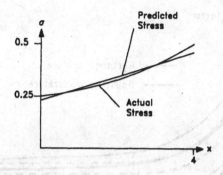

Figure 7. Theoretical and finite element predicted stresses for tapered rod.

Now the stress volume integral is given by

$$\Sigma = \int \left(\sigma / \sigma_{nom} \right)^m dV$$

and for $\sigma_{nom} = 0.5$

$$\Sigma = \int_0^4 \left(\frac{1/(4 - x/2)}{0.5} \right)^m (4 - x/2)dx$$

Hence

$$\Sigma = \frac{8}{m-2} (1 - 2^{2-m})$$

This enables the theoretical 50% load factors to be determined.

For a mean tensile strength, $\bar\sigma_{fv} = 0.50$, the 50% failure load factors from the finite element analysis are shown in fig 8 for the two integration rules and with different integration orders. Again, the same patterns can be seen; Gaussian quadrature converges quicker than the rectangular rule, Gaussian quadrature produces lower load factors and a Gaussian order of 4 is sufficient. Also the higher the value of the Weibull parameter, m, the greater the difference between the two methods.

However, the comparison shown in fig 8 is for the approximate finite element stress distribution. Also shown in fig 8 are the theoretical 50% load factors based on the actual stress distribution in the rod. The ratio between the correct theoretical load factors and the converged finite element approximate load factors are shown in Table 1, where R_{50} is defined as

$$R_{50} = \frac{50\% \text{ Load factor from finite element analysis and postprocessor}}{50\% \text{ Load factor from theoretical stress distribution}}$$

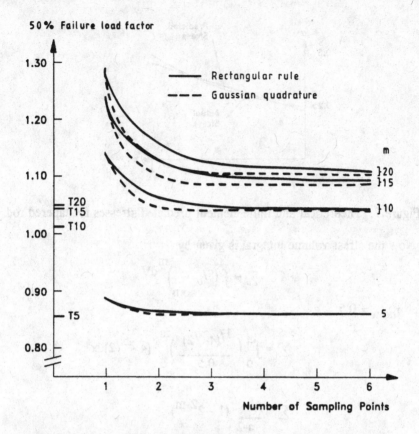

Figure 8. Variation of 50% failure load factor for tapered rod.

TABLE 1
Ratio of theoretical and finite element 50% load factors for tapered rod

m	R_{50}
5	1.008
10	1.027
15	1.041
20	1.050
∞	1.083

It can be seen that although there is an 8.3% error in the maximum stress calculated (i.e. when m = ∞) the averaging effect of the finite element method has meant that for lower values of the Weibull parameter m, the overall errors are reduced. This is because, as can be seen in fig 7, the finite element approximation produces areas where the stress prediction is higher than the theoretical value as well as the other way. In the limit, when m = ∞, only the maximum value counts and so there is no contribution to the integral from any point other than the maximum. At lower values of m, the areas where the finite element stress prediction is higher than the theoretical stress, have lowered R_{50}.

Combining the effect of the integration procedure in the stress volume integral and the approximate nature of the finite element method it can be seen that for low values of m the integration procedure used is not too important. For higher values of m the integration procedures give different results but four point Gaussian quadrature has produced a converged, although slightly inaccurate answer. This inaccuracy is inevitable if the finite element method is used but it is good that the errors are less when predicting 50% probability failure loads than that from the finite element method itself when predicting the maximum stress or stress concentration factor.

A NOTCHED BAR IN BENDING

A notched bar in bending has been suggested as a benchmark test for finite element postprocessors for brittle materials [10]. The geometry of the bar is given in fig 9 and this has been meshed as shown in fig 10. To obtain finer meshes each element in the base mesh has been subdivided into 2 x 2 and 3 x 3 elements. Also, a further test has been run with a 6 x 6 subdivision. The base mesh is called mesh 1, the mesh from subdividing each element in fig. 10 into 2 x 2 subelements is called mesh 1/2 etc.

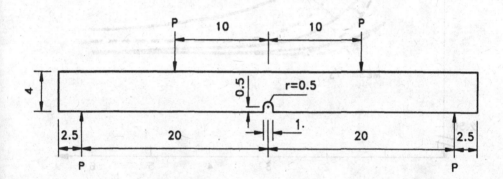

Figure 9. Notched bar in four-point bending.

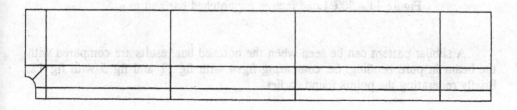

Figure 10. Finite element mesh for notched bar.

Again, the 50% load factors have been calculated for each mesh using both the Gaussian and rectangular rule and with a varying number of sampling points. The results are shown in figs 11 and 12 for m = 5 and 20 respectively.

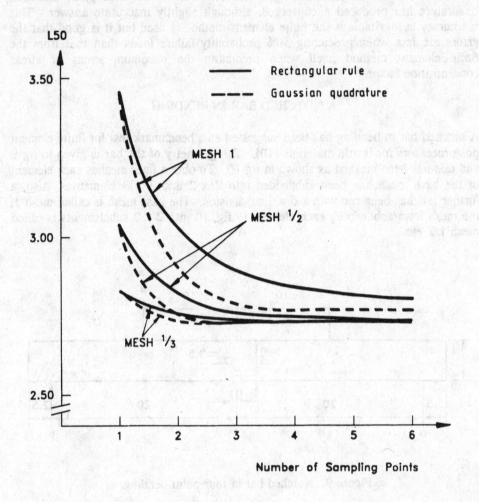

Figure 11. 50% Load factors for notched bar and m = 5.

A similar pattern can be seen when the notched bar results are compared with the beam in pure bending, i.e. comparing fig 4 with fig 11 and fig 5 with fig 12. Briefly reiterating the points found earlier:

1) Gaussian quadrature is superior to the rectangular rule for a given number of sampling points.

2) Convergence is quicker for a lower value of m.

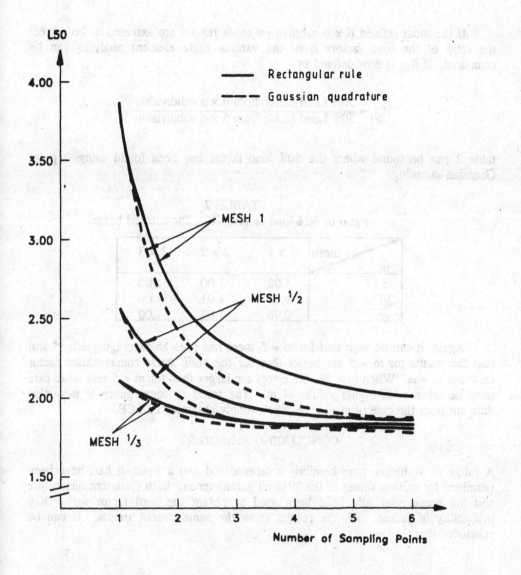

Figure 12. 50% Load factor for notched bar and m = 20.

3) Even with a coarse mesh and m = 5, using 4 point quadrature the results have converged. However, because of the finite element analysis approximations the error is about 2%. For m = 20 with the coarse mesh and with 4 point sampling the errors are about 8%.

4) 3 point sampling with the rectangular rule leads to considerably larger errors than using 4 point Gaussian quadrature.

5) The 50% load factor has again decreased with increasing m.

If the most refined 6 x 6 subdivision mesh results are assumed to be correct the ratio of the load factors from the various finite element analyses can be compared. If R_{50} is now defined as

$$R_{50} = \frac{50\% \text{ Load factor from } n \times n \text{ subdivision}}{50\% \text{ Load factor from } 6 \times 6 \text{ subdivision}}$$

table 2 can be found where the 50% load factor has been found using 4 point Gaussian sampling.

TABLE 2
Ratio of 50% load factors, R_{50}, for notched beam

m \ mesh	1 x 1	2 x 2	3 x 3
5	1.02	1.00	1.00
20	1.08	1.01	1.00
∞	0.96	1.02	1.00

Again, it can be seen that for m = 5 there has been an "averaging effect" and that the results for m = 5 are better than for the basic stress concentration factor i.e. when m = ∞. When m = 20, the errors are larger than for m = 5 and more care must be taken with higher values of m. The trend to convergence is now also different from the case of m = ∞, i.e. when simply obtaining an SCF.

CONCLUDING REMARKS

A range of problems, pure bending, a tapered rod and a notched bar, have been examined for various values of the Weibull parameter m. Both Gaussian quadrature and the rectangular rule have been used to obtain the load factor for a 50% probability of failure. All the results show the same overall trends. It can be concluded that

1. 4 point Gaussian quadrature is sufficient for postprocessing.

2. Convergence is quicker for lower values of m.

3. The errors in the postprocessing for the probabilistic assessment are generally similar to the errors arising from the finite element analysis. Postprocessing with 4 point Gaussian quadrature has not increased the errors.

REFERENCES

1. Sneddon, J.D. and Sinclair, C.D., Statistical Mapping and analysis of engineering ceramics data. Mechanics of Creep Brittle Materials - 1, Elsevier Applied Science Publishers, London 1989, 99-116.

2. Kittl, P. and Diaz, G., Weibull's fracture statistics, or probabilistic stength of materials: state of the art, Res Mechanica, 24, 1988, 99-207.

3. Jayatilaka, A de S., Fracture of Engineering Brittle Materials, Applied Science Publishers, London 1979.

4. Nemeth, N.N., Manderscheid, J.M. and Gyekenyesi, J.P., Design of Ceramic Components with the NASA/CARES Computer Program. NASA TM 102369, 1990.

5. Smart, J., The determination of failure probability using Weibull probability statistics and the finite element method, Res Mechanica, 31, 1990, 205-219.

6. Zienkiewicz, O.C. and Taylor, R.L., The Finite Element Method. 4th Edition, Volume 1, McGraw Hill, London, 1989.

7. Thiemeier T., Brückner-Foit, A., and Kölker, H., Influence of the fracture criterion on the failure prediction of ceramics loaded in biaxial flexure, J. Am. Ceram. Soc., 74, 1, 1991, 48-52.

8. Stanley, P., Fessler, H., and Sivill, A.D., An engineer's approach to the prediction of failure probability of brittle components, Proc. Brit. Ceram. Soc., 22, 1973, 453-487.

9. Smart J. Understanding and using finite elements. Short course notes, University of Manchester, 1990.

10. WELFEP. Benchmarks for postprocessors for brittle materials. 1990.

ON THE APPLICABILITY OF
THE SUPERPOSITION PRINCIPLE IN
CONCRETE CREEP

J. C. Walraven

Professor of Civil Engineering

Delft University of Technology

Delft, The Netherlands

J. H. Shen

Assistant of Civil Engineering

Darmstadt University of Technology

presently Krebs and Kiefer Consulting

Darmstadt, Germany

ABSTRACT

In this paper it is discussed how to choose and use a linear creep function to analyze the creep of concrete. Certain boundary conditions for a creep function are presented. With the help of an aging Kelvin-Voigt model, the physical significance of the boundary conditions can be proved. According to the principle of superposition, two integral types of the constitutive law are compared, and the generally correct form is pointed out, which is necessary to compute creep under influences of time-dependent variables such as temperature and humidity.

INTRODUCTION

Concrete is one of the most important materials and can be called the building material of this century. Constitutive laws for concrete, however, are quite complex. The complete stress-strain-time relation is rather non-linear

(Fig.1) [14,15]. The creep of concrete depends on its composition and furthermore on aging, temperature and humidity. So creep of concrete is an intricate phenomenon. With regard to the calculation of concrete creep a number of questions is still unanwsered.

In practice, if the stresses are within the range of low service stresses, i.e. less than $30 \sim 40$ % of the strength, the linear creep theory can be approximately used. That means that the creep strain ε is assumed to be linearly dependent on the applied stress σ:

$$\varepsilon(t,\tau) = \sigma \ \Phi(t,\tau) \tag{1}$$

where

$\Phi(t,\tau)$ = creep function;
t $\quad$ = time at the moment considered;
τ $\quad$ = time at the beginning of loading.

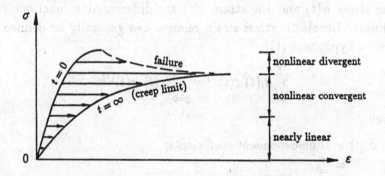

Fig.1 The stress-strain behaviour of concrete in reality

LINEARITY AND SUPERPOSITION

Mathematical linearity is defined to apply, if the following two equations are fulfilled, where $\mathcal{L}$ is a linear operator:

$$\mathcal{L}(\sigma_1 + \sigma_2) = \mathcal{L}(\sigma_1) + \mathcal{L}(\sigma_2)$$
$$\mathcal{L}(c\,\sigma) = c\,\mathcal{L}(\sigma) \tag{2}$$

where σ, σ_1 and σ_2 are variable stresses, and c is a constant.

The principle of superposition is actually an application of linearity. Assuming the stress-strain relationship to be linear, $\varepsilon = \mathcal{L}(\sigma)$, for different stresses σ_i, with $i = 1, \cdots, n$, we obtain the strain:

$$\varepsilon = \varepsilon_1 + \cdots + \varepsilon_n = \mathcal{L}(\sigma_1) + \cdots + \mathcal{L}(\sigma_n) = \mathcal{L}(\sigma_1 + \cdots + \sigma_n) = \mathcal{L}(\sigma) \quad (3)$$

The principle of superposition was at first formulated by Boltzmann in 1876. In order to calculate creep under variable stresses, the superposition principle is often used. If there are stresses σ_1 and σ_2, applied at τ_1 and τ_2, from eq.(1), the superposition can not be directly proven to apply [12], because the creep function $\Phi(t, \tau)$ is a time function with two variables:

$$\varepsilon(t, ?) = \varepsilon_1(t, \tau_1) + \varepsilon_2(t, \tau_2) = \sigma_1 \, \Phi(t, \tau_1) + \sigma_2 \, \Phi(t, \tau_2) = (\sigma_1 + \sigma_2) \, \Phi(t, ?) \quad (4)$$

DIFFERENTIAL TYPE OF CONSTITUTIVE LAW

If the stress $\sigma(t)$ and the strain $\varepsilon(t)$ are differentiable functions of time, the linear viscoelastic stress-strain relation can generally be defined by the following hypothesis [17]:

$$\sum_{i=0}^{m} a_i(t) \, \varepsilon(t)^{(i)} = \sum_{j=0}^{n} b_j(t) \, \sigma(t)^{(j)} \quad (5)$$

where

$a_i(t)$, $b_j(t)$ = time-dependent coefficients;
$(\;)^{(i)}$ $= \frac{d^i}{dt^i}$.

For a rheological model with serial and parallel combinations of springs and dashpots, the coefficients $a_i(t)$ and $b_j(t)$ consist of aging elastic moduli of springs and viscosities of dashpots. For several stress histories $\sigma_r(t)$, $r = 1, \cdots, l$ the superposition can be demonstrated:

$$\sum_{j=0}^{n} b_j(t) \left[\sum_{r=1}^{l} \sigma_r(t)^{(j)} \right] = \sum_{r=1}^{l} \sum_{j=0}^{n} b_j(t) \sigma_r(t)^{(j)}$$

$$= \sum_{r=1}^{l} \sum_{i=0}^{m} a_i(t) \varepsilon_r(t)^{(i)} = \sum_{i=0}^{m} a_i(t) \left[\sum_{r=1}^{l} \varepsilon_r(t)^{(i)} \right] \quad (6)$$

That means that the superposition principle can be used if the creep function in eq.(1) is a solution of the linear differential equation (5).

BOUNDARY CONDITIONS FOR CREEP FUNCTIONS

It would be ideal if the constitutive law could be formulated in a differential equation and the creep function could be found as a solution of this differential equation. Unfortunately this is almost impossible because of the complicated properties of concrete.

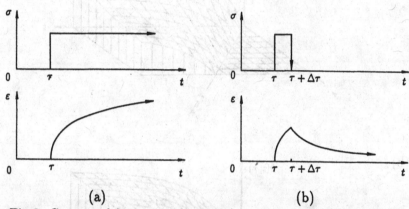

(a)   (b)

Fig.2   Creep under stepwise loading (a) or impulsive loading (b)

In practice, most suggested creep functions for concrete are empirical equations. However, not every function of two variables is acceptable as a creep function $\Phi(t, \tau)$. Certain mathematical and physical restrictions must be satisfied, i.e. for stepwise loading (Fig.2a):

- The creep must be positively proportional to the applied stress and, in the linear range, the creep of concrete should be limited:

$$0 = \Phi(t = \tau, \tau) \leq \Phi(t, \tau) \leq \Phi(t = \infty, \tau) < \infty \qquad (7)$$

- The creep increases and is convergent with time under constant stress:

$$\frac{\partial \Phi(t = \tau, \tau)}{\partial t} \geq \frac{\partial \Phi(t, \tau)}{\partial t} \geq \frac{\partial \Phi(t = \infty, \tau)}{\partial t} = 0 \qquad (8)$$

- The curvature of the creep function of concrete is not positive:

$$\frac{\partial^2 \Phi(t=\tau, \tau)}{\partial t^2} \leq \frac{\partial^2 \Phi(t, \tau)}{\partial t^2} \leq \frac{\partial^2 \Phi(t=\infty, \tau)}{\partial t^2} = 0 \qquad (9)$$

Fig.3 shows typical creep functions for concrete with and without aging (solidifying) effect under stepwise loading.

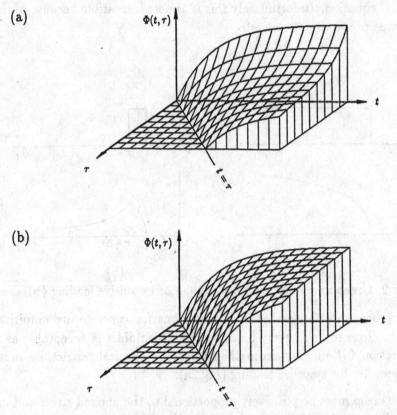

Fig.3 The creep function (a) with or (b) without aging effect

Assuming that a concrete prism is loaded at a time τ and totally unloaded after $\Delta\tau$, the sum of the creep deformation is, according to the principle of superposition (Fig.2b):

$$\Delta\varepsilon = \sigma\left[\Phi(t, \tau) - \Phi(t, \tau + \Delta\tau)\right] \qquad (10)$$

If $\Delta\tau$ is very small, we can write

$$\frac{\Phi(t, \tau + \Delta\tau) - \Phi(t, \tau)}{\Delta\tau} \rightarrow \frac{\partial\Phi(t, \tau)}{\partial\tau} \tag{11}$$

Also, for impulsive loading, certain boundary conditions can be derived:

• The creep remaining after total unloading is not negative, so

$$\Delta\varepsilon \geq 0 \quad \Rightarrow \quad \frac{\partial\Phi(t, \tau)}{\partial\tau} \leq 0 \tag{12}$$

• The creep doesn't increase after total unloading, so

$$\frac{\partial(\Delta\varepsilon)}{\partial t} \leq 0 \quad \Rightarrow \quad \frac{\partial^2\Phi(t, \tau)}{\partial t\partial\tau} \geq 0 \tag{13}$$

• The curvature of the creep function after total unloading must be positive:

$$\frac{\partial^2(\Delta\varepsilon)}{\partial t^2} \geq 0 \quad \Rightarrow \quad \frac{\partial^3\Phi(t, \tau)}{\partial t^2\partial\tau} \leq 0 \tag{14}$$

Now we consider two examples. The creep function of the first example is formulated as:

$$\Phi(t, \tau) = 1 - e^{-c(t-\tau)^m} \tag{15}$$

where c and m are two constant material parameters.
From the above conditions:

$$0 \leq \Phi(t, \tau) \leq \Phi(t = \infty, \tau) \quad \Rightarrow \quad c > 0 \tag{16}$$

$$\frac{\partial\Phi(t, \tau)}{\partial t} \geq 0 \quad \Rightarrow \quad m > 0 \tag{17}$$

$$\frac{\partial^2\Phi(t, \tau)}{\partial t^2} \leq 0 \quad \Rightarrow \quad m \leq 1 \tag{18}$$

That means that $c > 0$ and $0 < m \leq 1$ must be satisfied. For this reason, an application of this creep function, e. g. with $m > 1$, is unsuitable.
The second example is so called double power law [5]:

$$\Phi(t, \tau) = c\,\tau^{-m}\,(t - \tau)^n \tag{19}$$

with positive constants c, m and n. Differentiating eq.(19) by t and by τ, we have:

$$\frac{\partial^2 \Phi(t,\tau)}{\partial t \partial \tau} = -c\,n\,\tau^{-m-1}\,(t-\tau)^{n-2}\,[mt-(m+1-n)\tau] \geq 0 \qquad (20)$$

that is

$$t \leq \frac{m+1-n}{m}\tau \qquad (21)$$

The double power law will therefore voilate the boundary condition if $t > \frac{m+1-n}{m}\tau$. For $m = \frac{1}{3}$ and $n = \frac{1}{8}$, as used in [5], this is already the case for $t > 3.6\ \tau$. As a result the double power law is also not suitable for the use of the superposition principle and can lead to physical contradiction [2,9].

AGING KELVIN-VOIGT-MODEL

As an important example an aging Kelvin-Voigt model (Fig.4) is considered.

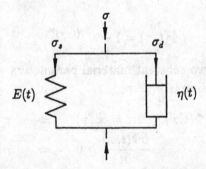

Fig.4 Kelvin-Voigt model with aging material parameters

Because the concrete is a solidifying material, the stress rate of the spring can be expressed as:

$$\dot{\sigma}_s = E(t)\,\dot{\varepsilon} \qquad (22)$$

where $\dot{\varepsilon} = \frac{\partial \varepsilon}{\partial t}$, $\dot{\sigma} = \frac{\partial \sigma}{\partial t}$, and the stress of the dashpot:

$$\sigma_d = \eta(t)\,\dot{\varepsilon} \tag{23}$$

so that the stress rate of the dashpot is equal to:

$$\dot{\sigma}_d = \eta(t)\,\ddot{\varepsilon} + \dot{\eta}(t)\,\dot{\varepsilon} \tag{24}$$

with $\ddot{\varepsilon} = \frac{\partial^2 \varepsilon}{\partial t \partial t}$.

Because of equilibrium of stresses, the differential equation for an aging Kelvin-Voigt model is:

$$\eta(t)\,\ddot{\varepsilon} + [E(t) + \dot{\eta}(t)]\,\dot{\varepsilon} = \dot{\sigma}_s + \dot{\sigma}_d = \dot{\sigma} \tag{25}$$

For creep under a constant stress $\sigma = \sigma_0$, applied at a time τ, the differential equation for the time $t > \tau$ becomes:

$$\ddot{\varepsilon} + K(t)\,\dot{\varepsilon} = 0 \tag{26}$$

where $K(t) = \frac{E(t) + \dot{\eta}(t)}{\eta(t)}$ is a material parameter which is dependent on t. The solution of the differential equation is:

$$\varepsilon = \sigma_0 \frac{1}{\eta(\tau)} \int_\tau^t e^{-\int_\tau^t K(t)dt} dt \tag{27}$$

for the initial conditions $t \to \tau$:

$$\varepsilon = 0, \quad \dot{\varepsilon} = \sigma_0 \frac{1}{\eta(\tau)} \tag{28}$$

As usually the creep strain can be formulated as:

$$\varepsilon = \sigma_0\,\Phi(t,\tau) \tag{29}$$

with the creep function:

$$\Phi(t,\tau) = A(\tau)\,[B(t) - B(\tau)] \tag{30}$$

where

$$A(\tau) \;=\; \frac{1}{\eta(\tau)}\, e^{G(\tau)} \tag{31}$$

$$B(t) - B(\tau) \;=\; \int_{\tau}^{t} e^{-G(t)}\, dt \tag{32}$$

$$G(t) - G(\tau) \;=\; \int_{\tau}^{t} K(t)\, dt \tag{33}$$

The modulus of elasticity of the spring and the viscosity of the dashpot can be calculated as, respectively,

$$E(t) \;=\; \frac{\dot{A}(t)}{A(t)^2\,\dot{B}(t)} \tag{34}$$

$$\eta(t) \;=\; \frac{1}{A(t)\dot{B}(t)} \tag{35}$$

From eq.(32) we get:

$$B(t) - B(\tau) \;\geq\; 0 \tag{36}$$

$$\dot{B}(t) \;\geq\; 0 \tag{37}$$

From eqs.(34) (35), $A(t) \geq 0$ and $\dot{A}(t) \geq 0$ are equivalent to $\eta(t) \geq 0$ and $E(t) \geq 0$ respectively. Because of eq.(30) the boundary conditions of $\Phi(t,\tau)$ have the following physical meaning:

$$\frac{\partial \Phi(t,\tau)}{\partial t} \geq 0 \qquad \equiv \qquad \eta(\tau) \geq 0 \tag{38}$$

$$\frac{\partial^2 \Phi(t,\tau)}{\partial t \partial \tau} \geq 0 \qquad \equiv \qquad E(\tau) \geq 0 \tag{39}$$

Substitution of eq.(29) into eq.(26) leads to an interesting property of the creep function according to the aging Kelvin-Voigt model:

$$\frac{\frac{\partial^2 \Phi(t,\tau)}{\partial t^2}}{\frac{\partial \Phi(t,\tau)}{\partial t}} \;=\; -K(t) \tag{40}$$

This includes that eq.(40) is independent of τ, if the material can be expressed by an aging Kelvin-Voigt model.

INTEGRAL TYPE OF CONSTITUTIVE LAW

The principle of superposition was introduced by Maslov (1941) [3, page 86] and McHenry [8] to analyse the creep of concrete.

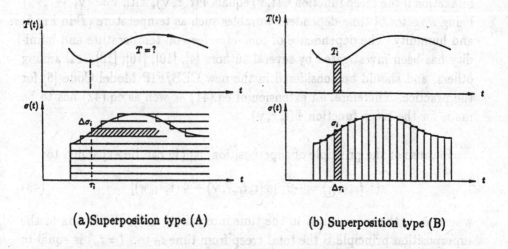

(a)Superposition type (A)   (b) Superposition type (B)

Fig.5 Horizontal and vertical decomposition of stress history

Fig.5 shows the superposition illustratively with variable stresses (stress increase or stress decrease with time). At the time t_0 we assume that the specimen is loaded with a stress $\sigma(t_0)$. The stress may be changed at the time t_1 to $\sigma(t_1) = \sigma(t_0) + \Delta\sigma$. According to the principle of superposition, type (A), the creep strain at the time $t \geq t_1 \geq t_0$ is the sum of the creep strains owing to $\sigma(t_0)$ and $\Delta\sigma$, respectively. For variable stresses at several times t_i for $i = 1, \cdots, n$, it leads to the following discrete equation (Fig.5a):

$$\varepsilon(t, t_0) = \sigma(t_0)\,\Phi(t, t_0) + \sum_{i=1}^{n} \Delta\sigma(t_i)\,\Phi(t, t_i) \tag{41}$$

If the stress $\sigma(t)$ is differentiable, eq.(41) may be written as an integral equation, in the following way:

$$\varepsilon(t, t_0) = \sigma(t_0)\ \Phi(t, t_0) + \int\limits_{t_0}^{t} \Phi(t, \tau)\ \frac{d\sigma(\tau)}{d\tau}\ d\tau \tag{42}$$

The eqs.(41) and (42), called constitutive laws, are well known for the calculation of creep or relaxation with variable stresses in practice [1], [6] and [7]. But the eqs.(41) as well as (42) are not suitable for the calculation of creep or relaxation if the creep function $\Phi(t, \tau)$ equals $\Phi(t, \tau, \mathbf{v})$, with $\mathbf{v} = \{v_1, \cdots, v_n\}$ being a vector of time-dependent variables such as temperature (T in Fig.5) and humidity. The dependence of concrete creep on temperature and humidity has been investigated by several authors [4], [10], [16], [13], [14] among others, and should be considered in the new CEB/FIP Model Code [6] for the practice. Therefore, an extension of eq.(41) as well as eq.(42) has to be made for the creep function $\Phi(t, \tau, \mathbf{v})$.

Because of the principle of superposition, eq.(1) can be expanded to:

$$\varepsilon(t, t_{i+1}, t_i) = -\bar{\sigma}_i\ [\Phi(t, t_{i+1}, \mathbf{v}) - \Phi(t, t_i, \mathbf{v})] \tag{43}$$

where, $\bar{\sigma}_i$ is the mean stress in the time interval (t_i, t_{i+1}). According to the superposition principle B the total creep from time t_0 to t ($= t_n$) is equal to the sum (Fig.5b):

$$\varepsilon(t, t_0) = \sum_{i=0}^{n-1} \varepsilon(t, t_{i+1}, t_i) = -\sum_{i=0}^{n-1} \bar{\sigma}_i\ [\Phi(t, t_{i+1}, \mathbf{v}) - \Phi(t, t_i, \mathbf{v})] \tag{44}$$

If the partial derivative of the creep function $\Phi(t, \tau, \mathbf{v})$ with respect to τ exists, while $n \rightarrow \infty$ and $\Delta t = (t_{i+1} - t_i) \rightarrow 0$, the discrete eq.(44) can be written as the integral:

$$\varepsilon(t, t_0) = -\int\limits_{t_0}^{t} \sigma\ \frac{\partial\Phi(t, \tau, \mathbf{v})}{\partial\tau}\ d\tau \tag{45}$$

The eqs.(44) and (45) are also the constitutive law based on the principle of superposition. If the creep function Φ is independent of $\mathbf{v}$, the eq.(44) and

(45) give the same results as the eq.(41) and (42).

As an example, we assume that σ_0, v_0 in the time interval (t_0, t_1) and σ_1, v_1 in the time interval (t_1, t) are constant; according to the superposition type (B) as well as eq.(44) we find for the creep at time t:

$$\varepsilon(t, t_0) = \sigma_0 \left[\Phi(t, t_0, v_0) - \Phi(t, t_1, v_0) \right] + \sigma_1 \, \Phi(t, t_1, v_1) \tag{46}$$

If the creep function Φ is independent of $\mathbf{v} = \{v_0, v_1\}$ or if $\mathbf{v}$ is constant, that is

$$\Phi(t, t_1, v_0) = \Phi(t, t_1, v_1) = \Phi(t, t_1) \tag{47}$$

eq.(46) may be written as

$$\varepsilon(t, t_0) = \sigma_0 \, \Phi(t, t_0) + (\sigma_1 - \sigma_0) \, \Phi(t, t_1) \tag{48}$$

This is just the result of the superposition (A) or eq.(41) with $n = 1$. Consequently, only if

$$\Phi(t, t_0) = \Phi(t, t_0, \mathbf{v})$$

Superposition (A) = Superposition (B)

The mathematical distinction between the type (A) equation (42) and the type (B) equation (45) for a complex creep function $\Phi(t, \tau, \mathbf{v})$ may also be discussed as follows:
Eq.(42) may be written as:

$$\varepsilon(t, t_0) = \sigma_0 \, \Phi(t, t_0, \mathbf{v}) + \int_{t_0}^{t} \Phi(t, \tau, \mathbf{v}) \, \frac{d\sigma(\tau)}{d\tau} \, d\tau \tag{49}$$

By partially integrating, we get

$$\varepsilon(t, t_0) = - \int_{t_0}^{t} \sigma(\tau) \, \frac{d\Phi(t, \tau, \mathbf{v})}{d\tau} \, d\tau$$

$$= - \int_{t_0}^{t} \sigma(\tau) \left(\frac{\partial \Phi}{\partial \tau} + \sum_{i=1}^{n} \frac{\partial \Phi}{\partial v_i} \, \frac{\partial v_i}{\partial \tau} \right) d\tau \tag{50}$$

Only if the following condition is fulfilled,

$$\sum_{i=1}^{n} \frac{\partial \Phi}{\partial v_i} \frac{\partial v_i}{\partial \tau} = 0 \tag{51}$$

that means that if $v = \{v_1, \cdots, v_n\}$ is constant or the creep function Φ is independent of v, then eq.(50) will be similar to eq.(45). Otherwise, the superposition type (A) equation (49) is not applicable. As a result it can be concluded that, if the influnces of temperature and humidily have to be considered when calculating creep and relaxation, only the superposition type (B) will lead to correct results.

The integral (45) can usually not be solved in a closed form. A numerical "step by step" method is developed in [15] and has been used to analyse the effect of imposed cyclic temperatures and strain histories for any linear creep function.

References

[1] ACI Committee 209: Prediction of Creep, Shrinkage, and Temperature Effects in Concrete Structures. Reported by Subcommittee II, Report No. ACI 209 R-82, ACI Publication SP-76, 1982.

[2] Alda, W.: Zum Schwingkriechen von Beton. Dissertation, TU Braunschweig, 1978.

[3] Bažant, Z.P.: Material Models for Structural Creep Analysis, Chapter 2, Fourth Rilem International Symposium on Creep and Shrinkage of Concrete: Mathematical Modeling, Northwestern University, Illinois, USA, 1986.

[4] Bažant, Z.P.: Concrete Creep at Variable Humidity: Constitutive Law and Mechanisms. Materials and Structures, RILEM, Vol.18, No.103, 1985.

[5] Bažant, Z.P., Osman, E.: Double Power Law for Basic Creep of Concrete, Materials and Structures, RILEM, Vol.9, No.49, 1976.

[6] CEB/FIP Model Code 1990, Bulletin D'Information No.195, CEB Co-
mite Euro-International du Beton. March 1990

[7] DIN 4227 Teil 1, 1988, sowie Erläuterungen zu DIN 4227 Spannbeton.
Deutscher Ausschüßfür Stahlbeton, Heft 320, Beuth Verlag GmbH, Berlin 1989.

[8] McHenry, D.: A New Aspect of Creep in Concrete and its Application
to Design. Proceedings ASTM, Vol.43, 1943.

[9] Müller, H.S.: Zur Vorhersage des Kriechens von Konstruktionsbeton.
Dissertation, Universität Karlsruhe, 1986.

[10] Neville, A.M., Dilger, W.H., Brooks, J.J.: Creep of Plain & Structural
Concrete, Construction Press, London and New York, 1983.

[11] Pfefferle, R.: Zur Theorie des Betonkriechens. Dissertation, Universität
Karlsruhe, 1971.

[12] Schade, D.: Einige eindimensionale Ansätze zur Berechnung des Kriechens und der Relaxation von Betontragwerken. Beton- und Stahlbeton-
bau, 3/1972.

[13] Shkoukani, H., Walraven, J.C.: Time Dependent Behaviour of Concrete
at Elevated Temperatures, Darmstadt Concrete, Vol.3, 1988.

[14] Shen, J.H.: Nonlinear Rheological Modeling for Concrete in Uniaxial
Compression. Darmstadt Concrete, Vol.5, 1990.

[15] Shen, J.H.: Lineare und nichtlineare Theorie des Kriechens und der
Relaxation von Beton unter Druckbeanspruchung. Dissertation (in preparation), Technische Hochschule Darmstadt, 1991.

[16] Springenschmid, R., Wagner, G.U., Schwarzkopf, M.: Temperaturspan-
nungen in Beton bei sommerlicher Erwärmung, Bauingenieur 53, 1978.

[17] Trost, H.: Auswirkungen des Superpositionsprinzips auf Kriech- und
Relaxationsprobleme bei Beton und Spannbeton. Beton- und Stahlbe-
tonbau, 10/1967.

CREEP FAILURE OF WELDMENTS IN THIN PLATES

M G NEWMAN and R E CRAINE
Faculty of Mathematical Studies,
University of Southampton
Southampton SO9 5NH, UK

ABSTRACT

A weldment is modelled as a thin Cosserat plate which, in addition to the parent material and weld metal, contains on each side of the weld a narrow heat affected zone (HAZ) and a narrow type IV region. Values for the strain rate in such a plate under uniaxial tension have been found previously. Here, a simple version of the Kachanov-Rabotnov damage equations is used and results are derived for the rupture time and position of rupture for the plate under tensile loading. Attention is focused on type IV cracking and the qualitative changes in the rupture caused by variations in the physical properties and dimensions of both the type IV region and HAZ.

INTRODUCTION

In power plant the design life of high temperature components operating in the creep regime is about twenty years. The integrity of such components is clearly of considerable importance, both for economic and safety reasons, and thus it is important to ensure that cracking and failure of components is minimised. The majority of practical problems associated with high temperatures are caused through cracking in or near welds, which are obvious regions of material inhomogeneity. These inhomogeneities are produced during the fabrication of the weld since the thermal gradients created during the localised melting cause changes in the parent material's metallurgical structure. The different zones that occur have physical properties that are usually markedly different from those of the parent material, their actual values depending upon the maximum temperature attained in the fusion zone and the local rates of cooling of the

structure, as well as the metallurgical properties of the parent material
[1,2,3]. Since some of the latter features are dependent on the welding
process, it is hoped that an enhanced understanding of how the properties
and sizes of the distinct regions influence type IV cracking will lead to
subsequent improvements in the welding process.

Metallurgical examination reveals that the weldment has four main
regions. The first region is the untransformed parent material. Adjacent
to the parent material is the narrow type IV region (or low temperature
HAZ) which contains metal which is weaker and more ductile than the parent.
The next region is the narrow high temperature HAZ, which is much stronger
and less ductile than the parent material. Finally, in the centre of the
weld is the weld metal itself which, for the widely utilised ferritic weld,
is slightly weaker than the parent metal. In this paper attention is
confined to this 'typical' ferritic weldment, with material properties
chosen so as to be appropriate for an investigation of type IV cracking.

A number of authors have considered the cracking that occurs in a
weldment, both experimentally and theoretically. Finite element models
have been used for pipe welds in [4,5,6] and experimental results support
the numerical predictions. The presence of the narrow material regions,
however, creates difficulties for finite element analyses and so it is not
certain that the numerical solutions obtained in [4,5,6] are accurate. In
a separate study Shakhmatov and Erofeev [7] deduce that the properties of
welded joints depend on the geometrical and mechanical properties of the
constituent materials.

Attempts to find a satisfactory mathematical model continued in the
1980's and Nicol [8] initiated a new approach, using the Cosserat theory of
plates and shells. Nicol formulated the plane strain creep problem for a
simple welded joint comprising two zones (the parent material and weld
metal) in a thin plate subjected to constant uniaxial tensile loading. The
Cosserat theory is more general than classical plate theories since it
permits the stress component normal to the plate to be non-zero, and it is
this extra degree of freedom that allows the thinning of the plate to be
included within a comparatively simple mathematical model.

A full derivation of the Cosserat theory for plates and shells can be
found in [9], where the work of many of the previous authors on the subject
is collected together and presented in a systematic way. Despite the
extent to which the general Cosserat theory has been developed very few
applications have been published, doubtless due to the apparent complexity

of the general theory. The major importance of Nicol's work was in
revealing how to apply the general theory to investigate the creep of a
thin welded plate. Nicol presented numerical results for a variety of weld
widths, and for weld metals that were both stronger and weaker than the
parent material, and his results were qualitatively consistent with
experiment. The analysis was extended by Nicol and Williams [10], who
obtained simple parametric equations describing the variation of creep
strain rate with material properties at certain key positions within the
weld. Nicol and Williams considered how these parametric equations could
be used to predict failure times in a weld but concluded that insufficient
experimental data was available to make firm predictions.

Hawkes [11] and Hawkes and Craine [12] further developed Nicol's model
by including both the narrow type IV region and HAZ. Numerical results
illustrating the effects on the creep strain rate caused by changing the
widths and strengths of the type IV, HAZ and weld regions and by varying
the creep index are presented in [11] and [12]. These results reveal that
in a weldment the interaction effects are extremely important, but very
complex. When narrow zones are present the value of the strain rate at a
point not only depends on the properties of the region in which it lies,
but can be significantly affected by the properties of adjacent, and even
non-adjacent, zones. In these situations, therefore, knowledge of the
homogeneous strengths of the constituent materials is not sufficient to
allow simple deductions to be made about the strain rates that occur in the
plate, or on its failure.

By including the additional metallurgical regions, and also
investigating pipes, Hawkes extended the range of applications of Nicol's
model to more useful situations. He also included in [11] a simpler
variant of Nicol's theory. Nicol generalised the multiaxial equations for
secondary creep by writing them in terms of variables that are utilised in
the Cosserat theory, whereas Hawkes in his modified theory used series
expansions and approximations emanating directly from the multiaxial
equations. The results obtained from both theories are identical when the
creep index n = 1, and they remain closely similar when n ≠ 1. In this
paper Hawkes' modified theory is used but a brief comparison with the other
more complicated theory is made.

The aim of the present work is to predict the position and time of
failure of a weldment, features not considered by Hawkes. A simple form of
the Kachanov-Rabotnov damage equations [13] is used to model the damage

associated with the creep process. The system of partial differential equations that results is still complicated, however, and a general analytical solution is not possible. For realistic values of the material parameters numerical solutions are presented for the rupture time and position of failure for a number of widths and strengths of the type IV region and HAZ. Arguments are put forward to support the qualitative features of these results.

GOVERNING EQUATIONS

A constant tensile stress σ is applied to a flat plate of thickness h. As discussed in the introduction it is assumed that the plate consists of distinct regions separated from each other by parallel plane surfaces that are normal to the plate's outer surfaces (see Figure 1). A rectangular Cartesian coordinate system $(x_1, x_2, x_3) \equiv (x, y, z)$ is introduced so that the central surface of the plate is at $z = 0$ and the top and bottom surfaces are at $z = h/2$ and $z = -h/2$ respectively. In addition, the x-axis is orientated so that the tensile stress σ is applied at infinity in the x-direction and the interfaces between the metallurgically distinct regions lie at $x = 0$, ℓ_1 and ℓ_2. The plane $x = \ell_3$, lying at the centre of the weld, is assumed to be a plane of symmetry.

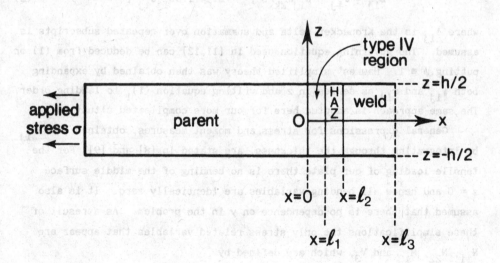

Figure 1. Cross-section of weldment.

The key feature in a Cosserat model is that the three-dimensional
plate is replaced by a two-dimensional surface within the plate, and that
at each point of this surface are associated directors which model in some
sense the thickness of the plate. The exact two-dimensional Cosserat model
for a plate of arbitrary thickness requires the existence of an infinite
number of directors at each point of the surface. When the number of
directors is restricted the theory becomes approximate, but for a fixed
finite number of directors the accuracy of the approximation increases as
the plate becomes thinner. Hence, a Cosserat model that includes a small
number of directors is more appropriate for thin plates.

In order to describe the damage of a weldment in the secondary creep
regime it is necessary to introduce constitutive equations connecting the
three-dimensional strain γ_{ij} and the three-dimensional stress σ_{ij}. In this
paper the equations used are those stated by Rabotnov [13]

$$\dot{\gamma}_{ij} = 1.5A\Sigma^{n-1}s_{ij}/\psi^p , \qquad \dot{\psi} = - B\Sigma^m/\psi^q , \tag{1}$$

where A, B, n, m, p and q are constants, a superposed dot denotes
differentiation with respect to time t and the Roman subscripts range over
the values 1,2,3. In equations (1) ψ represents the continuity and the
deviatoric stress tensor s_{ij} and stress measure Σ are defined by

$$s_{ij} = \sigma_{ij} - \delta_{ij}\sigma_{kk}/3 , \qquad \Sigma^2 = 1.5s_{ij}s_{ij} , \tag{2}$$

where δ_{ij} is the Kronecker delta and summation over repeated subscripts is
assumed. The governing equation used in [11,12] can be deduced from (1) on
putting $\psi \equiv 1$. Hawkes' simplified theory was then obtained by expanding
both γ_{ij} and σ_{ij} as series in z and writing equation $(1)_1$ to leading order.
The same approach is adopted here for our more complicated situation.

General expressions for stress and moment measures, obtained from σ_{ij}
by integrating through the thickness, are stated in [8] and [9]. For the
tensile loading of our plate there is no bending of the middle surface
z = 0 and hence all bending variables are identically zero. It is also
assumed that there is no dependence on y in the problem. As a result of
these simplifications the only stress related variables that appear are
N_{11}, N_{22}, M_{13} and V_3, which are defined by

$$N_{11}= \int_{-h/2}^{h/2} \sigma_{11}dz , \qquad N_{22}= \int_{-h/2}^{h/2} \sigma_{22}dz , \tag{3}$$

$$M_{13} = \int_{-h/2}^{h/2} z\sigma_{13}dz , \qquad V_3 = \int_{-h/2}^{h/2} \sigma_{33}dz. \tag{4}$$

Substituting series expansions for σ_{ij} into (3) and (4), the coefficients can be determined in terms of the quantities N_{11} etc. giving

$$\sigma_{11} = N_{11}/h, \quad \sigma_{22} = N_{22}/h, \quad \sigma_{13} = 12M_{13}z/h^3, \quad \sigma_{33} = V_3/h, \tag{5}$$

where the errors in each case are $O(z^2)$. Using equations (2) and (5) it can be shown that, to the same order,

$$\Sigma^2 = 0.75\{(N_{11}-V_3)/h\}^2 . \tag{6}$$

To specify the strain γ_{ij} a displacement vector $\underline{u}(x,y,t)$ and a director $\underline{\delta}(x,y,t)$ are associated with each point of the central surface of the plate, so that the three-dimensional displacement field $\underline{u}^*(x,y,z,t)$ is given approximately by

$$\underline{u}^*(x,y,z,t) = \underline{u}(x,y,t) + z\,\underline{\delta}(x,y,t) . \tag{7}$$

It then follows that for the problem considered in this paper, for which there is no dependence on y, the non-zero components of the strain γ_{ij}, to $O(z^2)$, are given by

$$\gamma_{11} = e_{11} , \qquad \gamma_{33} = \delta_3 , \qquad \gamma_{13} = \kappa_{13}z/2 , \tag{8}$$

where

$$e_{11} = \partial u_1/\partial x , \qquad \kappa_{13} = \partial\delta_3/\partial x . \tag{9}$$

The governing differential equations can now be determined by substituting the series for γ_{ij} and σ_{ij} into equations (1), and equating the leading order terms in z. Since $e_{22} = 0$ it is necessary for a non-trivial solution that

$$N_{22} = (N_{11}+V_3)/2 . \tag{10}$$

Eliminating N_{22} using (10), the remaining non-zero equations arising from $(1)_1$ can be rewritten

$$\dot{e}_{11} = 0.75A\Sigma^{n-1}(N_{11}-V_3)/(h\psi^p) = -\dot{\delta}_3 , \tag{11}$$

$$\partial\dot{\delta}_3/\partial x = 36A\Sigma^{n-1}M_{13}/(h^3\psi^p) , \qquad \dot{\psi} = -B\Sigma^m/\psi^q , \qquad (12)$$

where Σ can be expressed in terms of the other variables using (6). To complete the system of equations it is necessary to add the equilibrium equations for the plate (see [8], [9] or [11])

$$\partial N_{11}/\partial x = 0 , \qquad \partial M_{13}/\partial x = V_3 . \qquad (13)$$

Equations (11) to (13), together with (6), must now be solved subject to the boundary conditions

$$N_{11} \to N = \sigma h \quad \text{and} \quad M_{13} \to 0 \quad \text{as } x \to -\infty , \qquad (14)$$

$$N_{11}, M_{13}, \dot{\delta}_3 \text{ are all continuous at } x = 0, x = \ell_1 \text{ and } x = \ell_2, \qquad (15)$$

$$M_{13} = 0 \text{ at } x = \ell_3 , \qquad (16)$$

given that A, B, m, n, p and q take the piecewise constant values

$$(A,B,m,n,p,q) = \begin{cases} (A_a, B_a, m_a, n_a, p_a, q_a) & -\infty < x \leq 0 \\ (A_b, B_b, m_b, n_b, p_b, q_b) & 0 < x \leq \ell_1 \\ (A_c, B_c, m_c, n_c, p_c, q_c) & \ell_1 < x \leq \ell_2 \\ (A_d, B_d, m_d, n_d, p_d, q_d) & \ell_2 < x \leq \ell_3 \end{cases}, \qquad (17)$$

and that

$$\psi = 1 \text{ at } t = 0 \text{ for } -\infty < x \leq \ell_3 . \qquad (18)$$

From equation $(13)_1$ and the appropriate boundary conditions from (14) and (15) it follows that $N_{11} = N$ throughout the plate. The system (11), (12) and $(13)_2$ remains complex, however, and a full analytical solution has not been found. Some progress is possible in two simplified situations. Firstly, it was recently spotted that the system can be solved analytically for the strain rates at time $t = 0$ (when $\psi \equiv 1$) and, secondly, a small time series solution to the full system can be obtained. Both these solutions have been determined, although no details are presented here.

The system of governing differential equations (11), (12) and $(13)_2$ was non-dimensionalised and, for various values of the material parameters, solved numerically using finite differences. The two special solutions discussed in the previous paragraph were used as important, but partial, checks on the accuracy of the numerical results.

RESULTS

A comprehensive analysis of the system of equations remains difficult since there are many material parameters still unspecified. In this paper a number of assumptions are introduced which are consistent with an investigation of type IV cracking in a ferritic weldment. Experimental data suggests that it is reasonable to take $n_a = n_b = n_c = n_d = 4$, $m_a = m_b = m_c = m_d = 4$, $p_a = p_b = p_c = p_d = 4$ and $q_a = q_b = q_d = 4$. In the results presented it is also assumed that $q_c = 7$, although the numerical results obtained when $q_c = 10$ are little different. To specify the widths of the various regions it is convenient to introduce $x_b = \ell_1/h$, $x_c = (\ell_2 - \ell_1)/h$ and $x_d = (\ell_3 - \ell_2)/h$. In all the results displayed x_d, half the non-dimensional width of the weld region, equals 0.2 . The other widths x_b and x_c are varied. When specifying numerical values for A and B care must be taken with the units. With all n's and m's equal to 4 and time measured in hours, it is assumed that $A_a = 3.2 \times 10^{-15}$, $B_a = 3.8 \times 10^{-14}$, $B_b = 1.8 \times 10^{-11}$, $B_c = 1.5 \times 10^{-14}$ and $B_d = 2.2 \times 10^{-13}$. In all displayed results $A_d = 3A_a$ but different values are taken for the ratios A_b/A_a and A_c/A_a.

The finite difference scheme used has 50 nodes in each of the four regions and the results are stepped forward in time until ψ becomes zero at some point in the plate. This is the point at which the model indicates failure will occur, and for each numerical solution the point of failure x_r and the time t_r, in hours, for failure to occur have been calculated.

Figures 2 and 3 show plots for the strain rate $\dot{e}_{11}$ at time $t = 0$ for varying properties in the type IV region. In all these numerical results the width of the HAZ is fixed at $x_c = 0.2$ and its strength is given by $A_c = 0.1A_a$. Figure 2 shows results for $x_b = 0.1$ for A_b/A_a equal to 2 and 10, whereas Figure 3 displays results for $x_b = 0.2$ for the same two values of A_b/A_a.

It is evident from Figure 2 that the strong HAZ restricts the value of the strain rate in the type IV region to such an extent that the highest strain rates occur outside the latter region even though it contains the weakest material. The time to failure t_r and the position of failure x_r, associated with Figure 2, are stated in the first two lines of Table 1. To interpret the results it is useful to note that the simplified forms of equations (6), (11) and (12)$_2$ imply

$$\dot{\psi} \propto B\dot{e}_{11}/(A\psi^k) ,$$ (19)

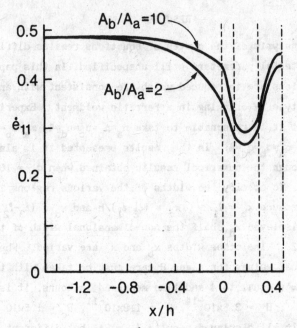

Figure 2. Plots of strain rate $\dot{e}_{11}$ against x/h at t = 0 for various strengths of type IV region.

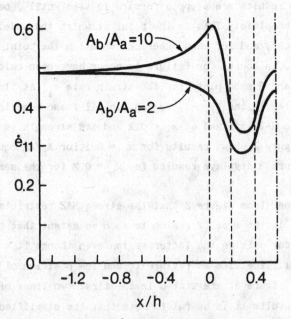

Figure 3. Plots of strain rate $\dot{e}_{11}$ against x/h at t = 0 for wider type IV region for various values of its strength.

TABLE 1

Position and time of failure for various widths and strengths of the
type IV region and HAZ

x_b	A_b/A_a	x_c	A_c/A_a	x_r	t_r (hours)
0.1	2	0.2	0.1	0	1906
0.1	10	0.2	0.1	0.01	8246
0.2	2	0.2	0.1	0.05	1296
0.2	10	0.2	0.1	0.06	4648
0.1	10	0.05	1.0	0.09	3387
0.1	10	0.05	0.1	0.06	4637
0.1	10	0.05	0.01	0.15	2805

where k = 3 in the HAZ and 0 elsewhere. Since the assumed value for B_b is
so much greater than the values of B in the other regions it is not
surprising that our plate usually ruptures in the type IV region. For
Figure 2 when A_b/A_a = 2 failure occurs at the parent-type IV interface,
where $\dot{e}_{11}$ has its maximum value. When A_b is increased by a factor of 5 the
strain rate increases at t = 0, but only slightly (as illustrated in Figure
2) and so equation (19) suggests $\dot{\psi}$ will decrease and t_r increase. The
results in Table 1 confirm this argument. In the latter situation, when
A_b/A_a = 10, the position of failure moves away from x = 0 since as t
increases the maximum strain rate occurs in the interior of the type IV
region, its value on the interfaces being restricted by the adjacent zones.

The strain rates at t = 0 for a wider type IV region, x_b = 0.2, are
displayed in Figure 3 and the corresponding failure parameters stated on
lines 3 and 4 of Table 1. As expected, with the wider type IV region the
strain rate in this weak region increases and the failure times are less
than the corresponding values associated with Figure 2. Both failures in
Figure 3 occur in the type IV region about half-way between the interface
x = 0 and the region's centre.

Changes to t_r and x_r caused by variations in the width and strength of
the HAZ are shown in Figure 4. Here x_b = 0.1, A_b/A_a = 10 and the non-
dimensional width of the HAZ is reduced to 0.05. The calculated values of
t_r and x_r are shown in the final three lines of Table 1. Narrowing the
strong HAZ obviously lessens its effect on reducing the strain rate in the
weak type IV region, and so the strain rate in the latter region is
increased and the value of t_r decreased. As A_c/A_a decreases from 1 to 0.1,
that is the HAZ becomes stronger, the strain rates are decreased and

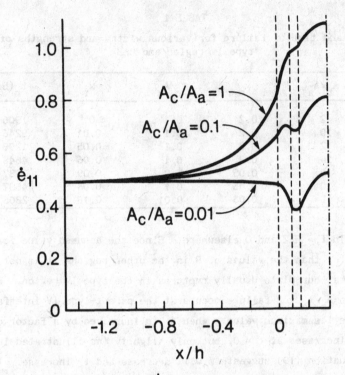

Figure 4. Plots of strain rate $\dot{e}_{11}$ against x/h at t = 0 for a narrow
HAZ of various strengths.

rupture times increase. Note, however, that for a very strong HAZ
(A_c/A_a = 0.01) the ratios B_c/A_c and B_b/A_b are comparable, and it is the
presence of the ψ^3 term in (19) which leads this time to failure in the
HAZ, at the position of maximum strain rate.

Comparison of Figures 2, 3 and 4, which show the strain rate in the
plate at t = 0, with the corresponding plots in [12] reveals that for fixed
zone widths and material strengths the strain rates are higher with Nicol's
original theory than when using Hawkes' modification as here. Hence it is
expected that the values of t_r obtained in this paper will be higher than
the corresponding values that would result from using the theory of Nicol
with damage included.

CONCLUSION

Using the Kachanov-Rabotnov damage method and the Cosserat theory of plates
results have been found for the position and time of rupture for a thin
welded plate containing multiple zones. As the material characteristics of
these zones are varied the results indicate that failure can occur at many

different places in the weldment. The position and time of failure are influenced by the material properties of both the region itself and its neighbouring regions.

Acknowledgements   MGN is grateful to the SERC and PowerGen for the award of a CASE studentship. The authors also thank Dr. J.A. Williams for many helpful discussions.

REFERENCES

1. Alberry, P.J. and Jones, W.K.C., Structure and hardness in creep resistant steels. Met. Sci. J., 1977, **14**, 557-566.

2. Roy, P. and Lauritzen, T., The relative strength of base metal and heat-affected zone in 2¾Cr-1Mo weldments – A microstructural evaluation. Welding Journal, 1986, **65**, 45s-47s.

3. Coleman, M.C., The structure of weldments and its relevance to high temperature failure. In Weldments, Physical Metallurgy and Failure Phenomena, Proc. of 5th Bolton Landing Conf., 1978.

4. Goodall, I.W. and Walters, D.J., Creep behaviour of butt-welded tubes. Inst. Mech. Engineers Conf. publication 13, 1973.

5. Walters, D.J., The stress analysis of cylindrical pipe welds under creep conditions. CEGB report RD/B/M3716, 1976.

6. Coleman, M.C., Parker, J.D. and Walters, D.J., The behaviour of ferritic weldments in thick section 1/2CrMoV pipe at elevated temperatures. CEGB report RD/M/1204/R81, 1981.

7. Shakhmatov, M.V. and Erofeev, V.V., The stress state and strength of welded joints with varying properties of the soft zone. Svar. Proiz., 1982, **3**, 6-7.

8. Nicol, D.A.C., Creep behaviour of butt-welded joints. Int. J. Engng Sci., 1985, **23**, 541-553.

9. Naghdi, P.M., The theory of shells and plates. In Handbuch der Physik, vol. VIa/2, 1972, reissued as Mechanics of Solids, vol. II, ed. C. Truesdell, Springer-Verlag, 1984.

10. Nicol, D.A.C. and Williams, J.A., The creep behaviour of cross-weld specimens under uniaxial loading. Res. Mechanica, 1985, **14**, 197-223.

11. Hawkes, T.D., Mathematical modelling of the creep of weldments using the Cosserat theory of plates and shells. Ph.D. thesis, University of Southampton, U.K., 1989.

12. Hawkes, T.D. and Craine, R.E., On the creep of weldments containing multiple zones in thin plates under uniaxial loading. Submitted to J. appl. Mech.

13. Rabotnov, Yu.N., Creep problems in structural members, North Holland, Amsterdam, 1969.

STRESS CALCULATION THE CERAMIC THERMAL BARRIER COATINGS FOR THE COOLED TURBINE BLADES

Y.A.TAMARIN
Institute of Aviation Materials
17, Radio st., 107006 Moscow, USSR
V.G.SOUNDYRIN, V.YU.KANAYEV
Central Institute of Aviation Motors,
2, Aviamotornaya st., 111250, Moscow, USSR

ABSTRACT

2D-finite element model for analysis of thermal and stress-strain state of ceramic thermal barrier coatings (TBC) on the cooled turbine blades allowing for the creep deformation in blade and coating is described. Coating life is estimated according to the criterion of exhaustion of ceramic coating adhesive strength.

INTRODUCTION

At present cooled blades with ceramic thermal barrier coatings (TBC) are getting wider use in design of aircraft advanced gas turbines. Having high heat resistance such coating work as a thermal insulator impeding heat penetration from gas to the blade metal.

At the same time under working conditions unfavorable thermal stresses (resulting from the distinction in mechanical properties of ceramic and blade materials) may take place in the ceramic layer causing TBC failure (Fig.1).

To carry out TBC stress analysis and its life estimation on the stage of design works an approach described herein can be used.

MATERIALS AND METHODS

Applied to the blade surface using electronic beam PVD method two-layer TBCs of ZrO_2/MeCrAlY type partially stabilized with Y_2O_3 are considered.

Ceramic layer of such TBCs has strongly pronounced columnar structure. The layer failure is caused mainly by the decreasing of adhesive strength on the boundary "ceramic-metal" while in operation with further spalling of the ceramic layer fragments. Spallation takes place after high temperature (1000 - 1150°C) exposure even with no stresses

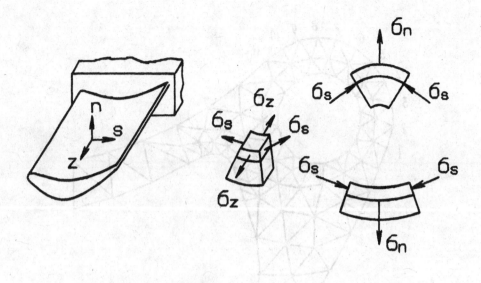

Figure.1. Thermal stresses in TBC.

in the specimen.This effect relates to the formation on the boundary
"ceramic-metal" of dense film of Al_2O_3 (or $Al_2O_3 \cdot Cr_2O_3$) oxides with no
pores;these oxides not only lower adhesive strength but also change
stress-strain state near this boundary,since their mechanical properties
differ from the properties of ceramic and metal TBC layers.

All processes resulted in decreasing of TBC ceramic layer adhesive
strength are indirectly corrected for by the adhesive strength integral
characteristic $\sigma_a \cdot \sigma_a (T, \tau)$ defined experimentally. σ_a is
breaking stress of separation of TBC ceramic layer from the surface of
specimen after exposure under temperature T for the time τ

TBC stresses (σ_s, σ_z) can be calculated from analysis of blade
and TBC thermal and stress-strain state taking into account their joint
deformation on various working regimes.For this purpose it is convinient
to carry out computer-aided numeric solution of 2D-problems of nonsta-
tionary (and stationary) heat transfer and thermoelasticity using finite
element method (FEM) on a uniform FEM grid for blade cross-section.

To correct for the blade and TBC creep deformations the authors
used creep flow theory and method of supplementary deformations.

Used FEM grid (Fig.2) includes triangular rectilinear and curvilinear finite elements (for blade) as well as quadrangular curvilinear
finite elements in the form of thin strips,two opposite sides of wich
are straight lines in the direction normal to the surface (for TBC
layers). Curvilinear sections of the FEM grid and sought functions
within each element are approximated by cubic functions in the local
coordinates.

The condition

$$\sigma_n < \sigma_a \tag{1}$$

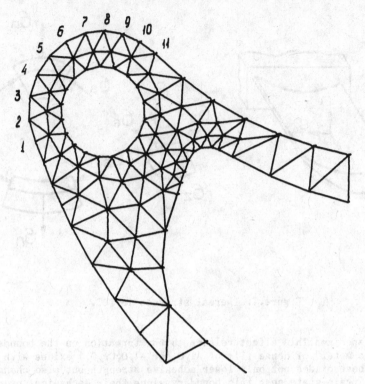

Figure 2. FEM grid fragment.

where σ_n is breaking stress of separation of TBC ceramic layer from the blade (see Fig.2), works as a criterion of TBC serviceability. These stresses appearing on the convex sections of the blade surface (when stresses σ_s in the ceramic layer are compressive) can be calculated according to the formula

$$\sigma_n \approx \frac{\Delta}{R}\sigma_s \tag{2}$$

where Δ - ceramic layer thickness;

R - radius of curvature of outer surface of the blade in its cross section.

Rated TBC life is defined by the time moment τ when condition (1) is broken.

RESULTS

Two-layer TBC of $ZrO_2 \cdot 8\%Y_2O_3/Ni \cdot 22\%Cr \cdot 11\%Al \cdot Y$ type with thicknesses of ceramic and metal layers of 100 μm and 60 μm respectively has been investigated.

Maximum inner TBC stresses appear in the cold state of the blade. When cooling down to the room temperature (20°C) after applying of TBC and high temperature annealing (1000°C) compressive residual stresses of ~ -400 MPa in TBC ceramic layer and tensile stresses of ~ +500 MPa in metal layer appear. For all this breaking stresses σ_n of separation of ceramic layer of the blade have their maximum of ~50 MPa at the zone of the input edge. On the working regimes thermal stresses in TBC ceramic layer at the zone of input edge are close to zero. After long run on these regimes accumulation of creep deformation takes place in the blade and TBC resulting in decreasing of residual stresses in the ceramic layer under room temperature (Fig.3).

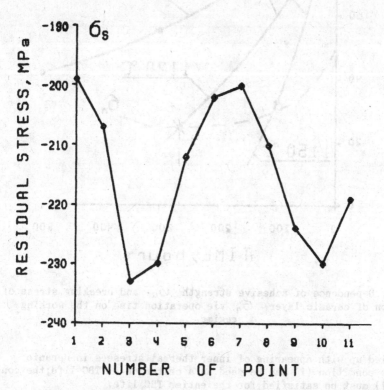

Figure 3. Residual stresses σ_S in TBC ceramic layer after 300 hours run.

On decreasing of stresses σ_S stresses σ_n in cold blade decreases as well. As shown on Fig.4 decrease rate of stresses σ_n of the blade having operation run is less than one of adhesive strength σ_a. At the time moment of τ =265 hours plots of $\sigma_n = \sigma_n(\tau)$ and $\sigma_a = \sigma_a(\tau)$ for the point 6 of the blade surface are cut (T=1120°C), thus adhesive strength condition (1) is broken.

Close value of time to failure of TBC of τ =260-280 hours was obtained during test of this blade. TBC failure began on the input edge of the blade.

CONCLUSIONS

The main cause of TBC failure is spallation of ceramic layer fragments from blade. The spallation is assisted by breaking stresses of separation

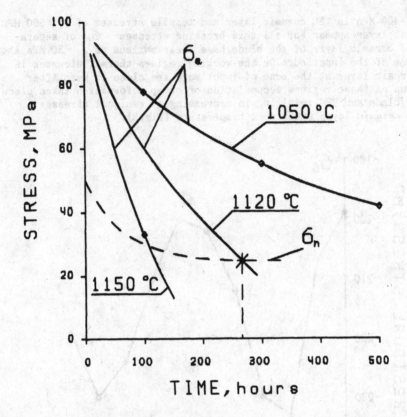

Figure 4. Dependence of adhesive strength σ_a and breaking stress of separation of ceramic layer σ_n vie operation time on the working regime.

σ_n tied up with appearing of inner thermal stresses in ceramic layer. The condition (1) can be used as a criterion of TBC life; the condition (1) must be satisfied for the entier TBC life.

MODELLING THE MECHANISMS AND MECHANICS OF INDENTATION CREEP WITH APPLICATION TO ZIRCONIA CERAMICS

J L HENSHALL, G M CARTER, K E EASTERLING and R M HOOPER
School of Engineering
University of Exeter
Exeter, Devon EX4 4QF, UK

and

W B LI
Department of Engineering Materials
University of Luleå
S-95187 Luleå, Sweden

ABSTRACT

Previous work [1] had shown that indentation creep occurred at low temperatures in zirconia ceramics. However, it was only possible to analyse the data in terms of a phenomenological approach, since it was concluded that dislocation glide was rate controlling at these low temperatures. A new theoretical analysis has since been developed [2] and the appropriate aspects of this modelling are presented, together with new data on single crystal calcia stabilised cubic zirconia and ceria stabilised tetragonal polycrystalline zirconia, to complement the previous results on yttria stabilised zirconias.

INTRODUCTION

The indentation technique, based on eg Vickers or Knoop indenters, is a well established method of characterising the hardness of a material [3]. It is used to provide both fundamental information regarding the deformation and fracture of ceramic materials, and as a useful guide to the operational response in many of the application areas where the primary loading is principally compressive. The technique can also be extended to study the creep resistance of a material subjected to the applied indentation pressure. In this case the hardness indenter maintains a constant load over a measured period of time under well controlled conditions, and changes in the indentation size are monitored. An important result of such testing is that all the materials studied, including diamond, are found to creep at temperatures well below half their melting points.

As discussed previously [1], the principal advantages of using microhardness indentation creep as a test method are:-
1. The high hydrostatic stress beneath the indenter inhibits fracture and thus allows the mechanics and mechanisms of the deformation processes to be investigated.
2. The small sample volume reduces the inherent variability problem with ceramics.
3. The applied stress state is similar to that developed in many practical situations.
4. The ease of specimen preparation, since only two relatively small parallel faces are required, one of which is well polished.
5. The test method is sufficiently rapid and low cost that it can be utilised to compare creep resistance throughout the development phase of new ceramics.

The indentation process induces complex deformation behaviour beneath the indenter. However, as described by Johnson [4], there are three regimes which can be indentified. Immediately adjacent to the indenter a heavily deformed zone is developed. This may include material which has sheared as a result of phase transformations and/or block slip, but it can be assumed that this material acts subsequently as a non-deforming 'plug' or extension of the indenter. This is enveloped by a plastic zone, which is in turn supported by an elastic 'hinterland' of material. It is time-dependent plastic deformation in the plastic zone which basically leads to the continuing penetration of the indenter. Previous models of this deformation [5-8] had assumed that conventional power-law creep was responsible for the time dependent deformation. However, it has now become clear that at the relatively low temperatures where zirconia ceramics are likely to be utilised, to take advantage of their high toughness, [9] the time-dependent deformation will occur as a result of dislocation glide [1,2]. This paper therefore reports further measurements of indentation creep in zirconia ceramics and the development of a dislocation glide-controlled model of indentation creep.

EXPERIMENTAL

The zirconias used in this study were yttria stabilised cubic single crystal (10% mole yttria), YCSZ, yttria stabilised cubic polycrystalline (6 mole % yttria), YCPZ, calcia stabilised cubic single crystal (14 mole % calcia), CaCSZ and ceria stabilised tetragonal polycrystalline zirconia (10.5 mole % ceria), CeTZP. The test specimens, approximately 5 mm x 5 mm x 3 mm, were cut with a diamond saw and the faces to be indented ground parallel, prior to final polishing with 1/4 micrometre diamond paste. Within the limits of conventional X-ray powder diffractometry, it was found that the materials

were either completely cubic phase, or completely tetragonal
phase, as appropriate. The single crystals were oriented by
the Laue back reflection technique such that the indented
faces were within 1 degree of (111). The YCPZ consisted of
equiaxed grains in the range of 1-5 micrometres, with a small
amount of closed porosity. The CeTZP had a typical grain size
of 2-3 μm, and contained small isolated pores and a small
number of larger (≅ 100 μm) irregularly shaped processing
defects.

All the indentations were made in air using a modified Leitz
Miniload microhardness tester incorporating a hot stage with a
Knoop diamond indenter at a load of 2.94 N, mounted on a
stable vibration free plinth. The indenter tip was placed in
contact with the specimen surface for a period of 30 minutes
to equilibrate the temperatures prior to actual testing. The
indenter was hydraulically damped such that full load was
applied after 18 seconds. At least 5 impressions were made at
dwell times of 10, 100, and 1000s, with a minimum of 3 at
10000s. Temperatures were monitored using a Type K
thermocouple and controlled to within 2 deg K of the nominal.
In all cases tests were performed initially at the highest
test temperature, and then at continually decreasing
temperatures. The indentational diagonal lengths were
measured optically with a precision of 0.2 micrometres. The
Knoop hardness is defined as the load divided by the projected
area of the indentation, and is thus the same as the mean
pressure. For the single crystal specimens the long axis of
the indenter was aligned along [110]. The (111) face was used
for the creep tests since the hardness anisotropy on this face
was found to be minimal [10].

Following the completion of the indentation creep tests, the
room temperature hardnesses of the specimens were remeasured
and found not to have changed. X-ray diffraction analysis
also showed no difference in peak positions prior to and after
testing. These factors both suggest that no phase changes or
intermediate phase precipitation occurred to affect the
results.

INDENTATION CREEP ANALYSIS

There are a number of possible mechanisms of plasticity, some
(or all) of which may contribute to indentation creep.
Following the creep equations given by Frost and Ashby [1],
hardness rate equations can be developed for the mechanisms,
shown schematically in Figure 1, of plasticity (dislocation
glide, 1,2), power low creep (climb plus glide, 3), power-law
breakdown (glide plus climb, 4), recovery (dislocation climb,
5), and diffusion (volume, grain boundary or dislocation pipe,
6,7).

Considering indentation creep due to dislocation glide in the
plastic zone. The shear strain rate $\dot{\gamma}$, is given by the
standard equation

$$\dot{\gamma} = \rho_m \, b \bar{v}$$

where ρ_m is the density of mobile dislocations, b is burger's vector and $\bar{v}$ is the mean dislocation velocity. The latter term, $\bar{v}$ is thermally activated according to

$$\bar{v} \propto \exp \left\{ \frac{-\Delta F_p}{KT} \left[1 - \frac{\sigma_s}{\hat{\tau}_p} \right] \right\}$$

where σ_s is the applied shear stress, ΔF_p is the activation energy of lattice-resistance controlled glide, $\hat{\tau}_p$ is the 0 K flow stress. Therefore as shown by Frost and Ashby [11] for conventional plasticity:-

$$\dot{\gamma} = \dot{\gamma}_p \left(\frac{\sigma_s}{\mu} \right)^2 \exp \left\{ \frac{-\Delta F_p}{KT} \left[1 - \frac{\sigma_s}{\hat{\tau}_p} \right]^{3/4} \right\}^{4/3} \tag{1}$$

where $\dot{\gamma}_p$ is a constant and μ is the shear modulus. The exponents 3/4 and 4/3 were derived from curve fitting to data obtained on a range of materials. A similar equation can be derived for glide limited by discrete obstacles, ie

$$\dot{\gamma} = \dot{\gamma}_0 \exp \left\{ -\frac{\Delta F}{KT} \left[1 - \frac{\sigma_s}{\hat{\tau}} \right] \right\}$$

where $\dot{\gamma}_0$, ΔF and $\hat{\tau}$ are the appropriate material constants. However, the zirconias studied here are all single phase solid solution strengthened materials, and thus equation (1) is more appropriate to describe the plasticity behaviour.

This can now be extended to indentation creep as follows. For the Knoop indenter the hardness H is given by

$$H = \frac{\alpha W}{1^2} \tag{2}$$

where W is load, 1 is long diagonal length and α is a geometrical constant = 14.23. Now since W is constant with time

$$\frac{\dot{a}}{a} = \frac{-\dot{H}}{2H} \tag{3}$$

The actual stress distribution around the indentation cannot be adequately described in detail at present. However, the magnitude of the stresses, in particular the average shear stress, σ_s, in the elastic-plastic zone will scale as the indentation pressure or hardness, ie

$$\sigma_s = C_1 H \tag{4}$$

C_1 is a proportionality constant which may vary from material class to material class, but in the present case it is assumed that $C_1 = 1/3$. When comparing the parameters for closely

similar materials, as with the range of zirconias studied,
then it is most unlikely that C_1 will vary from material to
material. For an incompressible material, the radial velocity
of an element at radius r in the elastic plastic zone is given
by $(a/r)^2\dot{a}$. Consequently the shear strain rate will be
$3a^2\dot{a}/(2r^3)$. Thus the mean strain rate throughout the elastic
plastic zone should be

$$\gamma = \frac{3}{2}\frac{a^2}{r^3}\dot{a} = \frac{12}{(1+k)^3}\frac{\dot{a}}{a} \tag{5}$$

where $k = c/a$ is the ratio of the plastic zone size, to the
indentation diagonal. It has been observed experimentally
that k is approximately constant in ceramics, but that there
is a critical temperature at which the value of k increases
markedly to a new constant value. Substituting from equations
(2)-(5) into (1) leads to

$$\dot{H}_1 = -\dot{\gamma}_p \frac{(1+k)^3}{6}\left(\frac{C_1}{\mu}\right)^2 H^3 \exp\left\{\frac{-\Delta F}{KT}P\left[1 - \frac{C_1 H}{\hat{\tau}_p}\right]^{3/4}\right\}^{4/3} \tag{6}$$

Similar analyses can be applied to the other mechanisms and
the results are given in Table 1, with full details presented
in reference 2. The net creep rate of a sample subject to an
indentation load W, at a temperature T is

$$\dot{H} = \text{greatest of } [\text{least of } (\dot{H}_1, \dot{H}_2), \dot{H}_3 \text{ or } \dot{H}_4, \dot{H}_5] + \text{greatest of } (\dot{H}_6, \dot{H}_7)$$

The current or instantaneous hardness value is thence
numerically calculated from this rate equation, if the initial
hardness is known.

The indentation process generates high local stresses,
particularly in ceramics, so that at low temperatures
($T < 0.3T_m$) it is found that glide plasticity is always the
dominant deformation mechanism [2], except for very fine
grained ceramics, ie d < 0.1 μm.

RESULTS

Figure 2 shows the variation of the short term (10s) hardness
with temperature for the four zirconias studied. The three
cubic stabilised materials all have similar hardnesses at room
temperature, but the variations with temperature are quite
different. The polycrystalline YCPZ decreases almost
uniformly with increasing temperature, whereas the single
crystal materials both appear to show a transition in
behaviour at ca 700 K for the yttria stabilised and 900 K for
the calcia stabilised. The ceria stabilised TZP is relatively
softer at room temperature but it appears to be of similar
hardness at ca 1100 K.

Figure 3 depicts the time dependent indentation behaviour for
these four materials at room temperature, 473 K and 573 K.

The lines drawn on the graphs are fitted using numerical
integration of equation (6) for the values given in Table 2.
The fit to the data for the yttria stabilised polycrystalline,
YCPZ and calcia stabilised single crystal is very close, but
the other results are not so closely matched. This is in part
due to a lack of refinement in the curve fitting procedures
used at present, and it is possible that relatively small
adjustments to the material parameters ($\cong$ 5-10%) would result
in improved matching of the theoretical curves and data, to be
similar to the YCPZ and CaCSZ.

TABLE 2

Material	ΔF_p (kJ/mole)	$\hat{\tau}_p$ (GPa)
•Yttria stabilised cubic single crystal (YCSZ)	570	5.3
Yttria stabilised cubic polycrystalline (YCPZ)	360	5.7
Calcia stabilised cubic single crystal (CaCSZ)	650	5.0
Ceria stabilised tetragonal polycrystalline	360	4.3

Values of activation energy, ΔF_p, and 0 K flow stress, $\hat{\tau}_p$ used
in calulating the theoretical curves in Figure 3.

DISCUSSION

All of the rate-equations of indentation creep presented in
Table 1 are derived on the basis that steady state conditions
have been established. The establishment of these steady
state conditions is generally relatively rapid (of order
seconds), although for some combinations of load and
temperature the incubation times may be quite prolonged [12].
No such incubation times were observed in the present work.
All of the rate equations, with the single exception of $\dot{H}_7$ for
single crystal diffusional creep, are independent of load.
Apart from the initial fitting to H_o, fundamental material
constants can be used in general. However, in the regime of
current interest, ie glide plasticity, there are few
alternative methods of determining the activation energy and
0 K flow stress in ceramics. Therefore the values of the
parameters have to be adjusted to fit the data. However, the
values which are determined by such a data fitting,
ie ΔF_p = 360-650 kJ/mol and $\hat{\tau}_p$ $\cong$ 4.3-5.7 GPa are physically
quite reasonable. This contrasts with the more conventional
log H vs log t linear plot which generally gives stress
exponents (>40) and unrealistically low activation energies
(as low as 36 kJ/mol) [1] for low temperature, 290-573 K,

indentation creep data. Therefore much greater confidence can be expressed in extrapolating the data to the longer times which are appropriate to operational service lifetimes. There is always, of course, the possibility of environmental effects influencing the behaviour, and hence hardness in the vicinity of the surface. This is also currently being investigated for zirconia and other materials [9,13] but care was taken to ensure that there was no environmental influence in the present results.

It is of interest to note that the single crystal hardnesses were higher than the polycrystalline values. Also the rate of decrease of hardness with both increasing time and temperature was greater for the polycrystalline zirconias. These observations are explained by the activation energies for the single crystal zirconias 570 and 650 kJ/mol, being greater than for the polycrystalline YCPZ and CeTZP, ca 360 kJ/mol. The 0 K flow stresses are all fairly similar, but, as observed, it might be expected that the CeTZP would have the lowest value since there are no excess anion vacancies required in the lattice.

CONCLUSIONS

The time and temperature dependence of the indentation hardness for single crystal yttria and calcia stabilised cubic zirconia, yttria stabilised polycrystalline cubic zirconia and ceria stabilised tetragonal polycrystalline zirconia have been measured between room temperature and 573 K, for times up to 10 000 s. The relative hardness decrease for the polycrystalline zirconias is greater than for the single crystals. Dislocation glide plasticity is the rate controlling deformation mechanism during the indentation creep. An analytical model of this process is developed and used to describe and interpret the data.

ACKNOWLEDGEMENTS

The authors would like to thank the UK S.E.R.C. and the Royal Swedish Academy of Science for financial support and Tenmat UK; Ceres Corp, USA and Hrand Djevahirdjian, Switzerland for the supply of materials.

REFERENCES

1. Henshall, J.L., Carter, G.M. and Hooper, R.M., Indentation creep in zirconia ceramics between 290 K and 1073 K. In Mechanics of Creep Brittle Materials - I, eds. A.C.F. Cocks and A.R.S. Ponter, Elsevier Applied Science, London, 1989, pp 117-128.

2. Li, W.B., Henshall, J.L., Hooper, R.M. and Easterling, K.E., The mechanisms of indentation creep. Acta Met. and Mat., in press.

3. Tabor, D., Indentation hardness and its measurement: some

cautionary comments. in <u>Microindentation Techniques in Materials Science and Engineering</u>, eds. P.J.Blau and B.R.Lawn, ASTM STP 889, American Society for Testing Materials, Philadelphia, 1986, pp. 129-159.

4. Johnson, K.L., The correlation of indentation experiments. <u>J. Mech. Phys. Sol.</u>, 1970, 18, 115-126.

5. Mulhearn, T.O. and Tabor, D. Creep and hardness of metals: A physical study. <u>J. Inst. Met.</u> 1960-61, 89, 7-12.

6. Atkins, A.G., Silverio, A. and Tabor, D., Indentation hardness and the creep of solids. <u>J. Inst. Met.</u>, 1966, 94 369-378.

7. Sherby, O.D. and Armstrong, P.E., Prediction of activation energies for creep and self diffusion from hot hardness data. <u>Metall. Trans.</u> 1971, 2, 3479-3484.

8. Roebuck, B. and Almond, E.A., Equivalence of indentation and compressive creep tests on a WC/Co hardmetal. <u>J. Mater. Sci. Let.</u>, 1982, 1, 519-522.

9. Thuraisingham, S.T. and Henshall, J.L., The influence of chemical environment on the indentation hardness and toughness of ceria-stabilised tetragonal polycrystalline zirconia. In <u>Proceedings of the 21st Biennial Conference of the Institute for Briquetting and Agglomeration</u>, ed. D.L.Roth, Hudson, USA, 19.

10. Guillou, M.-O., Carter, G.M., Hooper, R.M. and Henshall, J.L., Hardness and fracture anisotropy in single crystal zirconia. <u>J. Hard Materials</u>, 1990, 1, 65-78.

11. FROST, H.J. and ASHBY, M.F., <u>Deformation Mechanism Maps</u>, Pergamon Press, 1982.

12. Hooper, R.M. and Brookes, C.A., Incubation periods and indentation creep in lead. <u>J. Mater. Sci.</u>, 1984, 19, 4057-4060.

13. Carter, G.M., Henshall, J.L. and Wakeman, R.J., Influence of surfactants on the mechanical properties and comminution of wet-milled calcite. <u>Powd. Tech</u>. 1991, 65, 403-410.

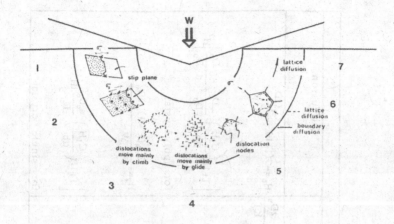

Figure 1 Schematic of the indentation "zones" and the deformation processes within the plastic zone.

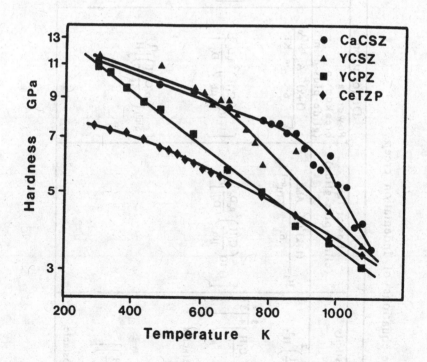

Figure 2 Variation of the Knoop Hardness (log scales) with temperature.

Table 1. Rate-equations of Indentation creep.

| | Dislocation | | | |
Glide plasticity	Power-law (climb-plus-glide)	Power-law breakdown (glide-plus-climb)	Climb	Diffusion
$\dot{H}_1 = -\dot{\gamma}_p \dfrac{(1+k)^3}{6}\left(\dfrac{C_1}{\mu}\right)^2 H^3$	$\dot{H}_3 = -\dfrac{(1+k)^3 C_1^n}{6}\dfrac{AbD_v}{KT\mu^{n-1}}$	$\dot{H}_4 = -\dfrac{(1+k)^3}{6}\dfrac{A}{\alpha'^n}\dfrac{b\mu D_v}{KT}$	$\dot{H}_5 = -\dfrac{40}{9}\dfrac{C_1^2(k^3-1)}{(\mu b)^2}\dfrac{\Omega D_v}{KT}$	$\dot{H}_6 = -7(1+k)^3\dfrac{C_1\Omega D_v}{KT d^2}$
$\exp\left[-\dfrac{\Delta F_p}{KT}\left[1-\left(\dfrac{C_1 H}{\hat{\tau}_p}\right)^{\frac{3}{4}}\right]^{\frac{4}{3}}\right]$	$\left[1+10\left(\dfrac{C_1 H}{\mu b}\right)^2\dfrac{a_c D_c}{D_v}\right] H^{n+1}$	$\left[1+10\left(\dfrac{C_1 H}{\mu b}\right)^2\dfrac{a_c D_c}{D_v}\right]\cdot$	$\left[C_2 H - \dfrac{8C_3\mu}{k^3-1}\right] H^3$	$\left[1+10\left(\dfrac{C_1 H}{\mu b}\right)^2\dfrac{a_c D_c}{D_v}+\dfrac{\pi\delta}{d}\dfrac{D_b}{D_v}\right] H^2$
for diffuse obstacles		$\left[\sinh\left(\dfrac{\alpha' C_1 H}{\mu}\right)\right]^n \cdot H$		for polycrystal
$\dot{H}_2 = -\dot{\gamma}_0\dfrac{(1+k)^3}{6}\cdot H\cdot$				$\dot{H}_7 = -\dfrac{2C_4(k^2-1)}{W}\dfrac{\Omega D_v}{KT}\cdot$
$\exp\left[-\dfrac{\Delta F}{KT}\left[1-\dfrac{C_1 H}{\hat{\tau}}\right]\right]$				$\left[1+10\left(\dfrac{C_1 H}{\mu b}\right)^2\dfrac{a_c D_c}{D_v}\right] H^3$
for discrete obstacle				for single crystal

GLOSSARY OF SYMBOLS USED IN TABLE 1

A = Dorn constant

a = radius of hydrostatic zone

a_c = cross-sectional area R dislocation core

b = Burger's vector = 3.22×10^{-10}m.

c = radius of elastic-plastic zone

C_1 = H/σ_s, taken to be 1/3

C_2 = H/σ_n, taken to be 1/3

$C_3 < 1$, and if the climb mechanism dominates $C_3 \approx 1$

C_4 = C_2

d = grain diameter

D_v = diffusion coefficient of lattice

D_b = diffusion coefficient of grain boundary

D_c = diffusion coefficent of dislocation core

k = c/a = 1.75

K = Boltzman constant

l = indentation diagonal length

n = exponent of power-law creep

T = temperature

T_m = melting temperature = 2843 K

W = load

α = constant for power-law breakdown

γ = shear strain

$\dot{\gamma}_o$ = pre-exponential of obstacle controlled glide

$\dot{\gamma}_p$ = pre exponential of lattice resistance controlled glide = 5×10^{11} s^{-1}

δ = thickness of grain boundary

μ = shear modulus = 71 G Pa

ρ_m = mobile dislocation density

σ_n = average local normal stress acting parallel to b

σ_s = shear stress

Ω = atomic volume

$\hat{\tau}$ = 0 K flow stress of obstacle controlled glide

$\hat{\tau}_p$ = 0 K flow stress of lattice resistance controlled glide

ΔF = activation energy of obstacle controlled glide

ΔF_p = activation energy of lattice resistance controlled glide

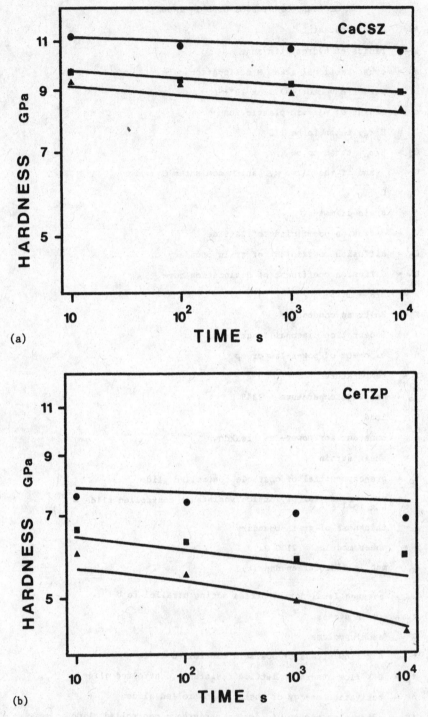

(a)

(b)

Figure 3   Variation of Knoop Hardness with time (log scales) at 293 K (●),
473 K (■) and 573 K (▲)for the four zirconias studied.   The curves
are derived using equation 6 in the text with values of
constants as given in Tables 1 and 2.

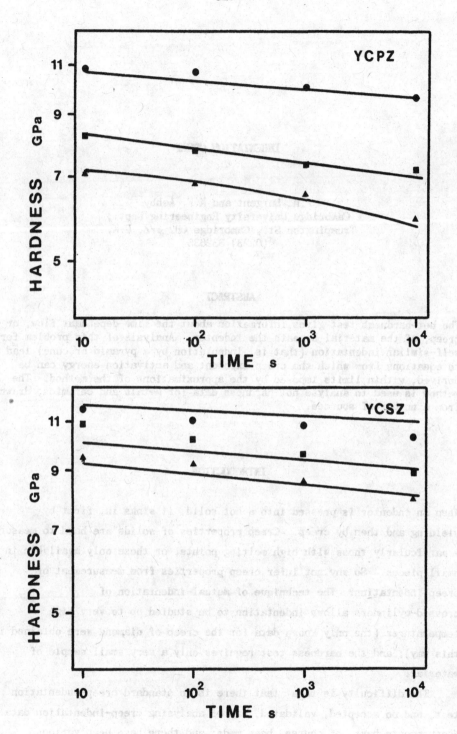

INDENTATION CREEP

P.M. Sargent and M.F. Ashby
Cambridge University Engineering Dept.,
Trumpington St., Cambridge CB2 1PZ, U.K.
(0223) 332635

ABSTRACT

The hot-hardness test gives information about the time-dependant flow, or creep, of the material beneath the indenter. Analysis of this problem for self-similar indentation (that is, indentation by a pyramid or cone) lead to equations from which the creep exponent and activation energy can be derived, within limits imposed by the approximations of the method. The method is used to analyse hot-hardness data for metals and ceramics, drawn from a number of sources.

INTRODUCTION

When an indenter is pressed into a hot solid, it sinks in, first by yielding and then by creep. Creep properties of solids are hard to measure - particularly those with high melting points, or those only available in small pieces. So why not infer creep properties from measurement of creep-indentation? The technique of mutual-indentation of crossed-cylinders allows indentation to be studied up to very high temperatures (the only known data for the creep of diamond were obtained in this way); and the hardness test requires only a very small sample of material.

The difficulty is this: that there is no standard creep-indentation test, and no accepted, validated, way of analysing creep-indentation data. Measurements have, of course, been made; and there have been various attempts to provide an analysis for the results. Early trys [1] made restrictive assumptions about the constitutive law for creep. Later attempts sought general solutions satisfying both equilibrium and compatability [2], or applied the reference-stress technique [3], or used dimensional analysis, calibrating the results against certain known limits (see below). All these, for practical purposes, give the same result. The dimensional analysis is the most transparent.

TABLE 1
Symbols, Definitions and Units

A	Projected area of indentation (mm^2)
$\dot{A}$	Rate of change of indentation area ($mm^2.s^{-1}$)
A_o	Area of indentation after initial yield (mm^2)
C_1, C_2, C_3	Indenter shape geometric constants (-)
C_4	Constant, function of C_1, C_2, C_3 and n (-)
$\dot{e}$	Strain rate (s^{-1})
$\dot{e}_o$	Strain rate at reference stress σ_o (s^{-1})
F	Force on indenter (kN)
H	Projected-area Hardness (GPa)
μ_{300}	Shear modulus at 300K (GPa)
n	Power Law exponent (-)
R	Gas constant ($kJ.mol^{-1}.K^{-1}$)
Q	Activation energy for creep ($kJ.mol^{-1}$)
σ, σ_s	Shear stress (GPa)
σ_o	Reference (shear) stress (GPa)
T_m	Melting or solidus temperature (K)
t	Time (s)
u	Indenter displacement (mm)
$\dot{u}$	Indenter displacement rate ($mm.s^{-1}$)

DIMENSIONAL ANALYSIS OF CREEP INDENTATION

An indenter is pressed into the flat surface of a solid, which creeps
(Fig.1). The constitutive law for creep (the constraints of which we would
like to extract from the measurements) is:

$$\dot{\varepsilon} = \dot{\varepsilon}_o \left[\frac{\sigma}{\sigma_o} \right]^n \tag{1}$$

Where

$$\dot{\varepsilon}_o = \kappa \, \exp - \frac{Q}{RT}$$

and $\dot{\varepsilon}$ is the strain rate, κ a kinetic constant, σ the deviatoric (or
shear) stress and σ_o is a reference stress at which the strain rate is
$\dot{\varepsilon}_o$.

The load F on the indenter is carried by a projected indent-area A, so
that the indentation pressure is $H = F/A$. It creates a stress field below
the indenter. For equilibrium, each component of stress in the material is
proportional to F/A (if F/A were doubled, for example, all the stresses
σ at each point beneath the indenter must double too) so that:

$$\sigma = C_1 \frac{F}{A} \tag{2}$$

where C_1 depends only on position but not on F or A. The stresses
cause the material beneath the indenter to creep and the indenter sinks in.
If the indenter is a pyramid or a cone, the stress and displacement fields
beneath it remain self-similar: they change only in scale, not in shape.
(It is possible to extend the analysis to spherical indenters, but the
stress and strain-rate fields do not remain self-similar). Consider a
representative material point below such an indenter, which is at a

distance which scales as $\sqrt{A}$ from the surface.

The displacement rate du/dt of the indenter is caused by a
strain-rate in the underlying material; let the strain-rate at this point
be $\dot{\varepsilon}$. For compatability, the displacement-rate divided by some relevant

length must be proportional to strain-rate. There is only one length in the problem: it is $\sqrt{A}$. (This is the treatment implicitly assumed by Roebuck and Almond [4]. Thus:

$$\dot{\epsilon} = \frac{C_2}{\sqrt{A}} \left[\frac{du}{dt}\right] \tag{3}$$

where C_2 characterises the plastic strain state induced by the indentation and depends on the indenter shape; for a given shape C_2 is also a constant at the representative material point. Inserting (2) and (3) into (1) gives:

$$\frac{du}{dt} = \dot{\epsilon}_0 \frac{\sqrt{A}}{C_2} \left[\frac{C_1}{\sigma_0}\frac{F}{A}\right]^n \tag{4}$$

For either a conical or a pyramidal indenter, the penetration is proportional to $\sqrt{A}$:

$$u = C_3 \sqrt{A} \tag{5}$$

where u is the current displacement and C_3 is a geometric constant depending only on the shape of the indenter, as shown in Figs 1 and 2. Differentiating (5) with respect to time and substituting in (4) gives

$$\frac{dA}{dt} = C_4 \dot{\epsilon}_0 A \left[\frac{F}{A\sigma_0}\right]^n \tag{6}$$

where $C_4 = 2.C_1{}^n / (C_2 C_3)$ is a constant, independent of load or temperature. Integrating with respect to time gives:

$$A = A_0 + (n C_4 \dot{\epsilon}_0)^{1/n} \frac{F}{\sigma_0} t^{1/n} \tag{7}$$

where A_0 is the area of the indentation caused by the initial high-strain-rate yielding, before the creep mechanism becomes dominant. The creep hardness $H = F/A$, is given by

$$H(t) = \left[1 - \frac{A_o}{A}\right] \frac{\sigma_o}{(n \, C_4 \, \dot{\epsilon}_o \, t)}^{1/n} \qquad (8)$$

Provided $A \gg A_o$, this simplifies to

$$H(t) = \frac{\sigma_o}{(n \, C_4 \, \dot{\epsilon}_o \, t)}^{1/n}$$

It is probable that σ_o has a slight temperature dependence caused by that of the modulus. To remove this we normalise by the shear modulus μ to give

$$\frac{H(t)}{\mu} = \frac{(\sigma_o/\mu)}{(n \, C_4 \, \dot{\epsilon}_o \, t)}^{1/n} \qquad (9)$$

This is the equation we use to analyse experimental data.

APPLICATION TO THE ANALYSIS OF EXPERIMENTAL DATA

In creep-indentation tests the load on the indenter is kept constant and the area of the indentation is measured as a function of time. This is done either by a continuous displacement measurement or, more usually, by making several different indentations, each for a different time under load. In any one test the temperature is kept constant.

The creep exponent n can be found in one of a number of ways. From eqn (9), the gradient of a plot of log (H/μ) against log (time) at constant temperature, T, is $-1/n$. If eqn (9) describes the data completely, the plot should be a family of parallel straight lines, one for each fixed temperature – and some (shown below) are like this. Curvature at short times means that the approximation $A \gg A_o$ may not be met; and straight lines which are not parallel implies that n depends on T. Alternatively the rate of indentation, rather than time, can be used. Differentiation of eqn (9) gives

$$\frac{dH}{H dt} = -\frac{1}{nt} \qquad (10)$$

or, using eqn (6) with F held constant

$$\frac{dH}{Hdt} = - C_4 \, \dot{\varepsilon}_0 \left[\frac{H}{\sigma_0}\right]^n \tag{11}$$

Then a plot of log (H) against log (-dH/Hdt) has a slope of 1/n and a
plot of log (-dH/Hdt) against log(t) superimposes all the data for a
material onto a single master curve with an intercept of 1/n at t = 1s
(or, equivalently, 1/100n at t = 100s) and a gradient of −1.

The <u>activation energy</u> of the creep process is found either by a plot
of ln(H/μ) against (T_m/T) at a constant time, with a gradient of
Q/nRT_m, or a plot of ln(time) against (T_m/T) at constant H/μ, which
has a gradient of Q/RT_m. The latter is perhaps the better method because
a constant H/μ implies a constant structural state.

Eqn (11) suggests that a log plot of a "stress-independent" rate
parameter

$$\ln \left[\frac{d(H/\mu)}{dt} \cdot \frac{1}{(H/\mu)^{n+1}}\right] \text{ against } T_m/T$$

should give a master-curve with slope Q/RT_m.

Finally, hardness data can be plotted approximately onto
deformation-mechanism maps [5], giving an idea of the <u>mechanism</u> of
deformation by noting that, at a given point on the creep indentation
curve, the strain rate (from eqns 3, 4 and 5) is

$$\dot{\varepsilon} \approx - \frac{1}{2} \frac{dH}{Hdt} \tag{12a}$$

(giving a shear strain rate of $\dot{\gamma} \approx \sqrt{3} \, \dot{\varepsilon}$); and the shear stress is of order

$$\sigma_s = \frac{H}{3\sqrt{3}} \tag{12b}$$

Results for Metals: Armco Iron and Copper

Figure 3a shows data of hardness, plotted as (H/μ), as a function of time,
t, at constant temperature, T, for Armco iron [6]. The data span for the
range T/T_m = 0.36 to 0.48 ; the upper end of this range is sufficiently
high that classical power-law creep behaviour is expected (see Frost and

Ashby [5] Figs 8.1 and 8.2). The data are tolerably well fitted by
straight lines, giving n = 10 - 13. (The alternative plot, log (-dH/Hdt)
against log (t), (Fig 3b) gives the same range for n). The activation
energy, Q, is derived from the plot of ln(t) against T_m/T at constant
H/μ, shown in Fig 3c. The slope gives Q = 272 - 327 kJ/mole (Q/RT_m = 18
- 22). The "stress-independent" rate master curve is shown in Fig 3d: the
agreement with theory is good.

TABLE 2

MODULI AND MELTING POINTS

	μ_{300} (GPa)	$-\dfrac{d\mu}{dT} \cdot \dfrac{T_m}{\mu_{300}}$	T_m (K)
Iron	64	0.81	1810
Copper	42.1	0.54	1356
Diamond	530.6	0.165	4000
Germanium	52	0.146	1211
Silicon	63.7	0.078	1687
Boron Nitride	240	0.2	4000

Data for iron, copper, germanium and silicon are from Frost and Ashby [5].
The melting points (at 1 atm) of diamond and boron nitride are estimates
based on moduli and electronic and thermodynamic data.

Some of the difficulties of the method are illustrated by the next
batch of data, for copper. Figure 4a shows indentation creep measurements
by Hill [7], Okada et al.[8] and Fairbanks et al. [9], covering the
temperature range T/T_m = 0.22 to 0.72. The lower end of this range is
outside the regime of classical power-law creep; and at the upper end,
scatter is considerable. The two plots to obtain the creep exponent

(Figs 4a and 4b) give values which range from n ≈ 8^{+}_{-} 4 at high
temperatures to n ≈ 60 at low. There is nothing unexpected in this: the
low-temperature data are in the power-law breakdown regime (see Frost and
Ashby [5], Fig 4.7). The high-temperature data are in the power-law creep
regime, and the limiting slope of Okada's data, n = 4.4, agrees well with

conventional creep experiments which give n = 4.8. The activation energy plot, shown in Fig 4c, shows similar scatter. At low temperatures, stress-activation reduces the apparent activation; at high, Okada's data give 170 kJ/mol $(Q/RT_m \approx 15)$ in reasonable agreement with conventional creep data (around 190 kJ/mol), but the data of Hill give a much larger value. The master-plot (Fig 4d) is not successful in bringing the data onto a single curve.

Results for Non-Metals: Silicon, Germanium, Diamond and Boron Nitride

The analysis presented here is valid for any geometrically-similar indentation geometry, which includes the case of mutually-indenting crossed-wedges. This was the geometry used to study creep in diamond [10]. The other materials were indented using diamond Vickers indenters [11-13]. The difference appears only in the factor C_3 which is 25.5 for Vickers pyramids and 16.0 for crossed octahedral-wedges (see Fig 2).

Figure 5 shows data for silicon, for T/T_m = 0.34 to 0.75, taken from the studies of Naylor [11], Morgan [12] and Okada [8]. The upper end of this range lies in the power-law creep regime (Frost and Ashby [5], Fig 9.3). The hardness measurements, plotted in Fig 5a against log (t), give a family of straight lines, parallel at the high temperatures, with a slope of about 4.8 (limited conventional creep data give n = 5) up to about $0.6T_m$, and with a slope of about 20 between 0.6 and $0.75T_m$. The values are confirmed by the plot of Fig 5b. The activation energies (Fig 5c: 84 to 618 kJ/mol, 6 RT_m to 44 RT_m) are mostly considerably lower than those of self-diffusion studies (around 490 kJ/mol, 35 RT_m), probably because the stresses are high (see discussion, below). The lower stress and higher temperature results give energies of between 27 and 44 RT_m. The master plot of Fig 5d brings the data approximately into coincidence with $Q \approx 13$ RT_m.

Data for germanium [8], [14], [15] from T/T_m = 0.48 to 0.97 are plotted in Figure 6. The mean creep exponent, n, is about 5 (range: 3.1 to 14.5), and the activation energy is 130 - 200 kJ/mol (12.7 to 20 RT_m), again lower than that expected from self-diffusion studies (287 kJ/mol). The master plot (Fig 6d) is successful in correlating the data, giving an activation energy of 13.7 RT_m.

Measurement of the creep indentation of Type I and Type II diamond [10] are harder to interpret. The data span the range T/T_m = 0.44 to 0.53. Figures 7a and 8a show the hardness data plotted against log (t); at

the higher temperatures, the lines are straight at all but the shortest
times, and give exponents n in the range 0.8 to 2.2. The activation-
energy analysis (Figs 7c and 8c) give values of Q in the range 800 to
2200 kJ/mol (24 RT_m to 66 RT_m).

Data for boron nitride, taken from Brookes [16], are plotted in Figure
9. The data lie in the temperature range $T/T_m \approx 0.12$ to 0.32, below that
in which creep is adequately described by the simple power-law of
equation (1). Not surprisingly, the values of n are high (14 to 50) and
the activation energies are very low (60 - 80 kJ/mol, 2 to 3 RT_m).

DISCUSSION AND CONCLUSIONS

It is clear from the analysis just presented, that creep-indentation
experiments can give fundamental information about creep. But it is
equally clear that there are pitfalls in interpretation to be avoided; that
the experimental technique, though improving fast, has still some way to go
before it can be thought of as reliable; and that the analysis, too, will
require further sophistication as experimental techniques improve. These
are worthwhile goals: the creep-identification method has the unique
capacity to give creep data for minerals often available only as single
crystals, for single phases in multiphase systems, and for materials which
creep only at very high temperatures and stresses.

The data analysis of section 3 showed that the simple theory, based on
a steady-state power-law creep equation, has the capacity to describe
creep-indentation data approximately. The fit is best, and the inferred
creep parameters n and Q are most accurate, when the indentation
temperature is above 0.5 T_m and the time is long (so that the indentation
pressure H is low). At the other extreme (low T/T_m, high H)
deformation involves all the complexity of the "power-law breakdown"
regime, or, in mechanistic terms, of stress-activated glide as well as
thermally activated, diffusion-controlled, creep. The data from this
extreme contains useful information too, but it is not yet clear how to
analyse it.

ACKNOWLEDGEMENTS

The authors wish to give thanks to Dr. Jan Czernuska, Dr. Brian Derby and Dr. Steve Roberts for helpful comments on drafts of this paper, to Dr. Andrew Hill, Dr. Atumasa Okada and Dr. Czernuska for supplying data in advance of publication, and to Prof. Chris Brookes, Dr. Hubert Pollack and Dr. Eric Almond for encouragement and help in locating indentation creep data. PMS thanks the Science and Engineering Research Council for the provision of an Advanced Fellowship.

REFERENCES

1. ATKINS, A.G., SILVERIO, A. and TABOR, D. (1966) J.Inst.Metals 94, 369-378.

2. MATTHEWS, J.R. (1980) Acta Met. 28, 311-318.

3. PONTER, A.R.S., PALMER, A.C., GOODMAN, D.J., ASHBY, M.F., EVANS, A.G. and HUTCHINSON, J.W. (1983) Cold Regions Science and Technology 8, 109.

4. ROEBUCK, B. and ALMOND, E.A. (1982) J.Mat. Sci. Lett. 1 519-521.

5. FROST, H.J. and ASHBY, M. (1982) in Deformation Mechanism Maps, Pergamon Press, Oxford, UK. Ch. 17: "Further Refinements on High Strain-Rates and High Pressures".

6. STEELE, R.K. and DONACHIE, M.J. (1965) Trans. ASM 58, 273-284.

7. HILL, A.D. (1984) "Modelling and Assessment of Diffusion Bonding", PhD. Thesis, University of Cambridge, UK, and personal communication.

8. OKADA, A., YAMAMOTO, Y. and TODA, R. (1987) J.Iron and Steel Inst. Japan 73 (9) 1186-1192 and personal communication.

9. FAIRBANKS, C.J., POLVANI, R.S., WEIDERHORN, S.M., HOCKEY, B.J., and LAWN, B.R. (1982) J.Mat. Sci. Lett. 1 391-393.

10. EVANS, T. and SYKES, J. (1974) Phil. Mag. 29, 135-147.

11. NAYLOR, M.G.S. (1982) "The Effects of Temperature on Hardness and Wear Processes in Engineering Ceramics", PhD. Thesis, University of Cambridge, UK.

12. MORGAN, J.E. (1977) "Indentation Hardness and Indentation Creep in Solids at Temperatures Below $0.5T_m$", PhD. Thesis, University of Exeter, UK.

13. BROOKES, C.A. (1981) in The Science of Hard Materials, Viswanadham, R.K., Rowcliffe, D.J. and Gurland, J. Eds., Publ. Plenum Press. Proc. 1st Int. Conf. Sci. Hard Matls., Aug. 1981, Jackson, Wyoming, USA.

14. GERK, A.P. (1975) Phil. Mag. 32, 355.

15. CZERNUSKA, J. (1987) "Indentation Creep of Intrinsic Germanium Single Crystal", personal communication.

16. BROOKES, C.A. (1984) Inst. Phys. Conf. Ser. No. 75: Chapter 3 (Proc. 2nd. Int. Conf. Science Hard Materials, Rhodes).

APPENDIX A1 - DATA MANIPULATION

The data analysis was greatly aided by Borland's "Reflex" data- handling package. Gradients of curves were calculated using a short program written in Pascal (Borland's, version 4.0) and the activation energy plots at constant H/μ were cross plotted by hand from the log (H/μ) v. log (t) plots using straight line interpolations between points.

Copies of the database formats, report document formats, all the data used in this paper and the Pascal program (including source code) are available from the authors on IBM PC floppy disk for a nominal fee of £10, Reflex is available from Borland Intl. (UK) Ltd., 8 Pavilions, Ruscombe Business Park, Twyford, Berkshire RG10 9NN, England.

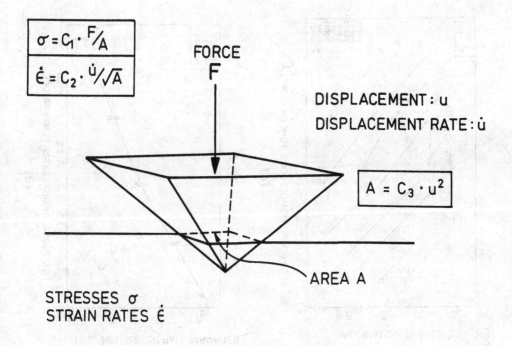

Figure 1. Diagram of a loaded pyramidal indenter.

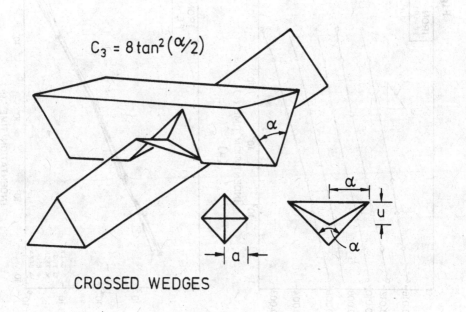

Figure 2. Diagram of a crossed-wedges indentation.

338

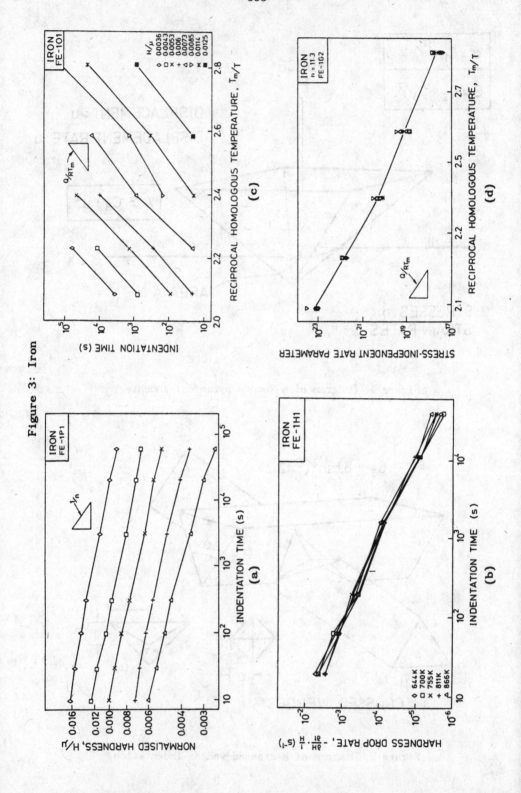

Figure 3: Iron

Figure 4: Copper

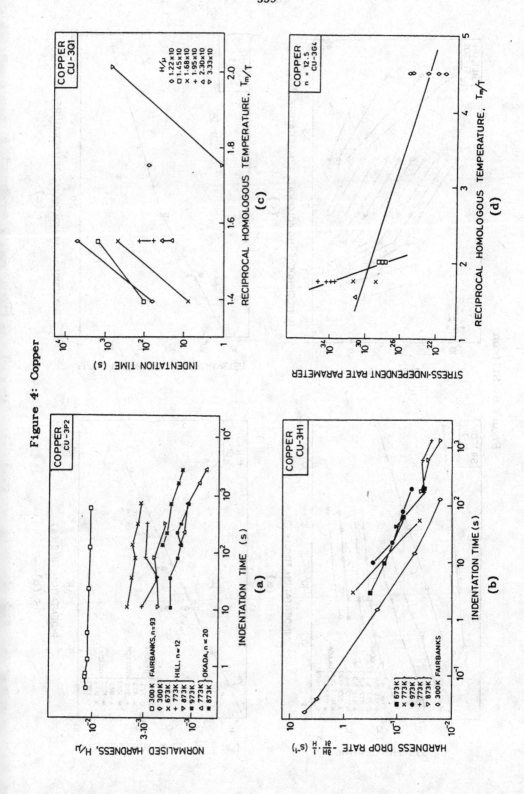

Figure 5: Silicon

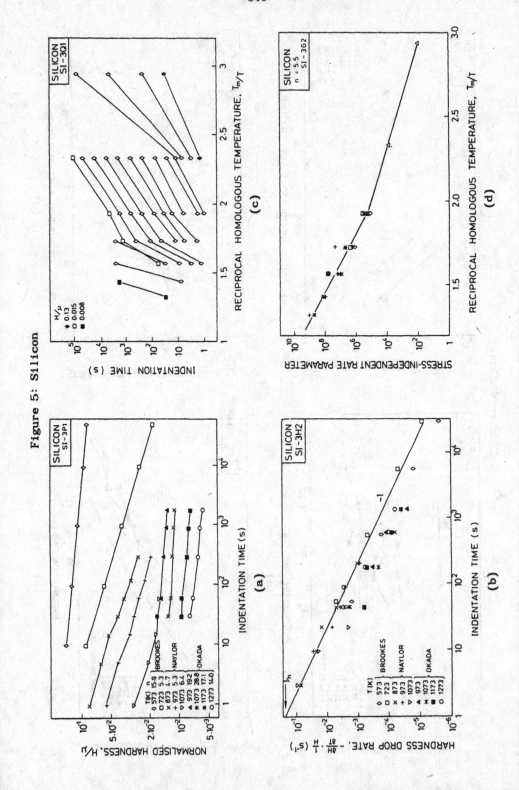

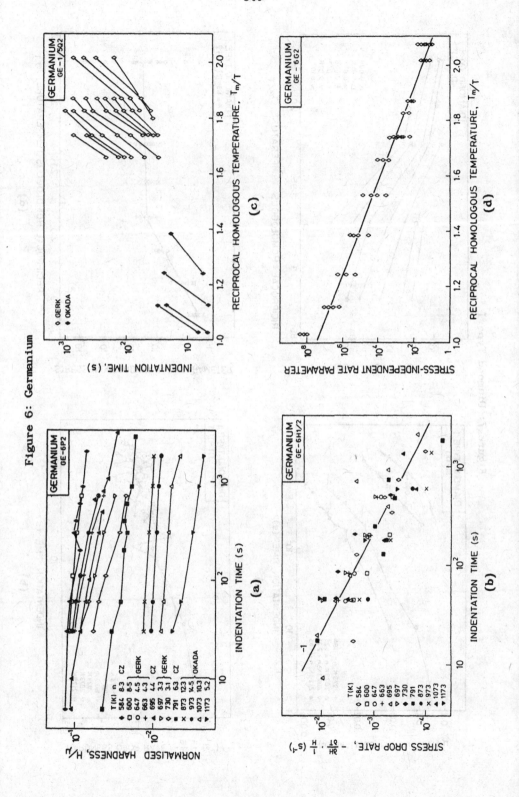

Figure 6: Germanium

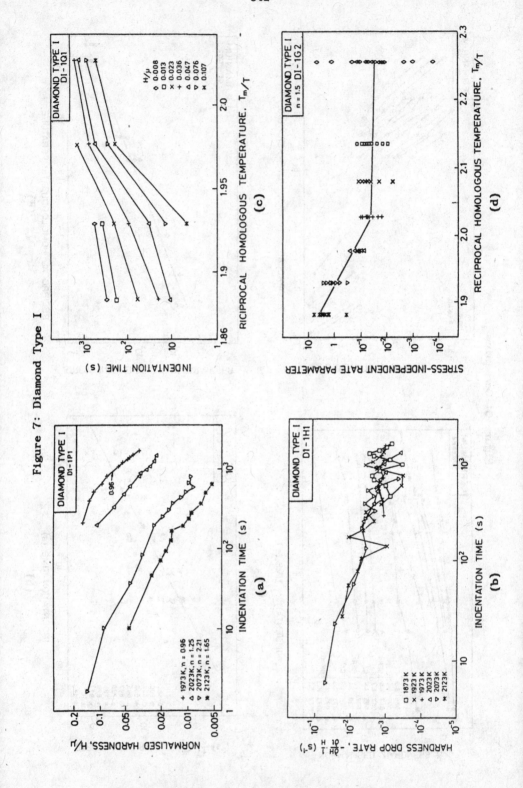

Figure 7: Diamond Type I

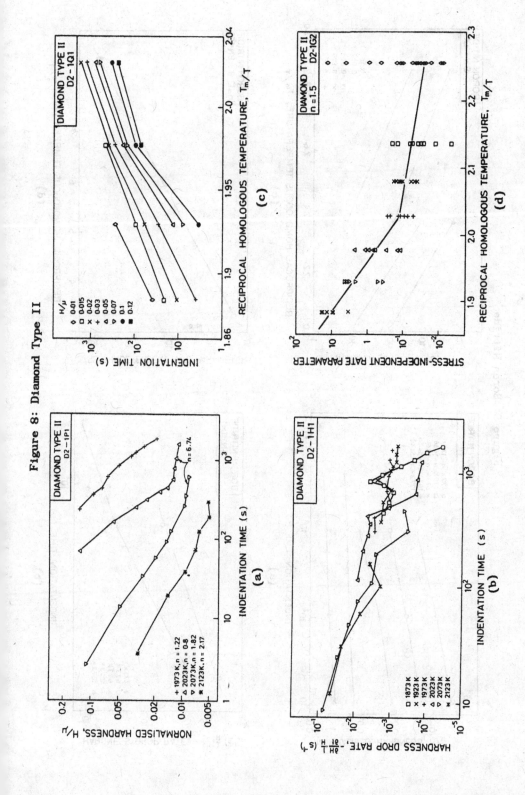

Figure 8: Diamond Type II

344

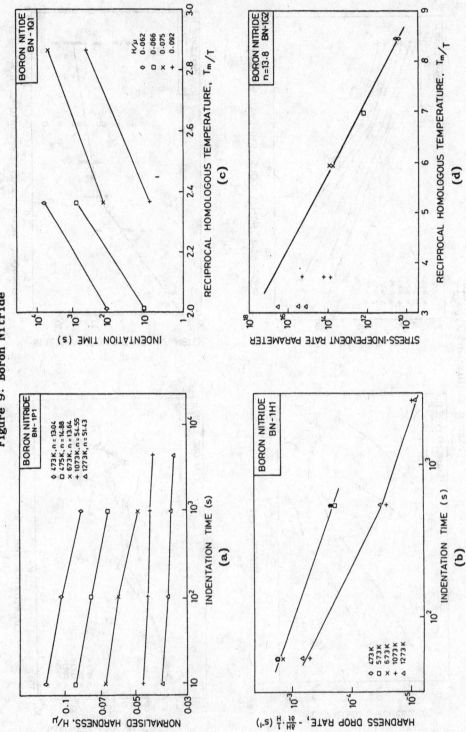

Figure 9: Boron Nitride

THE USE OF THE SOFT INDENTER TECHNIQUE TO INVESTIGATE IMPRESSION CREEP IN CERAMIC CRYSTALS

C. A. BROOKES, E. J. BROOKES AND G. XING
Department of Engineering Design and Manufacture
University of Hull, Hull, HU6 7RX, UK

ABSTRACT

Transmitted polarised light microscopy and dislocation etching techniques have been used to investigate the strain developed beneath the Knoop indenter and cones made from soft metals on (001) MgO surfaces. Time dependent plastic deformation, i.e. impression creep, of the ceramic crystal is observed for both cases. This phenomenon is shown to be due to an expansion of the dislocated volume – rather than an increase in the effective strain – such as to preserve its geometrical similarity with respect to the increased size of the indentation/impression.

INTRODUCTION

The strain developed in most ceramic materials during the formation of an indentation using conventional techniques, e.g. Vickers, Knoop and Rockwell, is generally sufficient to initiate a degree of fracture at homologous temperatures of $< 0.3\ T_m$. The resultant crack formation and distribution will depend on the nature of the stress field in particular regions of the deformed volume beneath the indenter but, even at room temperature in the hardest of solids [1], there is likely to be some dislocation movement and multiplication. Replacing the rigid pyramidal indenter with a cone made from a material which is significantly softer than the ceramic specimen is proving to be an effective method for studying the plastic deformation of these ceramics. Here the cone blunts and the mean pressure (P_m) applied to the surface is readily determined when the normal load is divided by the measured contact area on the flattened tip of the cone. This mean pressure tends to be about one third of the indentation hardness of the cone material [2] and by the use of alternative materials a wide and controlled range of P_m can be used to deform the specimen. Dislocation movement, identified by the use of etch pit techniques, is initiated at a threshold value of P_m and the corresponding critical resolved shear stress can be estimated by the application of a relevant model [3]. When materials of increasing hardness are used for the cones, and where otherwise the experimental conditions are the same, the dimensions of the dislocated volume remain roughly constant. Higher values of P_m induce greater dislocation activity within this constant volume and thus the effective strain increases with the hardness of the indenter material until, approaching that developed beneath conventional diamond indenters, cracks are initiated. The effective strain beneath the flattened cone can also be increased if the normal load is repeatedly applied and this can lead to significant workhardening, fatigue and fragmentation under lubricated sliding conditions [2].

The cumulative deformation associated with prolonging the 'dwell time', i.e. the time during which the normal load is continuously applied to the contacting cone, differs significantly from that associated with the increased strain and fatigue due to repeated loading. In the experiments described here it will be shown that there is time dependent plastic flow in ceramic crystals deformed under these conditions whilst, under a constant load, the applied mean pressure is decreasing concurrently. The reduction in P_m is generally determined by similar time dependent processes causing an increased contact area in the case of a soft indenter. This has been used to augment the scope of a method previously reported for estimating the critical resolved shear stress of magnesium oxide crystals at room temperature [3]. However, observations on the nature of the dislocated volumes formed in this way indicate a need for caution in applying creep mechanisms and models developed for uniaxial stress conditions where, generally, creep is manifest through a time dependent increase in strain.

In the interests of brevity, the work described in this paper will be limited to results obtained with magnesium oxide but it should be noted that we have obtained similar supportive evidence, albeit on a more limited scale, with a range of other ceramic crystals including nickel oxide [4,5], lithium fluoride [6], titanium carbide [2] and even diamond [7].

EXPERIMENTAL

All of this experimental work was carried out on the cube plane of magnesium oxide crystals deformed at room temperature - i.e. about 0.1 T_m.

Knoop Indentations

In the first series of experiments, a Leitz Miniload hardness machine was used with the long axis of a Knoop indenter aligned parallel to a <110> direction - this being the hardest direction and the one least likely to be affected by cracking [8]. A series of indentations were made using a standard dwell time of 12 seconds and with a normal load of either 300 gf or 500 gf to establish the characteristics of the dislocated volume. Subsequently, the dwell time was extended with a 300 gf load so that the indentation thus formed was of the same size as that produced by the 12 second indentations under 500 gf loads. Conventionally, this time dependent growth of the indentation under a constant load is termed indentation creep. For comparison, indentations were produced by 500 gf using the same prolonged dwell time of 3.5×10^5 seconds.

The nature and extent of the dislocated volume beneath these indentations can be most readily observed by viewing through the side of the transparent crystal, in a <010> direction, using transmitted polarised light to reveal the strain. Typical examples of the standard 300 gf and 500 gf, using this technique and for comparison with the extended dwell time 300 gf indentation, are shown in Figure 1. It should be noted that the mean pressure (Hk), calculated in the way described above, was 7.3 GPa, 7.2 GPa and 4.5 GPa respectively. It is apparent that the process of deformation which has enabled the time dependent growth of the 300 gf indentation over 3.5×10^5 seconds has resulted in much the same dislocated volume as that of the 500 gf standard time indentation with which it has been matched.

The use of transmitted polarised light microscopy to measure and compare dislocated volumes produced in other materials, such as diamond [9], is likely to prove extremely useful in future studies. However, it can be reinforced in such crystals as magnesium oxide by the so called serial polishing dislocation etch pitting technique. For this material we have used a dislocation etchant consisting of 4 parts NH_4Cl, 1 part H_2SO_4 and 1 part H_2O (immersed for approximately 4 minutes) and a chemical polish where the crystal is submerged in H_3PO_4 at $120°C$, leading to a removal rate of about 10 μm / second.

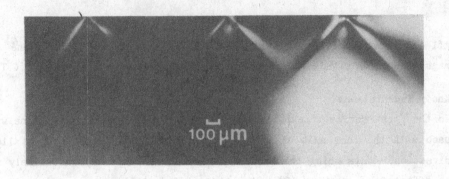

FIGURE 1. Side view in [100] of strain developed beneath Knoop indentations on (001) MgO surface. Normal load and dwell times: 300 gf, 12 s (left); 300 gf, 3.5×10^5 s (centre); 500 gf, 12 s (right).

In Figure 2 we show a sequence of micrographs illustrating the dislocation distribution developed by typical indentations produced in one of the series. On the original surface, dislocation lines which intersect the (001), lying on all six {110} slip planes, are responsible for a tight cluster of etch pits around the indentation (Figure 2(a)). Clearly, the two indentations of the same size have virtually identical dislocation rosettes developed around them - even though one was produced under standard conditions (B) whilst the other had been allowed to creep at a lower load (C). Here it should be noted that dislocation lines lying on the (110) and (1$\bar{1}$0) planes, which are normal to the (001), will intersect this surface with edge character - since they will be normal to the <110> direction of slip and their screw components will have emerged on to the surface. Consequently, rows of etch pits aligned in [110] and [1$\bar{1}$0] represent edge dislocations on (110) and (1$\bar{1}$0) respectively whilst dislocations aligned along <100> directions represent screw dislocations on (011), (01$\bar{1}$), (101) and (10$\bar{1}$) planes - which are all inclined at 45 degrees to the cube surface. The first two are briefly identified as {110}$_{90}$ and the second four as {110}$_{45}$ respectively. In addition, the indentations formed with a 500 gf normal load have developed a small degree of cracking on {110} planes due to dislocation interactions on adjacent but intersecting {110}$_{45}$ planes [8,10,11). These cracks are known to develop during the application of the normal load but do not appear to interfere significantly with subsequent time dependent plasticity. Nevertheless, in this series of experiments we have

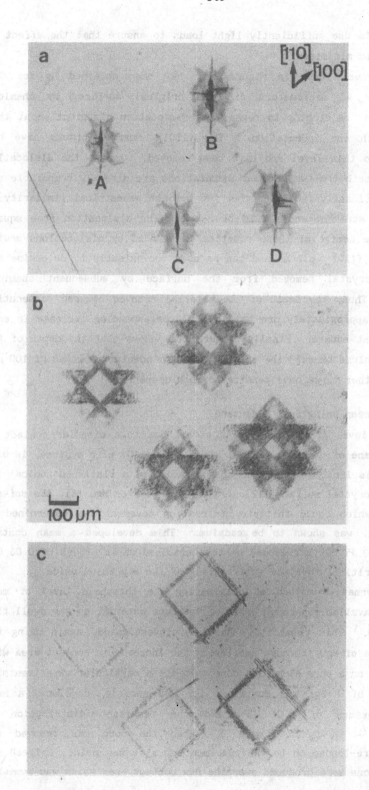

Figure 2. Distribution of dislocations produced by Knoop indentation on a (001) MgO surface (a); 40 μm (b) and 100 μm (c) below that surface. Standard (12 s) indentation at (A) and (B) with 300 gf and 500 gf respectively, extended dwell time (3.5 x 10⁵ s) at (C) and (D) with 300 gf and 500 gf respectively.

endeavoured to use sufficiently light loads to ensure that the effect of such cracks is minimised.

The surface shown in Figure 2(b) has been obtained by removing approximately 40 micrometers off the original surface, by chemical polishing, and re-etching to reveal the dislocation distribution at that level beneath the indentation. The $\{110\}_{90}$ surface cracks have not penetrated to this level and have been removed. Again, the dislocation rosettes beneath the two matched indentations are directly comparable and differ significantly from the other two but the geometrical similarity of them all is striking and should be noted. The dislocation free square region at the centre of these rosettes is bounded by dislocations moving on diverging $\{110\}_{45}$ planes and can be used conveniently to determine the amount of crystal removed from the surface by subsequent chemical polishing. Thus, the depth of the material removed between consecutive polishes is approximately one half of the corresponding increase in edge length of that square. Finally, Figure 2(c) shows that the depth of the dislocated volume beneath the standard 300 gf indentation is about 100 μm, whilst the other three have penetrated much deeper.

Aspects of Creep Using Soft Indenters

The maximum level of resolved shear stress due to a circular contact on the (001) plane of magnesium oxide, given $\{110\}<110>$ slip systems, is 0.35 P_m and this is located about 0.33 of the radius of a flattened conical tip beneath the crystal surface [12]. In earlier work on MgO [6], the softest metal cone which could initiate dislocation movement, as determined by etch pitting, was shown to be cadmium. This developed a mean contact pressure of 0.17 GPa corresponding to a resolved shear stress of 0.06 GPa – i.e. the critical resolved shear stress of the magnesium oxide.

An alternative method of determining the threshold level of mean pressure is available through creep of the cone material as the dwell time is increased. This possibility has been investigated, again using the (001) surface of MgO, through monitoring the increase in contact area with loading time on a pure aluminium cone. In these particular experiments, a normal load of 1 kgf was sustained for 300 seconds, developed a mean contact pressure of 0.25 GPa and the resultant distribution of dislocations is shown in Figure 3. When the cone was removed and immediately re-loaded on to the specimen but at a new point, a fresh set of dislocations were produced over the new contact area which was somewhat enlarged, due to creep of the aluminium, such that P_m was reduced to 0.20 GPa. Using a new cone of the same material, over a period of 10 x 10^4 seconds, the dislocated region stopped expanding when the mean contact pressure had reduced to 0.15 GPa. Furthermore, no dislocations were revealed in any part of the contact region when this particular flattened

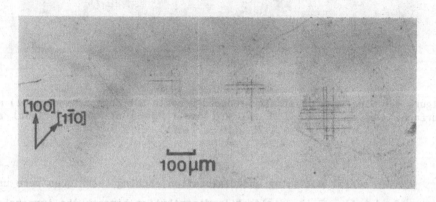

Figure 3. Dislocations produced just beneath (001) MgO surface by aluminium cone with dwell times of 300 s (left) and 10 x 10^4 s (right) under contact pressures of 0.25 GPa and 0.15 GPa respectively; and flattened cone after 600 s (centre) under 0.20 GPa.

cone was removed and immediately re-loaded. Clearly, the threshold mean pressure to initiate dislocation movement in magnesium oxide at room temperature is between 0.20 GPa and 0.15 GPa, as determined by this method, which compares well with the above work [6].

The strain produced in magnesium oxide crystals by aluminium indenters is insufficient to demonstrate, with sufficient clarity, that the depth of the dislocated volume increases with dwell time in similar fashion to the Knoop indentations shown in Figure 1. However, the phenomenon can be confirmed when the aluminium cone is replaced with one made from bronze. Figure 4 is a side view in a <100> direction, using transmitted polarised light, of two impressions made with a normal load of 1 kgf on the cube plane. The smaller impression was formed after 300 seconds, developing a mean pressure of 2.35 GPa, whilst the other one was the result of a 1.7 x 10^5 second dwell time and a mean pressure of 1.27 GPa. The geometrical similarity of the two impressions, i.e. the ratio of the contact area to the depth of the dislocated volume, is obvious.

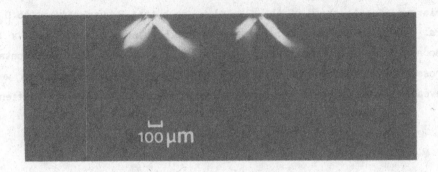

Figure 4.   Side view of strain produced beneath a bronze cone on (001) MgO
surface.   Normal load of 1 kgf and dwell times of 300 s (right) and
1.75 x 10^5 s (left).

Again, serial polishing and dislocation etching techniques were used
to verify details of the dislocation distribution beneath the impressions
in Figure 4.   As shown in Figure 5, which is of a region just below the
deformed surface, the diameter of the dislocation rosette is increased
with dwell time and this is consistent with the greater diameter of the
relevant impression and the depth of its dislocated volume.   Also, the
greater strain due to higher contact pressure with bronze, as compared to
aluminium, is reflected not only in a greater density of dislocations but
also some activity on $\{110\}_{90}$ slip planes.

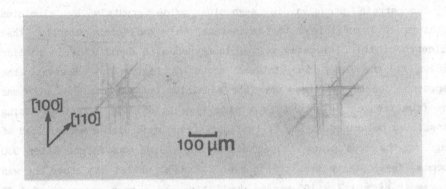

Figure 5.   Dislocations produced just beneath the two impressions shown in
Figure 4.   Contact pressures were 2.35 GPa (right) and 1.27 GPa (left).

DISCUSSION AND CONCLUSIONS

Further experimental data is necessary to develop a convincing model for impression creep in single crystals of ceramic materials at low homologous temperatures. However, there are a number of preliminary conclusions which can be supported by this and earlier work on magnesium oxide.

Arguably the most important observation is that dislocations are initiated in the harder crystal when the mean pressure due to point contact developed by softer solids exceeds a certain threshold level. When this threshold value of P_m is determined, either by using increasingly softer materials for the indenters or by allowing the indenter to creep, a reasonably accurate estimate of the critical resolved shear stress can be made. In practice, the threshold P_m may be one or two orders of magnitude less than the relevant indentation hardness of the ceramic crystal.

The dimensions of the dislocated zone, for a given normal load and dwell time, are essentially the same for soft and conventional indenters. But the density of dislocations within that constant volume, and therefore the effective or representative strain beneath the impression, increases with the hardness of the indenter. The effective strain within the dislocated volume can also be increased by repeated loading of a soft indenter on the same contact area or beneath the track formed by multiple traversals of a soft lubricated slider. In this manner, softer materials may induce workhardening and fracture in hard ceramic crystals by a process akin to fatigue.

The dislocated volume produced beneath conventional pyramidal and soft indenters increases with dwell time such as to preserve geometrical similarity. This process of time dependent plastic flow appears to be possible whilst the mean pressure exceeds the threshold level. It follows that, unlike conventional uniaxial creep conditions, both the magnitude of the maximum resolved shear stress and the level of strain decrease with time. Models based on a stable hydrostatic core produced beneath the indenter, within which there is no further dislocation movement, do not seem to be appropriate in these cases [13].

ACKNOWLEDGEMENTS

The authors would like to acknowledge the support of De Beers Industrial Diamond Distributors Ltd through research studentships to two of us (EJB and GX) and a grant to the Department.

REFERENCES

1. Brookes, C.A., Indentation hardness of diamond. In *Properties of Diamond*, ed: J.E. Field, Academic Press, London, 1979, 383-402.

2. Brookes, C.A., Shaw, M.P., and Tanner P.E., Non-metallic crystals undergoing cumulative work-hardening and wear due to softer lubricated metal sliding surfaces. *Proc. Roy. Soc., Lond.*, A409, 1987, 141-159.

3. Shaw, M.P., and Brookes, C.A., Dislocations produced in magnesium oxide crystals due to contact pressures developed by softer cones. *J. Mats. Sci.*, 24, 1989, 2727-2734.

4. Moxley, B., PhD Dissertation, University of Exeter, 1974.

5. Tanner, P.E., MSc Dissertation, University of Exeter, 1983.

6. Shaw, M.P. and Brookes, C.A., Cumulative deformation and fracture of sliding surfaces. *Wear*, 126, 1, 1988, 149-167.

7. Brookes, C.A., Brookes, E.J., Howes, V.R., Roberts, S.G., and Waddington, C.P., A comparison of the plastic deformation and creep of type I, type II and synthetic diamonds at 1100°C under conditions of point loading. *J. Hard Mats.*, 1, 1, 1990, 3-24.

8. Brookes, C.A., O'Neill, J.B., and Redfern, B.A.W., Anisotropy in the hardness of single crystals. *Proc. Roy. Soc., Lond.*, A322, 1971, 73-88.

9. Brookes, C.A., and Brookes, E.J., Diamond in perspective: a review of mechanical properties of natural diamond. Invited paper, Diamond Films '90 Conference, Crans Montana, Switzerland, September 1990. To be published in *J. Diamond and Related Materials*, 1, 1991.

10. Key, A.S., Li, J.C.M., and Chon, Y.T., *Acta Met.* 7, 1959, 694-696.

11. Chaudri, M.M., *Phil. Mag.*, A53, 1986, L55-L63.

12. Roberts, S.G., private communication.

13. Sargent, and Ashby, CUED-Mats/TR.145, Feb. 1989.

RELATIONSHIP BETWEEN CREEP AND FRACTURE OF ICE

M.A. RIST & S.A.F. MURRELL
Rock and Ice Physics Laboratory,
Department of Geological Sciences,
University College London,
Gower St, London, WCIE 6BT, UK.

ABSTRACT

After a brief introduction to the general brittle-ductile behaviour of polycrystalline materials we examine the flow and fracture properties of ice using recent experimental data. Results from uniaxial and triaxial compression tests, and from uniaxial tension tests, are presented that directly allow us to map out the range of various ice deformation mechanisms. In particular we examine the influence of microcracking on creep behaviour, the transition from flow to fracture, and the nature of uniaxial and triaxial brittle failure.

INTRODUCTION: GENERAL BRITTLE-DUCTILE BEHAVIOUR OF POLYCRYSTALS

Under an applied stress polycrystalline solids may deform elastically, plastically, by fracture or by cataclastic flow. The overall mechanical response of such solids may be greatly influenced by confining pressure, temperature, pore fluid pressure (if present) and strain rate (at high homologous temperatures). The range of dominance of various deformation and fracture mechanisms is mapped schematically in Figure 1, illustrating different fracture or flow transitions which are outlined below. (A more extensive overview of brittle-ductile transitions may be found in [1]).

Brittle cleavage (tensile fracture) characteristically occurs at low temperatures in

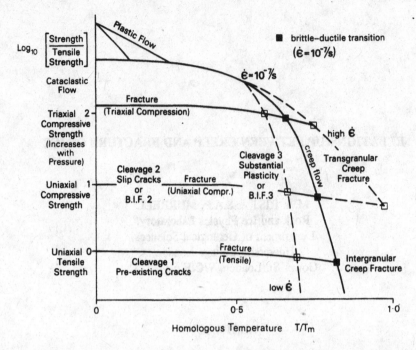

Figure 1. Schematic deformation and fracture mechanism map illustrating fracture-flow transitions for a polycrystalline solid.

solids for which the plastic flow stress is higher than the fracture stress. The measured tensile strength is controlled by microcracks [2] and on the basis of energy considerations is given by [3]:

$$\sigma_t = \{2E\Gamma/\pi(1-\nu^2)c\}^{\frac{1}{2}} \tag{1}$$

for a crack of length 2c in plane strain where E is Young's modulus, ν is Poisson's ratio and Γ is the work required to form unit area of fracture surface (the "fracture surface energy"). At low temperatures purely brittle failure is expected provided stresses are high enough to nucleate cracks (cleavage 2) or, if stresses are too low, brittle failure must occur by pre-existing flaws (cleavage 1). E and Γ are little influenced by temperature and so cleavage strength is expected to have a similarly weak temperature dependence. As temperature is increased there is typically a transition first from brittle cleavage 1 (or 2) to cleavage 3 which is preceded by substantial (1-10%) plastic strain that tends to cause crack blunting and raises the fracture surface energy, and then a further transition to creep flow.

Plastic flow and creep occur under stresses which are strongly dependent on strain rate, and also depend very strongly on temperature (especially at high temperature), since these processes are thermally activated. This is in marked contrast to cleavage fracture and generally leads to a sharp transition in deformation behaviour at a critical, strain-rate dependent, temperature [4]. Plastic flow takes place in response to a fixed non-zero state of shear or deviatoric stress (σ_s) and for polycrystalline ice over the range of strain rates encountered under laboratory and engineering conditions ($>10^{-10}s^{-1}$) power law creep is the dominant mechanism with strain rate given by:

$$\dot{\varepsilon}=A\sigma_s^n\exp(-Q/RT) \tag{2}$$

where n and A are material constants, Q is the activation energy, and R is the molar gas constant. For ice of grain size 1mm an estimate of the transition temperature from cleavage 1 to cleavage 3 is $T/T_m\approx0.63$ and for the transition from fracture to creep flow $T/T_m\approx0.88$ at a strain rate of $10^{-7}s^{-1}$.

For polycrystalline solids generally, hydrostatic confining pressures less than 0.001K (where K is the bulk modulus) cause negligible change in the plastic flow stress or creep rate although the former is increased, and the latter decreased, by higher pressures. (In terms of principal stresses $\sigma_1\geq\sigma_2\geq\sigma_3$ hydrostatic pressure is given by $[\sigma_1+\sigma_2+\sigma_3]/3$). Brittle cleavage and fracture, on the other hand, are generally strongly affected by confinement. In uniaxial compression truly brittle fracture occurs at a deviatoric stress which is an order of magnitude higher than the uniaxial tensile strength; either by a splitting mechanism [5] caused by buckling stresses or by a shearing mechanism [5,6,7,8]. With the introduction of a hydrostatic stress component the deviatoric fracture stress increases and the fracture process involves out-of-plane crack propagation which is stable, followed by crack linkage which eventually becomes unstable. A general theoretical treatment of the initiation of crack propagation in triaxially stressed solids is given by Murrell & Digby [7] and the initiation stress for open penny shaped cracks can be written:

$$\sigma_e=(\sigma_1-\sigma_3)/2=\{(2-\nu)^2\sigma_t[\sigma_t\,(4-\nu)+(\sigma_1+\sigma_3)]\}^{\frac{1}{2}} \tag{3}$$

and for closed cracks:

$$\sigma_e=(\sigma_1-\sigma_3)/2=\{2(2-\nu)(1+\sigma_c/\sigma_t)^{\frac{1}{2}}+\mu[(\sigma_1+\sigma_3)/2-\sigma_c]\}/(\mu^2+1)^{\frac{1}{2}} \tag{4}$$

where σ_c is the crack closure stress and μ is the coefficient of friction. These initiation conditions are not necessarily synonymous with macroscopic fracture and recent attempts have been made [5,8] to model crack linkage and instability processes in fracture.

EXPERIMENTAL METHODOLOGY

Creep-brittle processes in ice are currently being studied within the Rock and Ice Physics Laboratory at UCL using a purpose-built low temperature triaxial cell [9,10] in which a uniaxial stress (σ_1) and a hydrostatic pressure ($P=\sigma_2=\sigma_3$) are superimposed. The apparatus utilises specimen-based servo-control [11] to maintain a constant nominal strain rate during "strength" testing of cylindrical polycrystalline specimens 100mm long by 40mm diameter. The ice specimens used are isotropic, consisting of randomly oriented grains of uniform size (1-2mm). Specimens are protected from the nitrogen gas confining medium, which would otherwise damage the ice by exploiting surface flaws, using jackets made of soft indium metal. Current work is largely restricted to the ranges of temperature and pressure prevalent on earth (confining pressures up to 50MPa and temperatures above -60°C) whilst using high rates of strain (between $10^{-5}s^{-1}$ and $10^{-2}s^{-1}$ in compression) in order to promote crack growth and macroscopic fracture.

The apparatus allows the investigation of a wide range of complex failure mechanisms from purely ductile flow at higher temperatures and lower strain rates, to flow with distributed microcracking under intermediate conditions, to brittle failure by macroscopic fracture at higher strain rates and/or lower temperatures. Confining pressure acts to inhibit brittle failure and crack growth, and can also promote partial melting at high applied stresses and temperatures.

DEFORMATION BEHAVIOUR AT FIXED TEMPERATURE

The results from a series of constant strain rate uniaxial and triaxial compression tests conducted at -20°C are presented in Figure 2. The Figure shows the peak differential stress (σ_1-σ_3), or "strength", attained at various strain rates and confining pressures. The following modes of deformation are observed [12]:

(i) Uniaxial fracture (high $\dot{\varepsilon}$, zero confining pressure): axial splitting plus an accompanying broad shear band, the latter probably induced by frictional effects between ice and end caps.

(ii) Triaxial fracture (high $\dot{\varepsilon}$, low confining pressure): single predominant narrow shear fracture, generally independent of the specimen ends and tending to be inclined at an angle of ≈45° to the maximum principal stress. Few cracks apparent elsewhere in specimen.

(iii) Ductile deformation: in which no large scale macroscopic fracture develops and the specimen apparently deforms by flow, often with a large degree of distributed microcracking, depending on the strain rate and confining pressure. Cracks are generally

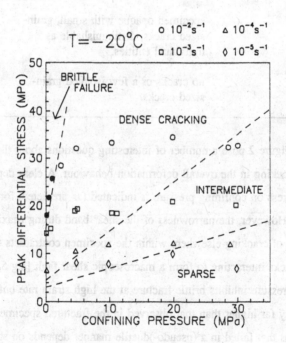

Figure 2. Ice strength and failure mode at -20°C. Crack densities refer to ductile specimens.

grain sized and aligned predominantly with their long axes parallel to the direction of the maximum principal compressive stress.

An attempt has been made to classify the extent of cracking activity in the pseudo-ductile specimens by visual inspection of the bulk ice after testing (all ductile tests were allowed to proceed to 5% strain). Following this, a particular nomenclature has been adopted, summarised in Table 1.

TABLE 1

Nomenclature used for describing ice specimens containing distributed microcracking after testing.

Crack Distribution	Definition
"Dense"	a myriad of small cracks throughout the specimen, too dense to be individually identified, many linking to form larger cracks.
"Intermediate"	specimen opaque with small, grain-sized cracks, distinguishable as individual entities.
"Sparse"	no cracks or a few isolated grain-sized cracks.

The data in Figure 2 pose a number of interesting questions about the role and influence of microcracking in the overall deformation behaviour. A clear dependence of the brittle fracture stress on confining pressure is indicated (as predicted, for example, by Equations 3 and 4). However, the narrowness of the shear band during triaxial fracture and the very low density of cracking elsewhere within the specimen contradicts the notion of large numbers of cracks interacting to form a macroscopic shear fault [e.g 5,13,14]. Higher imposed confining pressure inhibits brittle fracture at the high strain rate but induces an ultimate crack density far higher than that observed in the fractured specimens. The strength of specimens that failed in a (pseudo-)ductile manner depends on strain rate but is independent of confining pressure, at least above $\sigma_3=5$MPa. The maximum imposed confinement of 30MPa lowers the pressure melting point by about 3°C, but does not appear to have significantly affected the strength here.

The presence of cracks is expected to enhance bulk flow [15,16] although the influence of cracking activity on the ductile strength behaviour is by no means clear. It is apparent from Figure 2 that increasing confining pressure acts to inhibit the formation of cracks while higher differential stresses act to promote them; however, the presence of cracks does not necessarily lower the strength of the ice even at high volumetric densities. For example, the confined tests at the $10^{-4}s^{-1}$ strain rate span almost the whole range of cracking activity without any significant change in specimen strength. In fact, Sinha [15] has found that, under uniaxial conditions, cracks appear to be non-interacting up to strain rates of at least $10^{-4}s^{-1}$.

FLOW BEHAVIOUR - THE EFFECT OF TEMPERATURE

The strength of ice can be considered as a rate-sensitive yield stress and the differential failure stress for ductile deformation attained during constant strain rate tests has been shown to be similar to the fixed stress required to induce a similar minimum strain rate (secondary creep) in constant load creep tests [17,18]. Ductile strength data is therefore expected to follow a power law of the form given in Equation 2. Figure 3 shows a plot of $\log(\sigma_1-\sigma_3)$ versus $1/T$ covering temperatures -5°C to -45°C for ductile specimens with 5MPa confining pressure (small enough not to introduce significant pressure melting effects at high T but large enough for the strength to be pressure independent - see Figure 2). Stepwise regression of the data provides the following values (with standard deviations) for the material parameters: $Q=72\pm7kJmol^{-1}$, $n=4.3\pm0.2$, $\log A=14.9MPa^{-n}s^{-1}$ (correlation coefficient $r^2=0.98$). These values are compared with those from other recent studies of polycrystalline ice derived under a wide variety of test conditions in Table 2.

Our value for the activation energy is in close agreement with those derived by Barnes et al. [19] (from uniaxial creep tests), Budd & Jacka [20] (from an extensive range of borehole shear and laboratory compression tests) and Sinha [21] (from an examination of the delayed elastic strain of columnar ice). All these workers used comparatively low stresses (<3MPa) for which no cracking is expected and found n≈3 which is the widely accepted value reported in the literature under these conditions (for a review see [22]).

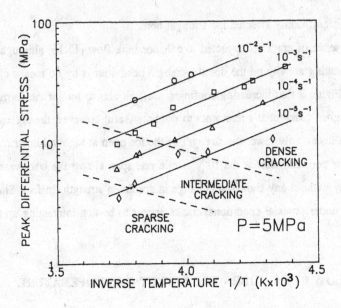

Figure 3. Ice strength (log scale) versus inverse temperature for specimens that failed in a ductile manner under a fixed confining pressure of 5MPa.

This value has been taken by many to imply that creep is controlled by the drag experienced by dislocations in the basal plane, although the attainment of secondary creep has also been considered to be largely controlled by processes occurring on non-basal planes [23]. At higher stresses Barnes et al. [19] found that a simple power law was no longer valid as n increased exponentially with stress. To estimate the change in the material parameters at higher stresses they cite the indentation hardness tests of Walker [24] which produced a similar value for Q of 73kJmol^{-1} at temperatures between -12°C and -25°C, but a higher value for n of around 4.4 for indentation pressures up to 50MPa, in close agreement with the value found here. Table 2 clearly shows the change in the value of the stress exponent from n=3 to n=4-5 as deviatoric stresses are increased above 2-3MPa.

It is generally considered that the increase in the stress exponent above n=3 is closely associated with the onset of microcracking. For example, Cole [25] observed a change during uniaxial strength tests at -5°C, from n=2.8 in the strain rate range 10^{-6}s^{-1} to

TABLE 2
Ductile flow parameters for polycrystalline ice derived under
various test conditions

Source	Temperature (°C)	Deviatoric Stress (MPa)	Strain Rate (s^{-1})	n	Q (kJmol^{-1})
Barnes et al. [19] Uniaxial creep tests	-8 to -45°C	0.1-3	10^{-9}-10^{-5}	3.1	75
Sinha [21] Uniaxial creep tests (columnar ice)	-5 to -45°C	0.5	small delayed elastic strains $\varepsilon < 2 \times 10^{-4}$	3.0	67
Budd & Jacka [20] Borehole shear data & Uniaxial comp. tests	-10 to -40°C	$\approx$0.02-1	$\approx 10^{-12}$-10^{-6}	$\approx$3	72
Cole [25] Uniaxial strength tests	-5°C	0.6-2 2-10	10^{-7}-10^{-6} 10^{-5}-10^{-3}	2.8 4.5	- -
Walker [24] Hardness tests	-12 to -25°C	7-50	t=10-10^{4}s	4.4	73
Jones [26] Triaxial strength tests	-11°C	Uniaxial: 2-11 Confined (P<70MPa): 2-20	10^{-6}-10^{-3} 10^{-6}-10^{-3}	5.0 4.0	- -
Durham et. al [27] Triaxial strength tests (P=50MPa)	-15 to -30°C -30 to -78°C	4-13 4-70	$\approx 10^{-6}$-10^{-4} $\approx 10^{-6}$-10^{-4}	4.0 4.0	91 61
Rist & Murrell Triaxial strength tests (P=5MPa)	-5 to -45°C	3-41	10^{-5}-10^{-2}	4.3	72

$10^{-7}s^{-1}$ with no apparent cracking (deformation being dominated by dynamic recrystallisation), to n=4.5 at higher strain rates where microcracking and extensive grain boundary deformation was observed. Jones [26] found n=5.0 from uniaxial strength tests but n=4.0 under confinement, implying that the change was due to the inhibition of cracking. This may not be the case, however, since his confined strength data show no change in the linear relationship between log(σ) and log($\dot{\varepsilon}$) (slope=n) from $10^{-7}s^{-1}$ (no cracking) to almost $10^{-2}s^{-1}$ where, presumably, cracking must have been extremely dense. Similarly, the strengths represented in Figure 3, used to calculate the flow parameters at

P=5MPa, cover the whole range of observed cracking activity from purely ductile deformation (no cracks in the specimen at -5°C, $10^{-5}s^{-1}$) to flow accompanied by "dense" microcracking with no apparent discontinuity.

During the hardness tests reported by Barnes et al. [19,24] cracking was inhibited by an effective hydrostatic pressure and so the high value for n must have been the result of some other mechanism enhanced by the high stresses - they suggest this reflects the increasing importance of non-basal glide as a deformation process. Duval et al. [23] also question the extent of the influence of cracking on the creep behaviour of polycrystalline ice, finding no apparent effect of microcracking on the strain to the minimum creep rate or the transition to tertiary flow.

Durham et al. [27] have found n=4 and Q=91kJmol^{-1} for post-failure "steady-state" flow in constant deformation rate triaxial tests over the temperature range -15°C to -30°C with P=50MPa. This confining pressure is sufficient to inhibit microcracking entirely but apparently leads to grain boundary softening which increases the activation energy at temperatures higher than -30°C [27,28]. Mellor and Testa [29] and Barnes et al. [19] observed a change in activation energy above -10°C which they associated with liquid at grain boundaries, a similar change not being observed in single crystals [30]. Below -10°C Barnes et al. considered only intergranular creep processes to be important. The 5MPa tests in Figure 3 cover the temperature range -5°C to -45°C but are too limited to identify any high temperature regime.

DEFORMATION MECHANISM MAP

Using our triaxial strength data we can sketch a deformation mechanism map for polycrystalline ice at high temperatures and strain rates (Figure 4). The flow data is plotted along with curves derived from the calculated creep parameters at P=5MPa and the observed crack density regimes are also marked. The uniaxial tensile fracture curve is drawn from preliminary tension tests we have conducted on our laboratory ice that indicate $\sigma_t \approx 1.4$MPa (mean value at -5°C to -15°C, $10^{-6}s^{-1}$ to $10^{-4}s^{-1}$) and is assumed to change little with temperature. We have also plotted uniaxial brittle compressive fracture

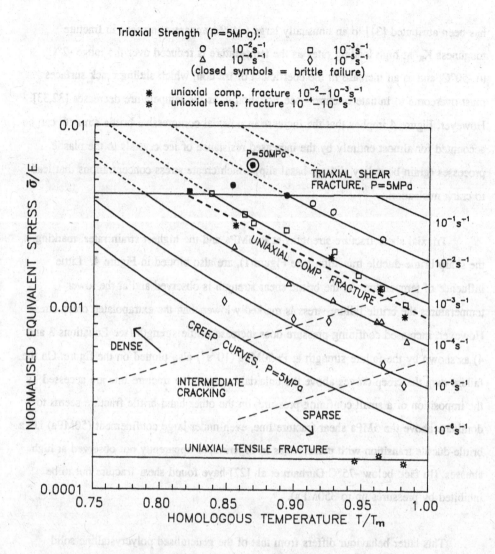

Figure 4. Deformation mechanism map for ice derived from experimental data.

strengths (derived at high strain rates 10^{-3}-10^{-2}s^{-1} where the brittle strength appears to be independent of deformation rate) and these are seen to show a very marked temperature dependence, similar to that displayed by the flow strengths. Note, however, that these fracture strengths lie well below the 10^{-2}s^{-1} creep curve.

The increase of uniaxial compressive fracture strength with decreasing temperature has been attributed [31] to an unusually large ($\approx$40%) apparent increase in fracture toughness K_{IC} at high loading rates as the temperature is reduced over the range -2°C to -50°C, and to an increase in the coefficient of friction, which sliding crack surfaces must overcome to initiate fracture - see Equation 4, as the temperature decreases [32,33]. However, Figure 4 implies that the increase in uniaxial compressive brittle strength can be accounted for almost entirely by the increased resistance of ice crystals to the plastic processes (grain boundary sliding, basal slip) which create stress concentrations that lead to crack nucleation.

Triaxial shear fracture strengths at P=5MPa and the highest strain rates, marking the sharp brittle-ductile transition (see Figure 1), are also plotted in Figure 4. Little influence of temperature on the brittle shear strength is observed and at the lower temperatures the brittle failure stress is markedly lower than the extrapolated creep stress. However, increased confining pressure does increase brittle strength (see Equations 3 and 4) as shown by the failure strength at P=50MPa (10^2s^{-1}) also plotted on the figure. On the failure map the creep curves above the uniaxial compressive fracture line are accessed by the imposition of a small confining pressure, on the other hand brittle fracture seems to dominate above the 5MPa shear fracture line, even under large confinement (50MPa) i.e. a brittle-ductile transition with pressure (as in Figure 2) is apparently not observed at high stresses. (In fact, below -75°C Durham et al. [27] have found shear fracture not to be inhibited by pressures up to 350MPa).

This latter behaviour differs from that of the generalised polycrystalline solid illustrated in Figure 1, where a transition to cataclastic, or crack enhanced, flow is expected at high confining pressures. However, the ice deformation map does illustrate clearly how the fracture strength in compression differs from, and is much greater than, that in tension for any given homologous temperature. Further, the triaxial compressive strength is always greater than the uniaxial compressive strength, the latter displaying a strong temperature dependence that implies a link with underlying plastic processes - emphasising the marked difference in the fracture processes occurring under uniaxial and triaxial stress conditions.

SUMMARY

The fracture and flow of polycrystalline ice and its dependence on temperature, strain rate and confining pressure has been examined using uniaxial and triaxial experimental test data. Typical creep behaviour involves a change from purely ductile flow at high temperatures and low strain rates to pseudo-ductile flow accompanied by increasingly dense distributed microcracking as temperature decreases and strain rate increases. The influence of this cracking on the overall deformation is unclear, but is seemingly negligible and does not alter the apparent activation energy. At high stresses under triaxial compression shear fracture predominates with strength dependent on pressure, but only weakly dependent on temperature. Tensile strength is much lower but also depends little on temperature. Conversely, uniaxial compressive fracture strength displays a strong temperature dependence, similar to that displayed by the flow strength. The underlying reason for this temperature dependence is unclear, as is the specific mechanism for triaxial shear fracture, and both require further investigation.

REFERENCES

1. Murrell, S.A.F., Brittle-to-ductile transitions in polycrystalline non-metallic materials. In Deformation Processes in Minerals, Ceramics and Rocks, eds D.J. Barber and P.G. Meredith, Unwin Hyman, London, 1990, pp109-37.

2. Griffith, A.A., The phenomena of rupture and flow in solids. Phil. Trans. Roy. Soc. Lond., 1920, A221, 163-98.

3. Murrell, S.A.F., The theory of the propagation of elliptical Griffith cracks under various conditions of plane strain or plane stress. Part I. Br. J. Appl. Phys., 1964, 15, 1195-210; Parts II and III, ibid., 1211-23.

4. Orowan, E., Fundamentals of brittle behaviour in metals. In Fatigue and Fracture of Metals, ed. W.M. Murray, Wiley, New York, 1952, pp. 139-167.

5. Ashby, M.F. and Hallam, S.D., The failure of brittle solids containing small cracks under compressive stress states. Acta Metall., 1986, 34, 497-510.

6. Murrell, S.A.F., The effect of triaxial stress systems on the strength of rocks at atmospheric temperatures. Geophys. J. Roy. Astron. Soc., 1965, 10, 231-81.

7. Murrell, S.A.F. and Digby, P.J., The theory of brittle fracture initiation under triaxial

stress conditions: I. Geophys. J. Roy. Astron. Soc., 1970, **19**, 309-34; II, ibid., 499-512.

8. Horii, H. and Nemat-Nasser, S., Brittle failure in compression: splitting, faulting and the brittle-ductile transition. Phil. Trans. Roy. Soc. Lond., 1986, **A319**, 337-74.

9. Sammonds, P.R., Murrell, S.A.F. and Rist, M.A., Fracture of multi-year sea ice under triaxial stresses: apparatus description and preliminary results. J. Offshore Mech. Arctic Eng., 1989, **111**(3), 258-263.

10. Sammonds, P.R., Murrell, S.A.F., Rist, M.A. and Butler, D., The design of a high-pressure, low temperature triaxial deformation cell for ice. Cold Reg. Sci. Technol., 1991, **19**(2), in press.

11. Rist, M.A., Sammonds, P.R. and Murrell, S.A.F., Strain rate control during deformation of ice: an assessment of the performance of a new servo-controlled triaxial testing system. Cold Reg. Sci. Technol., 1991, **19**(2), in press.

12. Rist, M.A., Murrell, S.A.F. and Sammonds, P.R., Experimental results on the failure of polycrystalline ice under triaxial stress conditions. In Proc. IAHR Ice Symp., Sapporo, Japan, 1988, pp118-27.

13. Hallbauer, D.K., Wagner, H. and Cook, N.G.W., Some observations concerning the microscopic and mechanical behaviour of quartzite specimens in stiff, triaxial compression tests. Int. J. Rock Mech. Min. Sci. & Geomech. Abstr., 1973, **10**, 713-726.

14. Costin, L.S., Damage mechanics in the post-failure regime. Mech. Mat., 1985, **4**, 149-60.

15. Sinha, N.K. Crack enhanced creep in polycrystalline material: strain-rate sensitive strength and deformation of ice, J. Mater. Sci., 1988, **23**, 4415-28.

16. Jordaan, I.J. and McKenna, R.F., Processes of deformation and fracture of ice in compression. In. Proc. IUTAM/IAHR Symp. Ice/Structure Interactions, St Johns, Canada, 1989, in press.

17. Mellor, M. and Cole, D.M., Deformation and failure of ice under constant stress or constant strain-rate. Cold. Reg. Sci. Technol., 1982, **5**, 201-19.

18. Mellor, M. and Cole, D.M., Stress/strain/time relations for ice under uniaxial compression. Cold Reg. Sci. Technol., 1983, **6**, 207-30.

19. Barnes, P., Tabor, D. and Walker, J.C.F., The friction and creep of polycrystalline ice. Proc. Roy. Soc. Lond., 1971, **A324**, 127-55.

20. Budd, W.F. and Jacka, T.H., A review of ice rheology for ice sheet modelling. Cold Reg. Sci. Technol., 1989, **16**, 107-144.

21. Sinha, N.K., Rheology of columnar-grained ice. Exper. Mech., 1978, **18**, 464-470.

22. Weertman, J., Creep deformation of ice. Annu. Rev. Earth Planet. Sci., 1983, **11**, 215-40.

23. Duval, P., Ashby, M.F. and Andermann, I., Rate-controlling processes in the creep of polycrystalline ice. J. Phys. Chem., 1983, **87**, 4066-74.

24. Walker, J.C.F., The mechanical properties of ice I_h. PhD Thesis, University of Cambridge, 1970.

25. Cole, D.M., Strain rate and grain size effects in ice. J. Glaciol., 1987, **33**, 274-280.

26. Jones, S.J., The confined compressive strength of polycrystalline ice. J. Glaciol., 1982, **28**, 171-7.

27. Durham, W.B., Heard, H.C. and Kirby, S.H., Experimental deformation of polycrystalline H_2O ice at high pressure and low temperature: preliminary results. J. Geophys. Res., 1983, **88**, B377-B392.

28. Kirby, S.H., Durham, W.B., Beeman, M.L., Heard, H.C. and Daley, M.A., Inelastic properties of ice I_h at low temperatures and high pressures. J. de Physique, 1987, **48**, Colloque C1 (3), 227-32.

29. Mellor, M. and Testa, R., Effect of temperature on the creep of ice. J. Glaciol., 1969, **8**, 131-45.

30. Jones, S.J. and Brunet, J.G., Deformation of ice single crystals close to the melting point. J. Glaciol., 1978, **21**, 445-56.

31. Schulson, E.M., The fracture of ice I_h. J. de Physique, 1987, **48**, Colloque C1 (3), 207-218.

32. Hallam, S.D., The role of fracture in limiting ice forces. In Proc. IAHR Symposium on Ice, 1986, Iowa, USA, Vol. 2, pp 387-319.

33. Schulson, E.M. The Brittle compressive fracture of ice. Acta. Metall., 1990, **38**, 1963-76.